Custom VLSI
Microelectronics

Custom VLSI Microelectronics

Stanley L. Hurst

PRENTICE HALL
New York   London   Toronto   Sydney   Tokyo   Singapore

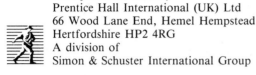

First published 1992 by
Prentice Hall International (UK) Ltd
66 Wood Lane End, Hemel Hempstead
Hertfordshire HP2 4RG
A division of
Simon & Schuster International Group

© Prentice Hall International (UK) Ltd, 1992

All rights reserved. No part of this publication may be
reproduced, stored in a retrieval system, or transmitted,
in any form, or by any means, electronic, mechanical,
photocopying, recording or otherwise, without prior
permission, in writing, from the publisher.
For permission within the United States of America
contact Prentice Hall Inc., Englewood Cliffs, NJ 07632.

Typeset in 10 on 12 point Times
by MCS Typesetters, Salisbury, Wiltshire, England

Printed and bound in Great Britain
at the University Press, Cambridge

Library of Congress Cataloging-in-Publication Data

Hurst, S. L. (Stanley Leonard)
  Custom VLSI microelectronics/Stanley L. Hurst.
    p.   cm.
  Includes bibliographical references and index.
  ISBN 0-13-194416-9
  1. Integrated circuits – Very large scale integration – Design
  and construction – Data processing. 2. Computer-aided design
  I. Title.
  TK7874.H883 1992   91-21743
  621.381′5 – dc20   CIP

British Library Cataloguing in Publication Data

Hurst, Stanley L.
  Custom VLSI microelectronics.
  I. Title
  621.3815

  ISBN 0-13-194416-9

---

1 2 3 4 5   96 95 94 93 92

Contents

Preface		xi
List of symbols and abbreviations		xiii
1	Introduction: the microelectronics evolution	1
	1.1 *Initial history*	1
	1.2 *The continuing evolution*	2
	1.3 *CAD developments*	7
	1.4 *Why custom microelectronics?*	11
	1.5 *Summary*	15
	1.6 *References*	15
2	Technologies and fabrication	18
	2.1 *Bipolar silicon technologies*	19
	2.1.1 Analogue ICs	21
	2.1.2 Transistor–transistor logic	28
	2.1.3 Emitter-coupled logic	30
	2.1.4 Integrated-injection logic technologies	32
	2.1.5 Collector-diffusion isolation	36
	2.1.6 Bipolar summary	38
	2.2 *Unipolar silicon technologies*	39
	2.2.1 nMOS insulated-gate devices	41
	2.2.2 pMOS insulated-gate devices	47
	2.2.3 CMOS technologies	49
	2.2.4 Silicon-on-insulator	54
	2.2.5 MOS summary	56
	2.3 *Memory circuits*	57
	2.4 *BiCMOS technology*	68
	2.5 *Gallium-arsenide technology*	70
	2.6 *A comparison of available technologies*	71
	2.7 *References*	75

3 Standard off-the-shelf ICs — 77

- **3.1** *Non-programmable SSI, MSI and LSI digital ICs* — 78
- **3.2** *Standard analogue ICs* — 80
- **3.3** *Microprocessors* — 81
- **3.4** *Memory* — 86
  - **3.4.1** Read-only memory (ROM) — 87
  - **3.4.2** Random-access memory (RAM) — 89
- **3.5** *Programmable logic devices* — 90
  - **3.5.1** Programmable logic arrays (PLAs) — 93
  - **3.5.2** Programmable array logic (PALs) — 96
  - **3.5.3** Programmable gate arrays (PGAs) — 100
  - **3.5.4** Programmable logic sequencers (PLS) — 100
  - **3.5.5** Other programmable standard products — 101
- **3.6** *Logic cell arrays (LCAs), sometimes termed programmable gate arrays (PGAs)* — 102
- **3.7** *Specialized off-the-shelf ICs (ASICs)* — 106
- **3.8** *Summary* — 107
- **3.9** *References* — 107

4 Custom microelectronic techniques — 110

- **4.1** *Full hand-crafted custom design* — 112
- **4.2** *Standard-cell techniques* — 116
  - **4.2.1** Fixed-cell architectures — 117
  - **4.2.2** Soft-cell architectures — 123
  - **4.2.3** Macrocells — 123
  - **4.2.4** Analogue cells — 127
- **4.3** *Gate-array techniques* — 128
  - **4.3.1** Choice of general-purpose cells for digital applications — 132
  - **4.3.2** Routing considerations — 145
  - **4.3.3** Channel-less architectures — 153
  - **4.3.4** Analogue and mixed analogue/digital arrays — 157
- **4.4** *Maskless fabrication techniques* — 161
  - **4.4.1** E-beam direct-write-on-wafer — 163
  - **4.4.2** Other technologies — 166
- **4.5** *Summary and technical comparisons* — 169
- **4.6** *References* — 175

5 Computer-aided design — 179

- **5.1** *IC design software* — 183
  - **5.1.1** Behavioural- and architectural-level synthesis — 184
  - **5.1.2** Functional- and gate-level synthesis — 191

	5.1.3 Physical design software	193
	5.1.4 Test programme formulation	208
	5.1.5 Analogue and mixed analogue/digital synthesis	209
	5.1.6 Programmable logic devices	210
5.2	*IC simulation software*	215
	5.2.1 Behavioural-, architectural- and functional-level modelling	216
	5.2.2 Gate- and switch-level modelling	218
	5.2.3 Silicon-level simulation	225
	5.2.4 Back-annotation and post-layout simulation	230
	5.2.5 Analogue and mixed analogue/digital simulation	232
	5.2.6 Simulation accelerators and system simulation	235
5.3	*Silicon compilers*	239
5.4	*CAD hardware availability*	242
	5.4.1 Workstations	243
	5.4.2 Personal computers (PCs)	244
	5.4.3 Networking	247
5.5	*CAD software availability*	250
	5.5.1 Design and layout software	253
	5.5.2 Simulation software	253
	5.5.3 PLD design software	254
	5.5.4 Printed-circuit board (PCB) design software	255
	5.5.5 Housekeeping software	256
	5.5.6 Data formats and standards	256
5.6	*CAD costs*	260
5.7	*Summary*	262
5.8	*References*	265

6	Test-pattern generation and design-for-testability	271
6.1	*General introduction*	271
6.2	*Basic testing concepts*	277
	6.2.1 Digital circuit test	277
	6.2.2 Analogue circuit test	281
	6.2.3 Mixed analogue/digital circuit test	282
6.3	*Digital test-pattern generation*	283
	6.3.1 Controllability and observability	284
	6.3.2 Fault-effect models	288
	6.3.3 Test-pattern generation	295
6.4	*Test-pattern generation for memory and programmable logic devices*	303
	6.4.1 Random-access memories	303
	6.4.2 Fixed memories (ROM and PROM)	305
	6.4.3 Programmable logic arrays	306
6.5	*Microprocessor testing*	309

Contents

6.6	*Design-for-testability (DFT) techniques*	310
	6.6.1 Ad-hoc design methods	310
	6.6.2 Structured DFT methods	314
	6.6.3 Self-test DFT methods	324
6.7	*PLA design-for-testability techniques*	334
6.8	*I/O testing and boundary scan*	340
	6.8.1 I/O testing	342
	6.8.2 Boundary scan and board testing	342
6.9	*Further testing concepts*	347
	6.9.1 Input compaction	349
	6.9.2 Output compaction	350
	6.9.3 Syndrome and transition counting	352
	6.9.4 Further PLA and BIST testing concepts	355
	6.9.5 Arithmetic and spectral coefficients	357
	6.9.6 On-line checking	367
	6.9.7 Cellular arrays and special macros	370
	6.9.8 Cellular automaton logic-block observation (CALBO)	372
6.10	*The silicon area overheads of DFT*	376
6.11	*Summary*	380
6.12	*References*	381

7 The choice of design style: technical and managerial considerations — 389

7.1	*The microelectronic choices*	393
	7.1.1 Digital-only products	393
	7.1.2 Analogue-only products	403
	7.1.3 Mixed analogue/digital products	405
7.2	*Packaging*	405
7.3	*Time to market*	410
7.4	*Financial considerations*	413
	7.4.1 NRE costs	413
	7.4.2 CAD costs	415
	7.4.3 Learning costs	416
	7.4.4 Testing costs	417
	7.4.5 Overall product costings	419
7.5	*Summary*	424
7.6	*References*	424

8 Conclusions — 426

8.1	*The present status*	426
8.2	*Future developments*	427
8.3	*Final summary*	432
8.4	*References*	433

Appendix A: The elements and their properties 434

Appendix B: Fabrication and yield 440

Appendix C: The principal equations relating to bipolar transistor performance 446

Appendix D: The principal equations relating to unipolar (MOS) transistor performance 453

Index 461

Preface

In its short life span microelectronics has become the most complex and astounding of our every-day technologies, embracing as it does physics, chemistry, materials, thermodynamics and micromechanical engineering, as well as electrical and electronic engineering and computer science. No one person can hope to be expert in all these aspects.

And yet, in spite of all this complexity and sophistication we often take the end-products for granted. Nowadays, our homes contain thousands of transistors in domestic appliances and communications and entertainment equipment, to add to the vast range of applications in industrial, military, space and commercial products. To make all this possible has required the simultaneous evolution not only of the ability to make the microelectronic circuits themselves, but also the ability to design them in the first place and test them during and after production.

This is not a text on detailed silicon design and fabrication methods. Instead, it is principally concerned with an increasingly important branch of microelectronics, namely custom microelectronics, whereby specific circuit designs can be realized rapidly and economically in small-quantity production quantities. The terminology ASIC ('application-specific IC') has been widely used up to now, but the more accurate terminology USIC ('user-specific integrated circuit') is now used increasingly. Custom circuits have only become viable with the maturity of both the circuit fabrication methods and the computer-aided-design resources necessary for their design and test, thus releasing the circuit designer from personal involvement with a vast range of design and fabrication detail.

However, having said this, it is the hallmark of a good engineer to be aware of all the aspects involved even though he or she may not need – or indeed not be able – to do anything about them. This text adopts this approach, and will attempt to give a comprehensive overview of all aspects of custom microelectronics, including the very important but difficult managerial decisions of when or when not to use it.

The concept of custom microelectronics is not new; it has been around for over two decades, but probably did not achieve very great prominence until the early-1980s. The introduction of the microprocessor as a readily available standard part delayed the wider adoption of custom microelectronics, and still is a major force in the how-shall-we-make-it product design decisions. Nevertheless, custom ICs offer their particular technical or economic advantages, and every product designer should be aware of the strengths as well as the weaknesses of both custom and other design styles. The new generation of VLSI programmable ICs is another up-and-coming player in this area. These are some of the matters which we will consider.

It is assumed that readers of this text will be familiar with the principles of electronic devices, semiconductor physics, basic electronic circuits and systems design as covered in many standard teaching texts. For readers requiring a more detailed treatment than is covered here, then specific texts dealing solely, for example, with device physics and fabrication, with computer-aided-design, or with design-for-test should be consulted. Regrettably no one text (and no one author!) can cover all these aspects in depth, but it is hoped that the references contained in this text may point to such additional in-depth information.

On a personal note, may I thank all my acquaintances over the years in industry and academia who have collectively added to my awareness of this subject area, particularly former research colleagues at the University of Bath and academic colleagues in the Electronics Discipline of the Open University. Also, on the testing side, very many years of pleasant co-operation must be acknowledged with the VLSI Design and Test Group of the Department of Computer Science at the University of Victoria, BC, Canada. To them and very many others throughout the world, my thanks.

Stanley L. Hurst
Bristol
England

List of symbols and abbreviations

ALU	arithmetic logic unit
ASIC	application-specific integrated circuit (*see also* CSIC and USIC)
ASM	algorithmic state machine
ASTTL	advanced Schottky transistor–transistor logic
A-to-D	analogue-to-digital
ATE	automatic test equipment
ATPG	automatic test-pattern generation
BiCMOS	bipolar/CMOS (fabrication)
BFL	buffered FET logic
BILBO	built-in logic block observation
BJT	bipolar junction transistor
C	capacitor
CAD	computer-aided design (*see also* CAE and CAM)
CAE	computer-aided engineering
CAM	computer-aided manufacture
CCD	charge-coupled device
CDI	collector-diffusion-isolation
CIF	Caltech Intermediate Form
CML	current-mode logic
CMOS	complementary MOS (*see also* nMOS and pMOS)
CPU	central processing unit
CSIC	custom- or customer-specific integrated circuit (*see also* USIC)
DA	design automation
DFT	design-for-testability
DIL or DIP	dual-in-line package or dual-in-line plastic (package)
D-MESFET	depletion-mode metal-semiconductor field-effect transistor (GaAs device, *see also* E-MESFET)
DMOS	dynamic MOS
DRAM	dynamic read-only memory (*see also* SRAM)
DRC	design rules checking
D-to-A	digital-to-analogue
E^{2}PROM	*see* EEPROM

EAROM	electrically alterable ROM
ECAD	electronic computer-aided design
ECL	emitter-coupled logic
EDA	electronic design automation
EEPROM	electrically erasable ROM
E_g	energy band gap of an atom
E-MESFET	enhancement-mode field-effect transistor (*see also* D-MESFET)
EPROM	erasable programmable ROM
ERC	electrical rules checking
eV	electron-volt
FAST™	advanced Schottky TTL technology originated by Fairchild
FET	field-effect transistor
FPGA	field-programmable gate array
FPLA	field-programmable logic array
FPLS	field-programmable logic sequencer
FTD	full technical data (*see also* LTD)
GaAs	gallium-arsenide
GDSII	graphics design system, version II (a mask pattern standard)
Ge	germanium
GHz	gigahertz
GPIB	general-purpose instrumentation bus, electrically identical to Hewlett-Packard's instrumentation bus HPIB
HCMOS	high-speed CMOS
HDL	hardware description language
HILO	a commercial simulator for digital circuits
IC	integrated-circuit
IGFET	insulated-gate FET
I^2L	integrated-injection logic
I^3L	oxide-isolated integrated-injection logic
I/O	input/output
ISL	integrated Schottky logic
JFET	junction FET
K	1024 (as used in defining memory size)
kHz	kilohertz
L	inductor, or length in MOS technology geometry
LAN	local area network
λ	lambda, a unit of length in MOS design layout
L/W	length/width ratio, used in MOS design layout
LCA	logic cell array
LOCMOS	local-oxidation CMOS
LSI	large scale integration (*see also* SSI, MSI, VLSI)
LSTTL	low-power Schottky TTL
LTD	limited technical data (*see also* FTD)
MESFET	metal-semiconductor FET

MgCMOS	metal-gate CMOS (*see also* SgCMOS)
MHz	megahertz
mil	one-thousandth of an inch
MIPS	million instructions per second
MOS	metal-oxide–silicon
MOST	metal-oxide–silicon transistor
ms	millisecond (10^{-3} s)
MSI	medium scale integration (*see also* SSI, LSI, VLSI)
nMOS	n-channel metal-oxide–silicon (circuit)
NRE	non-recurring expenditure or engineering
ns	nanosecond (10^{-9} s)
OEM	original-equipment manufacturer
op.amp.	operational amplifier
PAL	programmable array logic
PCB	printed-circuit board
PDM	process device monitor (*see also* PED and PMC)
PED	process evaluation device (*see also* PDM and PMC)
PG	pattern generation
PGA	pin grid array (packaging)
	or
	programmable gate array
PLA	programmable logic array
PLD	programmable logic device
PLS	programmable logic sequencer
PMC	process monitoring circuit (*see also* PDM and PED)
pMOS	p-channel metal-oxide–silicon (circuit)
PMUX	programmable multiplexer
ps	picosecond (10^{-12} s)
PROM	programmable ROM
QIL	quad-in-line (package)
R	resistor
RAM	random-access memory (*see also* DRAM and SRAM)
RAS	random-access scan
ROM	read-only memory
RTL	resistor-transistor logic
	or
	register transfer language
SBC	standard buried collector (fabrication)
SDFL	Schottky-diode FET logic
SgCMOS	silicon-gate CMOS (*see also* MgCMOS)
Si	silicon
S-a-0	stuck-at-0
S-a-1	stuck-at-1
SiO_2	silicon dioxide

Si₃N₄	silicon nitride
SMD	surface mount device
SOI	silicon-on-insulator
SOS	silicon-on-sapphire
SOSMOS	silicon-on-sapphire CMOS
SPICE	simulation program, integrated circuit emphasis
SRAM	static read-only memory (*see also* DRAM)
SSI	small scale integration (*see also* MSI, LSI, VLSI)
STL	Schottky transistor logic
STTL	Schottky transistor–transistor logic
TPG	test-pattern generation (*see also* ATPG)
TTL	transistor–transistor logic
UCA	uncommitted component array
ULA™	uncommitted logic array
ULSI	ultra-large-scale integration
μP	microprocessor
μs	microsecond (10^{-6} s)
USIC	user-specific integrated circuit, used in this text in preference to ASIC or CSIC
V_{BB}	d.c. base supply voltage
V_{CC}	d.c. collector supply voltage
V_{DD}	d.c. drain supply voltage
V_{EE}	d.c. emitter supply voltage
V_{GG}	d.c. gate supply voltage
V_{SS}	d.c. source supply voltage
VDU	visual display unit
VHSIC	very high-speed integrated circuit
VLSI	very large scale integration (*see also* SSI, MSI, LSI)
VMOS	a form of MOS fabrication
W	width (in MOS technology geometry)
WSI	wafer scale integration

1 Introduction: the microelectronics evolution

The development of microelectronics spans little more than one-quarter of a century – less than one-half of a person's life expectancy – but, during this short period it has become the most pervasive technology that has yet been developed. It touches all aspects of our lives: transportation, communications, entertainment, medical matters, comfort and safety, and yet many professional engineers have still to be involved in the design of products or systems using microelectronics.

In this introductory chapter we will survey the developments which have led to the present levels of microelectronic expertise. As noted in the Preface it is assumed that the reader is already familiar with basic electronic components, particularly bipolar and MOS transistors, and also with basic electronic circuits and semiconductor physics. We will not deal with these important fundamentals in detail, but instead will attempt, throughout this text, to cover the subject areas in a way that is relevant to a design engineer who uses microelectronics as the means of producing new and innovative company products.

1.1 Initial history

The development of the first transistor, in 1948, was the start of the microelectronics evolution [1], although it was not until about the mid-1950s that suitable discrete devices became available for use in industrial products. Early commercial devices were germanium junction transistors, where the collector and emitter regions were diffused from opposite sides into the base region to form the pnp or npn construction. Although germanium has higher electron and hole mobility than silicon, 0.39 and 0.19 $m^2 V^{-1} s^{-1}$, respectively, for Ge, compared with 0.14 and 0.05 $m^2 V^{-1} s^{-1}$ for silicon, thus making it easier to achieve high-frequency performance, germanium transistors suffer from several inherent disadvantages, including the following:

1. poorer temperature characteristics than silicon, which results in unacceptably high leakage currents at elevated temperatures;
2. the inability to protect the exposed transition perimeter (the 'metallurgical

junction') between p-regions and n-regions which is very sensitive to pollution from oxygen or water vapour, as can readily be done by an impervious silicon dioxide layer in silicon technology; and consequently

3. the need to provide very effective hermetically sealed packages to give long-term reliability.

Details of germanium versus silicon and other III/V semiconductor compounds may be found in many standard physics and fabrication texts [2]–[5].

The gradual development of silicon technology, and in particular the development of planar fabrication methods wherein all the fabrication steps are made on one surface (plane) of the silicon wafer, see Chapter 2, removed the shortcomings of germanium and opened the way to integrating complete circuits on silicon as we now know them. Silicon was initially more difficult to process than germanium [6], which accounted in part for germanium being the first commercially used technology.

The first person to perceive the possibility of a fully interconnected monolithic circuit, rather than the fabrication of discrete devices which were then interconnected by wires or other separate means, was probably G. W. A. Dummer of the Royal Radar Establishment (now the Royal Signals and Radar Establishment, RSRE) at Great Malvern, England [7]–[9]; this was in 1952. However, the bulk of the development work was pioneered in the United States, particularly by Jack Kilby of Texas Instruments, followed by other companies such as Fairchild and Sprague. By 1962 small-scale integrated (SSI) packages with four logic gates per package were becoming widely available.

1.2 The continuing evolution

Since the early-1960s, the history of silicon technology has been one of continuing and rapid evolution based largely upon planar techniques. During the following two decades chip complexity increased from the initial small-scale integrated (SSI) capability through medium-scale integration (MSI) and large-scale integration (LSI) to the present status of very-large-scale integration (VLSI) capability. This is illustrated in Figure 1.1 [10], [11]. Since the silicon area required for transistors is less than that required for resistors and capacitors, the impact of this evolution on electronic circuit design was to encourage the use of active rather than passive devices. This change of emphasis has in its turn given greatly improved circuit performance, since more overall gain is readily available for new circuit configurations. The operational amplifier and the microprocessor (see Chapter 3) are both good examples of circuit design techniques which silicon technology has made possible.

The increase in capability shown in Figure 1.1 combines two effects, namely the growth in possible chip area and the decrease in possible on-chip feature size. The improvement in these two parameters is shown in Figure 1.2(a). At the same time,

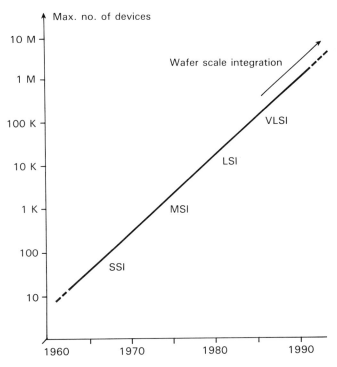

Figure 1.1 The classic Moore's Law graph, showing the increase in maximum possible capability per single IC chip since 1960. SSI is usually taken as tens of transistors, MSI hundreds of transistors, LSI thousands of transistors and VLSI tens of thousands of transistors.

performance and costs have improved, as indicated in Figure 1.2(b). These developments are still continuing, with ultra-large-scale integration (ULSI) and wafer-scale integration (WSI) now being considered [12], [13]. The starting point of fabrication, however, remains the growing of pure silicon, which has increased from the original pulled-crystal growth capability which gave wafers of only one- or two-inch diameter to the present wafer sizes of six-inch (150 cm) and, more recently, eight-inch (200 cm) diameter. Crystal defects and discontinuities in wafers which affect the subsequent formation of a defect-free epitaxial layer have also been reduced.

Overall characteristics such as those shown in Figure 1.2 usually represent the best technology for each parameter; the fastest available speed per gate at any given time, for example, is likely to be bipolar, whereas maximum chip area will be MOS. Nevertheless, these overall characteristics represent the trend in capability of all the various silicon technologies.

The two technologies which have been continuously developed over this period are of course bipolar and MOS (metal-oxide–silicon) [2]–[5], [13]. (As will

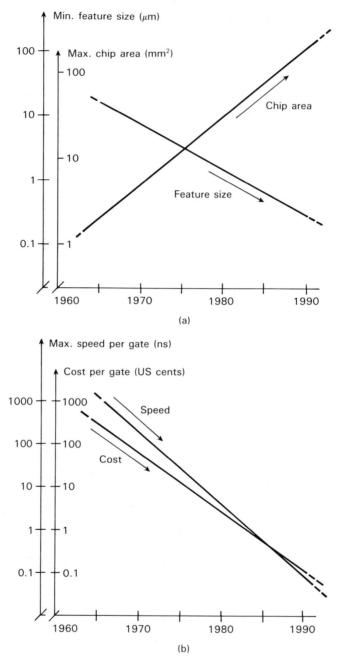

Figure 1.2 The improvement in specific parameters during the past three decades: (a) chip area and minimum on-chip feature size; (b) maximum speed and cost per logic gate.

be noted in Chapter 2, the terminology 'MOS' is something of a misnomer since the original metal-gate construction has now almost entirely been superseded by polysilicon-gate construction. The terminology 'POS' has never been adopted however.) Chapter 2 will consider the fabrication details of the families in each of these categories and their comparative performance. Very briefly, the significant features of each technology may be summarized as follows:

1. *Bipolar*
   - faster than MOS, but usually taking more power;
   - a higher transconductance g_m than is generally available from MOS (see Appendix D), thus giving better signal amplification performance;
   - fewer fabrication steps required compared with high-performance complementary MOS/CMOS fabrication techniques.

2. *MOS*
   - generally much higher impedance levels than in bipolar, thus reducing on-chip power dissipation but limiting the drive capability and increasing the time necessary to charge and discharge circuit capacitances;
   - device performance very strongly dependent upon the device geometry, particularly the length L and width W of the transistors (see Appendix D), thus making it readily possible to tailor device performance during the chip layout stage, a facility completely absent in bipolar design.

Bipolar has always been, and will always be, the fastest technology; MOS, and particularly CMOS, has always been the technology with the better on-chip packing density and lowest power consumption, although the latter may not remain dominant if CMOS operating speeds increase and if present developments in small-geometry bipolar circuits are pursued [14], [15].

However, except where absolute maximum speed capability is required, MOS has been and will remain the dominant technology for memory circuits, both read-only memory (ROM) and random-access memory (RAM) [16], [17]. Memory chips with over four million transistors are already available, with feature sizes on the silicon of under 1 μm. These capabilities and the decreasing cost per bit are shown in Figure 1.3. Further developments to reduce transistor channel length to about one-quarter of a micron have been announced, which is approaching the limits possible with optical lithography [11]–[13].

Outside those mainstream technologies are others which have been actively pursued. They include the following:

1. BiCMOS: which attempts to merge the advantages of bipolar and MOS on the one chip without incurring an intolerable increase in fabrication complexity and hence cost [18], [19].
2. Silicon-on-insulator technology: which attempts to replace the bulk silicon semiconductor substrate with an insulating substrate; sapphire has been

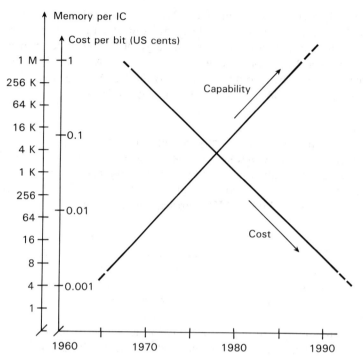

Figure 1.3 The increase in memory capability over three decades, driven largely by nMOS fabrication capability and with capacity approximately doubling every two years.

considered most frequently, giving silicon-on-sapphire (SOS) devices which have the attribute of possessing high radiation hardness [11], [20].
3. The use of III/V compounds instead of silicon, particularly gallium arsenide (GaAs) which offers a higher speed and higher temperature performance than silicon [21], [22].
4. Monolithic microwave integrated circuits, frequently using GaAs, but where great emphasis is placed upon transmission-line interconnect in order to handle signals without reflection in the GHz region [23], [24].

Even more recent research has been performed on GaAs-on-silicon, which attempts to overcome the cost and difficulties of growing perfect GaAs crystals from which the wafers are made by using proven silicon technology for the bulk substrate and forming, by epitaxial growth, a GaAs working layer on this silicon substrate [25]. Unfortunately, the lattice constant of GaAs is 0.563 nm compared with 0.543 nm for silicon; this 4% mismatch in crystal dimensions means that it is not possible to grow GaAs on Si directly without an unacceptably high percentage of crystal dislocations, and in order to reduce this defect density, some intermediate processing using other compounds is required to 'interleave' the two semiconductor materials.

We shall have no occasion to refer to the last two areas again in this text, but further mention of (1), (2) and (3), particularly (1), will be found in subsequent chapters. None yet compete on financial terms with conventional bipolar and MOS technologies, and are very unlikely to do so in the foreseeable future in view of the vast expenditure in research and development which has gone into these mainstream technologies.

1.3 CAD developments

The ability to be able to fabricate chips containing tens or hundreds of thousands of transistors would be meaningless without the ability to design the circuit in the first place. In the very early days of MSI and LSI where possibly five to seven different layouts were involved, the chip design was done by hand on large drawing boards, usually 100–200 times full size. The mask layouts were cut by hand in dimensionally stable plastic sheets (rubyliths) which were visually checked both individually and collectively by laying one on top of the other, being viewed on a light box which provided uniform underneath illumination.

This design procedure was highly inefficient and costly, taking many man-years of design effort as chips grew in size, and with many errors in the fine detail of the layout. Right-first-time designs were almost impossible to achieve, and automation was clearly necessary. However, up to the early 1960s, computer systems were inadequate, in terms of computational power, graphics capability and data storage, to handle all the information and geometric rules involved in translating a circuit requirement into a chip design. Peripherals such as large visual display units and large plotters still had to be developed.

As well as difficulties in the physical layout design, in the early days there were no means of simulating the circuit layout to check for correct functionality and performance before manufacture. Breadboard techniques could no longer be relied upon to simulate the circuit behaviour, since the performance of the devices on the chip and the interconnect capacitances were not the same as on a breadboard model. As chip complexities increased, these problems all escalated. It was therefore necessary for the computer industry to mature alongside the increasing maturity of the microelectronics fabrication. 'Correct-by-construction' would be the objective of the supporting computer-aided-design (CAD) developments, although even now this has not been a 100% achievement in all circumstances.

The first generation of CAD tools for IC design appeared in the late-1960s and early-1970s. These were aimed directly at the chip layout design; they were graphics machines with large storage capability which could manipulate geometric shapes and sizes on VDU screens, and provide design rule checking (DRC) of the resulting layouts to ensure that all the geometric design rules of minimum widths, spacings, overlaps, etc., were correct. These early machines did not, however, provide any electrical rule checking (ERC) to ensure that circuit design rules were correct.

The two main commercial contenders in this market were Applicon and Calma, although some very large industrial organizations also developed their own in-house systems based on mainframe computers. These machines grew in capability, and were sometimes referred to as 'polygon-pushers' because of their specific duties. Their drawback was that they remained graphics engines only, not being capable of other design activities such as simulation. Thus, although they grew in size their demise eventually came because of their limitations

This 'dinosaur' era, as it has been unfairly called, came to a close in the late-1970s with the introduction of workstations as we now know them. The particular contenders in the marketplace were Daisy, Mentor and Valid. These products were smaller, cheaper and more relevant for CAD purposes than those previously available, with software to provide not only graphics capabilities but also schematic capture to enable a circuit design to be entered into the CAD system, and some simulation capability. They were the first major step along the path of providing a truly hierarchical suite of CAD software, which would link all the design phases from

- initial circuit design and simulation,
- the required IC layout design,
- post-layout simulation ('back-annotation'),
- test vector generation,

to final manufacture and test.

The principal reason for the emergence of workstations was the availability of the microprocessor. Early microprocessors, such as the Intel 4040, were largely designed by hand, and it is therefore significant that their increasing commercial availability, at an economic price, was the means of producing CAD tools for the more rapid and successful design of the next generation of microprocessors and other LSI and VLSI circuits. This hand-in-hand evolution is still continuing: new CAD tools lead to better or more comprehensive chips, which in turn power more comprehensive CAD tools. However, the fabrication potential has until now always managed to keep ahead of the capabilities of the CAD to use its full potential, as indicated in Figure 1.4.

A later entry into this CAD market was of course the personal computer, the PC. Early PCs did not have the graphics or simulation capability to be very useful, being largely used for schematic capture purposes only, in order to enter a circuit design into a database. Subsequent layout and simulation activities could be transferred to a more powerful workstation or mainframe computer. However, the capabilities of the PC have continuously increased, particularly with the new versions introduced in the 1988–89 period, so that there is now a blurring of

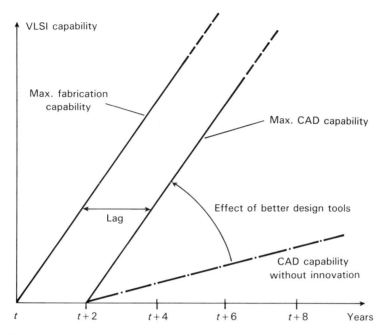

Figure 1.4 A graphical representation of the time lag that is present between maximum fabrication capability and the maximum design capability of the CAD resources. (Source: based upon [26].)

capability (and price) between workstations and PCs. Figure 1.5 illustrates these two contenders in the IC–CAD field.

Although vast strides have been made in the development and adoption of CAD for IC design purposes since the early-1970s (see Figure 1.6, for example), there still remains room for further evolution. It is generally considered that the software rather than the hardware is in need of most improvement; for example, a 'silicon compiler' which would allow a designer to enter a circuit description into the CAD system, and which would then automatically produce an efficient correct-by-design chip layout without designer intervention, is still not fully available. Indeed, the interpretation of the word 'silicon compiler' still gives rise to debate [26]–[29].

A more in-depth consideration of these CAD hardware and software resources will be the subject of Chapter 5. In particular, the desirability for these resources to be fully hierarchical by nature in order to handle all levels from initial design to final test, and if possible to be technology independent, will be stressed. Further information may be found elsewhere [26], [30]–[32].

10 *Introduction: the microelectronics evolution*

Figure 1.5 The workstation and personal computer hardware platforms for IC CAD activities. (Courtesy: (a) Mentor Graphics Corporation, OR; (b) Daisy/Cadnetix Inc., CA.)

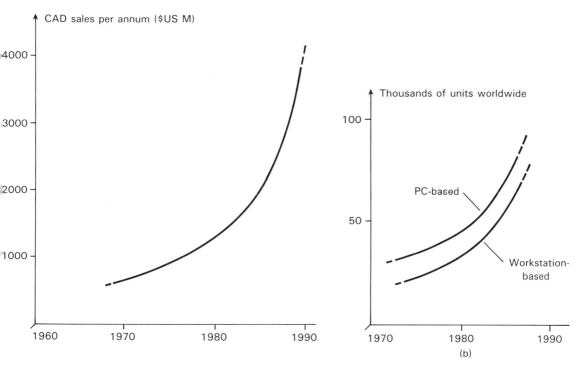

Figure 1.6 The increase in CAD resources since the early 1970s: (a) world-wide sales of CAD systems; (b) the total number of workstations and PCs in use for CAD purposes.

1.4 Why custom microelectronics?

Every new product designed by an original-equipment-manufacturer (OEM) is novel in some way or other. If it is an electronic product, then the circuit requirements will be unique to that product, and, provided that the product specification is within the capabilities of the existing state-of-the-art, these requirements can always be made from an appropriate assembly of standard off-the-shelf parts. So why custom microelectronics?

The fundamental and only reason for the introduction of custom microelectronics is to satisfy the need of the product designer who would like – or requires – one IC package to do exactly what he or she wants, rather than having to assemble together a miscellany of standard packages to do the job. This gives the following advantages:

- fewer packages to assembly in the final product,
- a smaller final product,
- better product reliability due to fewer package connections,

and also possibly some sales advantages over competitors. In simple terms, this is the *raison d'etre* for custom microelectronics, although for very specialized applications, such as in portable or satellite equipment, the size or weight advantages may be mandatory. As we will see in Chapter 4, other technical advantages, such as improved overall performance, may also accrue.

It has always been possible to commission an IC design which contains exactly what the product designer requires. This is known as 'full-custom' design, with the chip design being undertaken in every detail so as to produce the fastest and smallest circuit possible within the existing fabrication resources. The problems with this approach are cost and time to design. Final fabrication costs are small compared with these design costs. Before the advent of good CAD tools a full-custom LSI chip could take two or three years to design and finally manufacture, with a low guarantee of correct circuit operation without one or more design iterations. Hence, this design route was only relevant for military or other non-commercial applications when cost was not a prime consideration; for the commercial market more viable means of production were needed.

It was, however, clear at a very early stage that it was not always necessary to design every individual IC from scratch. This was evident in the development of the range of vendors' off-the-shelf 7400-series TTL circuits, where proven basic logic configurations were used in the expanding product range, with a confidential in-house library of circuit designs being continuously built up and improved. The concept of a *library of standard circuit designs* being available for custom design activities was therefore obvious.

Another early concept of the 1960s was that if a number of components or basic circuits were made available on a standard chip which could be mass-produced as a *stock production-line item* up to but not including the final interconnections between all these on-chip elements, then all that would be needed to produce a unique chip would be the final interconnection design and its fabrication. In essence, all the discrete devices or logic gates with which a designer was familiar could be on the chip, with only the interconnections as the principal custom design activity.

Both the above custom approaches are directly targeted at saving design time, and therefore costs, by using already-designed data or already-designed chip layouts. It was only by using such pre-designed resources that custom ICs became economically viable for the OEM designer in a competitive market, but to gain these advantages, supporting CAD was also vitally necessary.

Figure 1.7 indicates the interlocking relationships which were involved in this successful evolution of custom microelectronics [33]–[35]. This currently leaves us with the following three principal categories of custom microelectronics, namely:

(a) full-custom ICs, where the chip layout is fully optimized ('hand-crafted') for its particular duty, and does not rely heavily upon pre-designed circuit elements – this expensive design methodology is dying out with the increase in capabilities of the two further categories listed below;

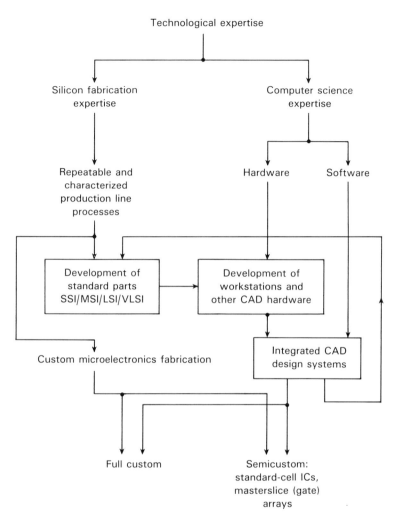

Figure 1.7 The relationships between silicon technology and computer science which have influenced the developments and viability of custom microelectronics. Full custom, however, remains a minority interest except where very large production volumes of a circuit are anticipated.

(b) standard-cell ICs, where the chip layout is compiled from pre-designed standard circuit elements ('building-blocks') held in the cell library of a CAD system – recent developments have allowed the CAD system to trim or 'fine-tune' these cell designs in order to optimize their specification for the specific application; and
(c) the pre-fabricated but uncommitted array IC usually, but possibly incorrectly,

(a)

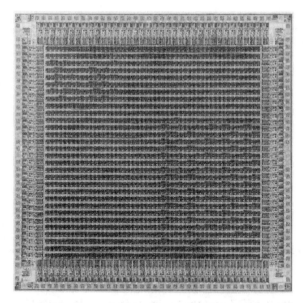

(b)

Figure 1.8 The chip architecture of two commercial custom chips, both from the same vendor for comparison purposes: (a) a 1.5 μm CMOS standard-cell design, with large standard blocks (macros) of read-only memory (ROM) and random-access memory (RAM) around the perimeter of the chip and 6000 gates of digital logic towards the centre; (b) a 1.5 μm CMOS gate-array design, with 3140 logic gates arranged in equispaced rows. (Courtesy of NEC Electronics, Mountain View, CA.)

referred to as a 'gate array', where only the final interconnection design on a fixed chip layout has to be custom designed and made.

Figure 1.8 illustrates the last two categories with the regular structure which typifies the latter category being evident.

1.5 Summary

Thus the coming of age of both the semiconductor industry and digital computer has given us not only standard off-the-shelf microelectronic products of increasing complexity, but also the means whereby custom ICs may be designed and fabricated in relatively small quantities at an economic overall cost. The following chapters will deal in greater detail with the technical aspects of this subject, and, finally, an attempt will be made to quantify the many parameters involved in the cost equations.

Custom microelectronics has now been given wide publicity for over a decade. Many surveys of products and studies of potential user have been made in this period, and will continue to be made [27], [37]–[40]. However, with one or two noticeable exceptions, most of the products and the custom circuits developed up to now have been entirely digital, with no analogue capability. Analogue has been a slow starter in this field, although an increasing number of vendors are beginning to address this area [41]. Nevertheless, analogue capability and its adoption at present remains the greatest weakness, and perhaps the greatest need, in the custom microelectronics field, one that we shall have occasion to address specifically in later chapters.

As a final observation appropriate to this general introductory chapter, there are a number of available encyclopaedic reference books which give exceedingly good overviews of technological subjects, for example Ref. [42]. Readers are recommended to these publications which have very high standards of both acumen and accuracy.

1.6 References

(Note, the surveys marked with an asterisk are subject to periodic revision, usually annually or bi-anually. Readers may check these publications for the latest listed data.)

1. Kilby, J. S. 'Invention of the integrated circuit', *IEEE Trans. Electron Devices*, Vol. ED.23, July 1976, pp. 648–54.
2. Sparkes, J. J., *Semiconductor Devices: How they work*, Van Nostrand Reinhold, Wokingham, UK, 1987.
3. Till, W. C. and Luxton, J. T., *Integrated Circuits: Materials, Devices and Fabrication*, Prentice Hall, Englewood Cliffs, NJ, 1982.
4. Neudeck, G. W. and Pierret, R. F., *Modular Series on Solid State Devices*, Vols. 1–4, Addison-Wesley, Reading, MA, 1988/1989.

5. Sze, S. M., *Physics of Semiconductor Devices*, John Wiley, NY, 1981.
6. Thomas, R. C., 'Czochralski silicon', *IEEE Trans. Electron Devices*, Vol. ED.31, November 1984, pp. 1547–9.
7. Noyce, R. N., 'Microelectronics', *Scientific American*, Vol. 237, September 1977, pp. 63–9.
8. Dettmer, R., 'Prophet of the integrated circuit', *IEEE Electronics and Power*, Vol. 30, 1984, pp. 279–81.
9. Roberts, D. H., 'Silicon integrated circuits: a personal view of the first 25 years', *ibid.*, pp. 282–4.
10. Moore, G., 'VLSI: some fundamental challenges', *IEEE Spectrum*, Vol. 16, April 1979, pp. 30–7.
11. Dillinger, T. E., *VLSI Engineering*, Prentice-Hall, Englewood Cliffs, NJ, 1988.
12. Meindl, J. D., 'Ultra-large scale integration', *IEEE Trans. Electron Devices*, Vol. ED.31, 1984, pp. 1555–61.
13. Saucier, G. and Trile, J. (eds), *Wafer Scale Integration*, North-Holland, Amsterdam, 1986.
14. Andrews, W., 'New-generation ECL challenges GaAs in speed and power', *Computer Design*, Vol. 28, January 1, 1989, pp. 38, 39.
15. Vora, M., 'Fairchild's radical process for building bipolar VLSI', *Electronics*, Vol. 53, September 4, 1986, pp. 55–9.
16. Pugh, E. W., Critchlow, D. L., Henle, R. A. and Russell, L. A., 'Solid state memory development at IBM', *IBM Journal of Research and Development*, Vol. 25, 1981, pp. 585–602.
17. Dennard, R. H., 'Evolution of the MOSFET dynamic RAM – a personal view', *IEEE Trans. Electron Devices*, Vol. ED.31, 1984, pp. 1549–55.
18. Dettmer, R., 'BiCMOS – getting the best of both worlds', *IEEE Electronics and Power*, Vol. 33, 1987, pp. 499–501.
19. Cole, B. C., 'Is BiCMOS the next technology driver?', *Electronics*, Vol. 61, 4 February 1988, pp. 55–7.
20. Maly, W., *Atlas of IC Technologies: An introduction to VLSI processes*, Benjamin/Cummings Publishing, Menlo Park, CA, 1987.
21. Morgan, D. V., 'Gallium arsenide: a new generation of integrated circuits', *IEEE Review*, Vol. 35, September 1988, pp. 315–19.
22. Greiling, P., 'The historical development of GaAs FET digital IC technology', *IEEE Trans. Microwave Theory and Techniques*, Vol. MTT32, 1984, pp. 1144–56.
23. Howe, H., 'Microwave integrated circuits: a historical perspective', *ibid.*, 1984, pp. 991–6.
24. McQuiddy, D. N., Wassel, J. W., Lagrange, J. B. and Weisseman, W. R., 'Monolithic microwave integrated circuits: A historical perspective', *ibid.*, 1984, pp. 997–1008.
25. Dettmer, R., 'GaAs on Si: the ultimate wafer?', *IEEE Review*, Vol. 35, April 1989, pp. 136, 137.
26. Russell, G. (ed.), *Computer Aided Tools for VLSI System Design*, IEEE Peter Peregrinus, London, 1987.
27. Editorial Staff, 'Directory of silicon compilers', *VLSI System Design*, Vol. 9, March 1988, pp. 52–6.*
28. Ayres, R. F., *VLSI Silicon Compilation and the Art of Automatic Microchip Design*, Prentice-Hall, Englewood Cliffs, NJ, 1983.
29. Horbst, E. (ed.), *VLSI Design of Digital Systems*, North Holland, Amsterdam, 1985.
30. Editorial staff, *Semiconductor IC Yearbook*, Elsevier Scientific Publications, Oxford, UK, 1988.*

31. Rubin, S. V., *Computer Aids for VLSI Design*, Addison-Wesley, Reading, MA, 1987.
32. Russell, G., Kinniment, D. J., Chester, E. G. and McLauchlan, M. R., *CAD for VLSI*, Van Nostrand Reinhold, Wokingham, UK, 1985.
33. Freeman, W. J. and Freund, V. J., 'A history of semicustom design at IBM', *Semicustom Design Guide*, Summer 1986, pp. 14–22.
34. Rappaport, A. S., 'Technical evolution of the semi-custom design process', *ibid*, pp. 40–2.
35. Open University, *The Management Perspective: Microelectronics Matters*, PT504 Microelectronics for Industry, The Open University, Milton Keynes, UK, 1987.
36. Hinder, V. L. and Rappaport, A. S., 'Employing semicustom: a study of users and potential users', *VLSI System Design*, Vol. 8, May 1987, pp. 6–25.
37. Editorial Staff, 'Gate-array directory', *EDN*, Vol. 32, 25 June 1987, pp. 134–201.*
38. Editorial Staff, 'Survey of gate array and cell libraries', *VLSI Systems Design*, Vol. 8, November 1987, pp. 76–101.*
39. Editorial Staff, 'Buyers' guide to ASICs and ASIC design tools', *Computer Design*, Vol. 26, August 1987, pp. 67–136.*
39. Editorial Staff, 'Survey of CAE systems', *VLSI Systems Design*, Vol. 8, June 1987, pp. 51–77.*
40. Editorial Staff, 'Survey of IC layout systems', *ibid.*, Vol. 8, December 1987, pp. 63–77.*
41. Editorial Staff, 'Survey of analogue semicustom ICs', *ibid.*, Vol. 7, May 1987, pp. 89–106.*
42. Muroga, S., 'Very large scale integrated design', in Vol. 14 of the *Encyclopedia of Physical Science and Technology*, Academic Press, NY, 1987.

2 Technologies and fabrication

The semiconductor technologies used in custom microelectronic circuits are essentially the same as those used in standard off-the-shelf ICs. Indeed, it would be strange if this was otherwise since the custom IC market is still small compared with the global demand for standard ICs, and hence the cost of semiconductor developments is mainly recouped from sales of standard parts. There may have been a few exceptions to this, for example very large, high-tech companies engaged in the computer or instrumentation industry who require components at the leading edge of technology, but in general the normal customer for user-specific ICs (USICs) will be using proven technology in the ICs which he or she commissions from a vendor. (In this text we will use the terminology 'user-specific IC' (USIC) for a custom IC rather than the term 'application-specific IC' (ASIC), which has been widely used up to now. The reason is that 'ASIC' is becoming increasingly used for off-the-shelf VLSI ICs made for specific purposes, see Chapter 3, Section 3.7, which are truly application-specific but available to all. The alternative designation 'customer-specific' with the abbreviation 'CSIC' has also been suggested, but this – although technically very explicit – does not sound quite as nice as USIC.)

Chapter 1 looked back upon the broad developments in germanium, silicon and other semiconductor technologies, from which it is evident that silicon is the dominant technology both now and for the foreseeable future. Germanium technology, with its poor temperature characteristics (compared with silicon) due to its lower energy bandgap (0.72 eV for Ge compared with 1.12 eV for Si), will not be mentioned further.

The depth of treatment in the following sections is intended to be appropriate for readers for whom custom microelectronics may be a part of their system design activity, and who should therefore be aware of the methods of fabrication and the characteristics of the various alternative technologies. Full details of the physics and chemistry of fabrication and basic device design is not our purpose; for this in-depth information, which is necessary for those engaged in the manufacture of the ICs or design at the silicon level, reference should be made to more detailed texts in specific areas [1]–[8].

2.1 Bipolar silicon technologies

By definition, bipolar technology involves both hole and electron flow in the action of the active devices, unlike unipolar (MOS) technology which is based upon the controlled flow of majority carriers only [1], [3], [4], [7], [8].

The three principal devices involved in bipolar technology are resistors, diodes and npn or pnp junction transistors. The basic fabrication process involved is a planar process, involving the diffusion or implantation of areas of doped silicon into a silicon substrate so as to form appropriate p-type or n-type areas. 'Planar' means that all the stages of fabrication are performed on one surface (plane) of the silicon wafer.

Figure 2.1 shows the basic fabrication method of creating doping areas in silicon. The substrate may be either n-type or p-type material. Between the two types of silicon a pn junction diode is created, which if the applied voltages are appropriate will conduct across the junction in one direction or act as a non-conductor (barrier) between the two regions.

Resistors and diodes may be formed by making connections as shown in Figure 2.2(a) and (b). Junction transistors require a further window and diffusion stage to form the third of the pnp or npn regions, as indicted in Figure 2.2(c). In the case of the npn transistor, the flow of electrons from emitter to collector is determined by the positive voltage applied between the base and emitter, which injects holes into the base region.

These are the basic fabrication methods used in all bipolar technologies. One further major step may be involved in modern bipolar technologies and in most MOS technologies, this being an expitaxial process that is, an epitaxial process which is performed on the basic silicon substrate [1], [2]. This is the growing of an epitaxial layer ('epi. layer') of very precisely doped silicon on top of the starting

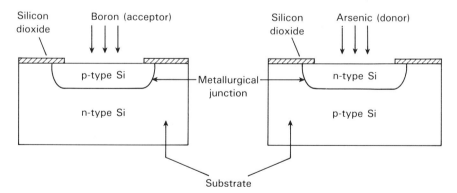

Figure 2.1 The planar fabrication process, with the diffusion of areas of doped silicon into the substrate made through a 'window' in the protective silicon dioxide layer so as to form appropriate p-type or n-type areas.

Technologies and fabrication

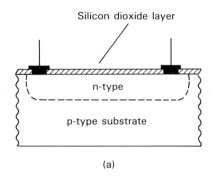

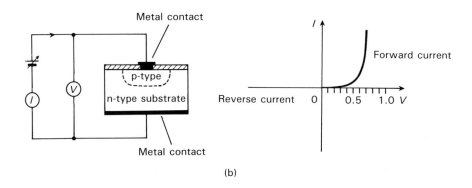

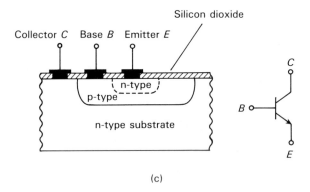

Figure 2.2 Basic concepts of integrated resistor, diode and bipolar junction transistor planar construction: (a) resistor – the length, area and diffusion density determine the effective resistance; (b) simple pn junction diode, with connections made to each region; (c) npn transistor, with connections made to all three regions.

Bipolar silicon technologies

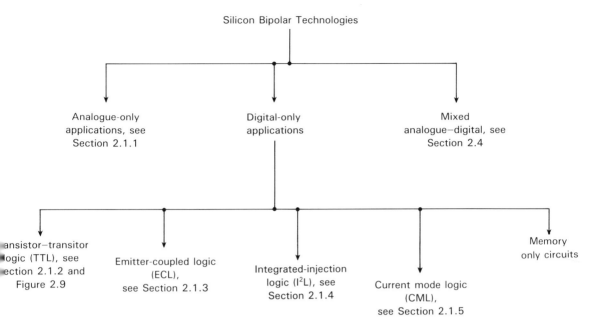

Figure 2.3 Bipolar silicon technologies. The mixed analogue–digital area is, however, subject to considerable evolution at present. Memory technologies will be reviewed separately in Section 2.3.

silicon substrate; it is in this thin, precise epi. layer, usually n-type silicon on a p-type substrate, that all the principal steps of device fabrication and isolation are done rather than in the bulk substrate. This feature will be noted in most of the following illustrations, and is generally necessary where maximum performance is required. The epi. layer should also contain fewer crystal structure defects than the bulk substrate, hence improving the production yield as well as providing more accurately controllable doping levels.

The present range of silicon bipolar technologies is given in Figure 2.3. The use of discrete transistors and other devices has not been included, although for many high-power and/or high-voltage applications discrete semiconductors rather than monolithic circuits still remain appropriate.

2.1.1 Analogue ICs

Analogue bipolar circuits employ both npn and pnp transistors, together with junction diodes, resistors and sometimes small-value capacitors. Precise fabrication and layout details depend very largely upon the individual IC manufacturers, and although not generally publicized in detail are based upon the fundamentals introduced in Figure 2.2

There are, however, two distinct methods of achieving isolation between the devices fabricated on a wafer. The first, and oldest, method is the junction-isolated process, which relies upon the presence of a reverse-biased pn junction barrier between adjacent devices to ensure isolation. The second, more recent, method is to use some method of silicon dioxide isolation, which provides an insulating barrier between devices.

Figure 2.4 illustrates the cross-sectional arrangement of a bipolar transistor with junction isolation. This process is sometimes termed the 'SBC (Standard Buried Collector) process', since it requires a buried area of n^+ silicon to be first made below each transistor on a p-type substrate, which will form a low-resistance sub-collector contact area for each transistor. The seven process fabrication steps are as follows; appropriate masks are used to define each window, the silicon dioxide protecting layer at the surface being re-formed between steps when necessary:

1. Open window (mask 1) in the SiO_2 (silicon dioxide) protecting layer on the lightly-doped p-type substrate and diffuse in an n^+-type buried sub-collector.
2. Remove remaining SiO_2 and grow a lightly doped n-type silicon epitaxial layer over the whole wafer.
3. Re-grow SiO_2 and open new window (mask 2) to diffuse in a deep p^+-type trench ('moat') around the perimeter of the device to form an isolated island.
4. Re-grow SiO_2 and open new window (mask 3) to diffuse in the p-type base area, leaving a thin n-type collector width in the epi. layer between base and buried sub-collector.
5. Re-grow SiO_2 and open new windows (mask 4) to diffuse in the n^+-type transistor emitter area and also the collector contact.

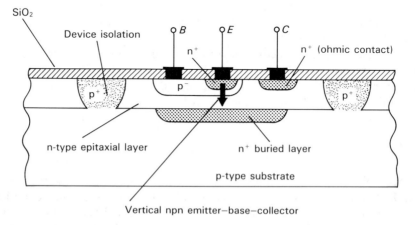

Figure 2.4 The cross-section fabrication details of an SBC bipolar npn junction transistor with junction-isolation (not to scale).

6. Re-grow SiO₂ and open new windows (mask 5) for contacts.
7. Cover with interconnection metal and etch away (mask 6) to form final required interconnect pattern.

With the substrate of this structure connected to the most negative voltage of the circuit, the pn junction around each device formed by the p-type moat is effectively reverse-biased, and therefore provides the device isolation. The n^+ buried layer also prevents the formation of parasitic vertical pnp transistors which may otherwise occur between base, collector and substrate regions.

Whilst still very widely used, this SBC process has drawbacks in that the active area of the transistor is only the area below the emitter, which is within the base area, and the base area is itself within the collector area. Also, due to lateral diffusion, the minimum width of the p^+ isolation moat will be about twice the depth of the epitaxial layer, so that in total the active area of the transistor is often less than 5% of the total device area. Figure 2.5 illustrates this shortcoming, the active area under the emitter being only 2.67% of the total device area.

The use of silicon dioxide isolation instead of pn junction isolation is an example of MOS fabrication developments being applied to bipolar technology. This allows a decrease in transistor area and hence an increase in device density and speed.

The cross-section of a typical oxide-isolated npn transistor fabrication is shown in Figure 2.6. The first fabrication steps of making the buried n^+-region and the growth of the epitaxial layer remain as previously described. The next three steps, however, create the SiO₂ isolation boundaries as follows:

1. A thin SiO₂ layer followed by a thick protecting Si₃N₄ (silicon nitride) layer is grown over the whole surface, the Si₃N₄ then being lightly oxidized. (If Si₃N₄ is formed directly on the epitaxial surface it causes surface damage due to different thermal expansions. Hence the need for the thin intervening SiO₂ layer.)
2. Windows where the oxide-isolation boundaries are required are cut through the Si₃N₄ and SiO₂, followed by an etch which is allowed to dissolve away about one-half the thickness of the exposed epitaxial layer.
3. A boron implant to form a p^+-region is then made on the surface of all these etched boundaries, followed by a long, high-temperature cycle to grow a thick SiO₂ layer in these boundaries.

Since SiO₂ occupies about twice the volume of the silicon from which it is produced, the effect of the last step is to cause the SiO₂ to grow deeper into and fill the cuts in the epitaxial layer. The SiO₂ growth finally reaches the silicon substrate, with the p^+ implant being driven ahead of it.

The subsequent processing stages window and implant the n^+- and p-regions into the epitaxial layer after removal of the thick Si₃N₄ layer. However, unlike the SBC process where the emitter region is within the base and collector regions, the

Minimum feature size	5 μm
Worst-case alignment tolerance between levels	2 μm
Epitaxial-layer thickness	10 μm
Collector-base junction depth	5 μm
Emitter-base junction depth	3 μm
Minimum emitter-to-collector spacing at surface	5 μm
Minimum base-to-isolation spacing at surface	5 μm
Minimum collector contact n^+ diffusion to isolation spacing	5 μm
Minimum collector contact n^+ diffusion to base spacing	5 μm
Buried-layer diffusion (both up and down)	2 μm
Buried layer to isolation spacing	5 μm
Lateral diffusion = vertical diffusion	

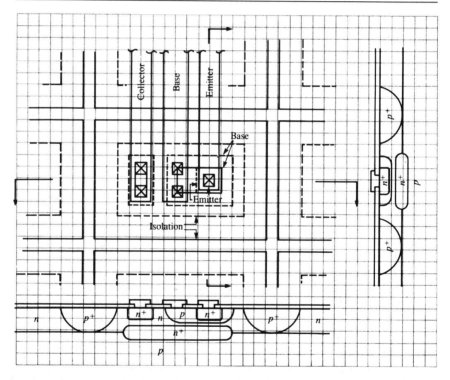

Figure 2.5 An example minimum area bipolar transistor layout on a 5 μm grid with a given feature size of 5 μm. The solid lines in the top plan view represent the edges of the fabrication mask patterns, with the dotted lines representing the final spread of emitter, base and collector diffusions assuming equal vertical and lateral diffusion distances. Note that the isolation area occupies over 60% of the total transistor area, illustrating the pressure to improve the means of isolation in all fabrication processes. (Reprinted with permission from [2].)

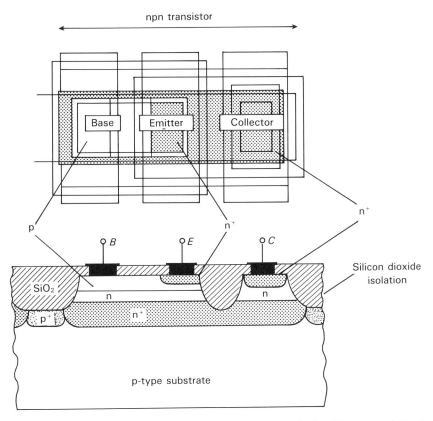

Figure 2.6 The cross-section fabrication details of a typical oxide-isolated bipolar junction transistor [9]. The base and emitter regions may be interchanged from as shown here by some manufacturers (not to scale).

base, emitter and collector regions of the oxide-isolated junction transistor are 'side-by-side'. Also, in the detailed fabrication steps the emitter, base and contact regions are 'self-aligned', that is, they are positioned by windows made in the SiO₂ rather than relying upon the accuracy and positioning of separate masks of the exact areas required – this is a carry-over from CMOS processing and will be mentioned later. The total effect of this method of fabrication is to make the overall transistor size considerably smaller than with the SBC process, giving higher on-chip packing density and better performance.

Different manufacturers have different variations of this oxide-isolated process. Polysilicide – a low-resistance form of polysilicon – may be used instead of metal for the connection to the base, emitter and collector regions, another technique which has been taken from CMOS technology.

Yet another way of providing device isolation is now available for high-performance bipolar structures. This is 'deep-trench' isolation, as illustrated in

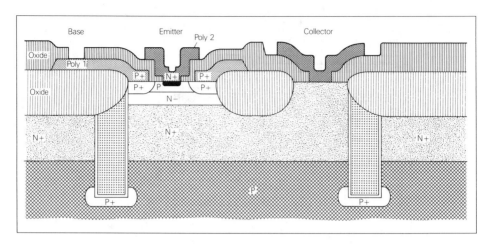

Figure 2.7 The HE bipolar fabrication process of Plessey Semiconductors, with 1 μm double-polysilicon and trench isolation. (Courtesy of Plessey Research and Technology, UK.)

Figure 2.7. The deep isolating trenches which reach down to the substrate are cut by a reactive-ion etching process, which produces a sharp etch with minimum side-spreading. The surfaces of these trenches are then oxidized to produce the isolating SiO_2, and the remaining volume is filled with polysilicon. This isolation process is done after an Si_3N_4 protective layer has been applied and windowed, as in the oxide-isolated process described above.

The above illustrations have each been npn devices. The alternative polarity pnp devices are available, but do not inherently have the same maximum performance as npn since the majority carriers are holes rather than the electrons in a pnp device [7]. In the SBC process the pnp transistors are frequently lateral devices, with the pnp action lying along the plane of the device rather than vertically as in Figure 2.4; in oxide-isolated and deep-trench fabrication the n^+ buried layer forms part of the base rather than the collector.

The remaining devices required for analogue circuit design, namely diodes, resistors, and possibly capacitors, are all available in silicon technology. Diodes are frequently formed by using one of the pn junctions of an npn transistor, either the collector–base junction with possibly the base shorted to the emitter, or the base–emitter junction with possibly the collector shorted to the base. Both have different characteristics due to the different doping densities of the transistor regions, the fastest being the base–emitter junction with the collector shorted to the base [1]. In general, therefore, the silicon area occupied by a diode is the same as that of a transistor.

Resistors and capacitors are formed by appropriate areas in the structure. Resistors may be made in the polysilicon layer, with values from 20 Ω to 20 kΩ with good matching but poor absolute-value tolerances. Alternatively, a p-type diffusion can be used. (In MOS technology a permanently biased on field-effect transistor may be used as a resistor, (see Section 2.2), which occupies far less silicon area than bipolar technology methods.) Capacitors may be formed using polysilicon and metal as the two capacitor plates, or, alternatively, the depletion layer capacitance of a reverse-biased pn junction may be employed. The latter is voltage sensitive, and can only provide very small capacitance values, generally less than 1 pF. In general, the use of capacitors is avoided if at all possible. Figure 2.8 illustrates the range of bipolar components which are available from a typical manufacturer for general-purpose analogue circuit use, using in this case a polysilicon-emitter, oxide-isolation process.

At present bipolar technology is dominant for analogue applications because of its superior performance compared with MOS technologies. For this reason we have introduced the basic bipolar fabrication details here under analogue considerations. Further fabrication details may be found in many more specialized texts [1], [2], [3], [9], [10]; specific custom microelectronic products using bipolar technology will be covered later.

However, the recent developments in bipolar technology, with emphasis upon reduced device size and increasing packing density, may in the future present a strong challenge to the present dominance of CMOS for LSI/VLSI digital applications. Therefore, the bipolar fabrication concepts introduced here may need to be much more widely appreciated than is the case at present.

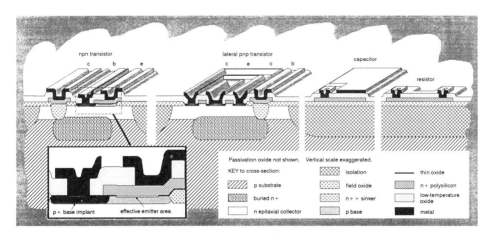

Figure 2.8 Typical structures available in a commercial 2 μm geometry bipolar process. The lateral pnp transistors have much lower performance than the vertical npn transistors, typically f_T = 25 MHz compared with 5 GHz. (Courtesy of STC Components, UK.)

28   *Technologies and fabrication*

2.1.2 *Transistor–transistor logic*

Transistor–transistor logic (TTL) represents the original and still most widely used technology for digital-only applications, being the standard off-the-shelf product for original-equipment-manufacturer (OEM) use from late 1950s. It has, however, evolved from its original 'standard' form into many variants which provide enhanced performance capabilities. Figure 2.9 indicates these developments.

Figure 2.10(a) shows the original standard TTL logic configuration, employing a multiple-emitter transistor as the input stage and a totem-pole output stage giving active pull-up to V_{CC} and active pull-down to 0 V for the logic 1 and logic 0 output states respectively. The principal disadvantage with this standard circuit is its high power consumption, about 10 mW per gate, with a sharp current spike when switching from one output logic level to the other because of a momentary conducting path from V_{CC} to 0 V whilst the two output transistors are changing state [1], [7].

Variations of this circuit to provide higher speed or lower power consumption

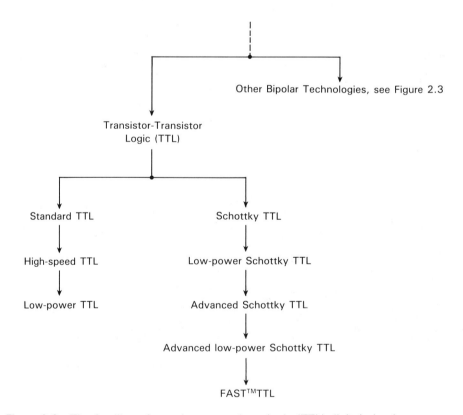

Figure 2.9 The families of transistor–transistor logic (TTL) digital circuits.

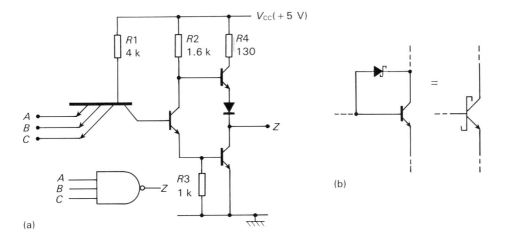
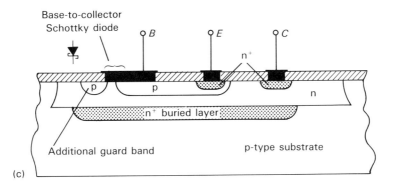
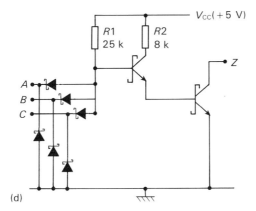

Figure 2.10 TTL circuit details: (a) standard SSI TTL, with a propagation delay of about 10 ns and a maximum frequency of operation of about 25 MHz; (b) the Schottky diode used for the prevention of collector saturation; (c) the fabrication details of (b); (d) typical low-power Schottky TTL NAND gate; cf. the resistor values with those of (a).

by varying the circuit resistor values were produced. However, the principal modification that improved efficiency came with the adoption of Schottky diodes to eliminate transistor saturation and hence increase switching speeds. These variants are collectively known as Schottky-clamped TTL.

A Schottky diode is a rectifying junction formed between aluminium and lightly doped n-type silicon [1], [3], [7]. The aluminium may be considered to act as weak p-type dopant at the surface of the silicon, resulting in a junction with pn properties. However, conduction is almost entirely due to electron emission from the silicon into the aluminium, which gives rise to junction characteristics which differ from normal pn junctions, particularly in the negligible storage time and low forward volt-drop, typically 0.35 V compared with 0.6–0.7 V for normal pn junctions. However, its very low reverse breakdown voltage precludes this form of junction diode from being used as a rectifying device. Note that a junction between aluminium and more heavily doped n-type silicon becomes a normal ohmic connection and not a pn junction, since the p-type dopant effect of the aluminium is now swamped by the higher n-type dopant.

The application of Schottky diodes in bipolar transistor circuits is now very widespread, both as separate diodes for isolation and unidirectional purposes and between base and collector of npn transistors where they can be incorporated with very low silicon area overheads in order to prevent collector saturation. Figure 2.10(b) shows the latter, with Figure 2.10(c) showing typical fabrication details. As the collector voltage falls and approaches the saturation value, the Schottky diode between base and collector will commence to conduct, preventing the collector potential from dropping further to its $V_{CE(sat)}$ value, excess input current being diverted from the transistor base to the collector as soon as the Schottky diode conducts.

A typically low-power Schottky TTL circuit is given in Figure 2.10(d), which shows Schottky diodes used as the gate inputs rather than the multiple-emitters shown in Figure 2.10(a).

The other TTL families listed in Figure 2.9 generally follow the principles shown in Figure 2.10, with individual manufacturers' variations. FAST™ TTL, for example, is a second-generation, low-power Schottky family, which employs oxide-isolation to increase the speed and packing density of the digital circuits.

2.1.3 *Emitter-coupled logic*

Emitter-coupled logic (ECL) represents the fastest available silicon-based technology for digital circuits, surpassed only by modern gallium-arsenide (GaAs) developments, see Section 2.5. However, hand-in-hand with this very-high-speed capability goes high power dissipation, which limits the number of gates per IC, and small voltage swings between logic 0 and logic 1 which cause interfacing complexity to other items. It is not therefore currently regarded as a VLSI technology and does not feature very significantly in the custom microelectronics area.

Figure 2.11 shows the circuit topology of a typical commercial ECL gate. Transistors $T1$ and $T2$ form the main component, an emitter-coupled ('long-tail pair') circuit, where constant current is switched from one side to the other depending upon whether an input voltage is less than or greater than the reference voltage on the base of $T2$. The resistor values $R1$, $R2$ and R_E are chosen such that all transistors are working in the non-saturated (active) mode, so that maximum switching speed results. Voltage swings are also small, typically 0.8 V, so that charging and discharging times of circuit capacitances are fast, these times also being minimized by the low circuit resistance values. Emitter-followers $T3$ and $T4$ provide low output impedance drive, and adjust the output voltage levels so as to be appropriate to drive further similar circuits.

Because of its small voltage swings and high power dissipation, which requires

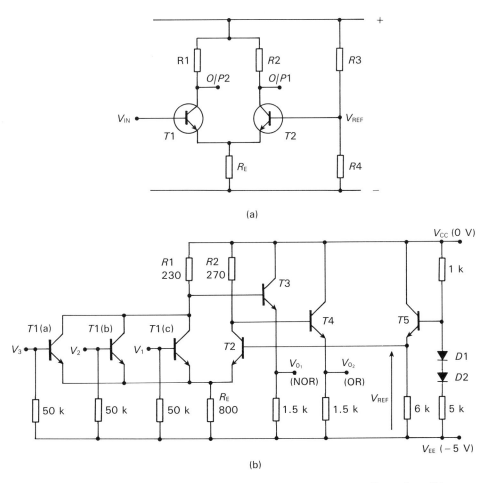

Figure 2.11 Emitter-coupled logic (ECL): (a) the basic circuit configuration; (b) a commercial 3-input gate with NOR and OR outputs.

very complex cooling of large circuits, ECL is not widely used except in the very fastest computers and specialized high-speed applications.

2.1.4 Integrated-injection logic technologies

The original integrated-injection logic (I^2L), alternatively referred to as 'merged-transistor logic' (MTL), was first developed in the early-1970s in an attempt to develop a bipolar logic family with reduced power consumption and good packing density. The full range of technology developments is indicated in Figure 2.12. It has never been used for standard off-the-shelf logic ICs, but has been adopted for memory applications and some commercial custom IC products.

The fabrication techniques used in I^2L are similar to those already covered, but the circuit configurations are completely different [1], [3], [11], [12]. Figure 2.13(a) shows the basic I^2L gate configuration. Transistor $T1$ is a lateral pnp device and $T2$ is vertical multiple-collector device, the fabrication of these two transistors being merged into one, as shown in Figure 2.13(b). Junction isolation is used for isolation between separate gates.

The action of I^2L is as follows: $T1$, termed the 'injector', provides a constant

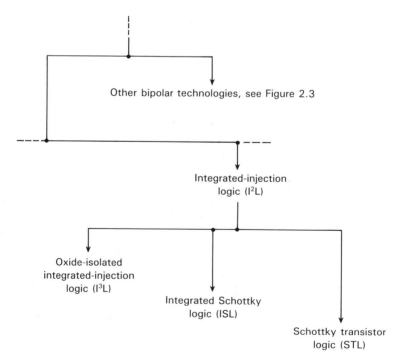

Figure 2.12 The development of integrated-injection logic (I^2L).

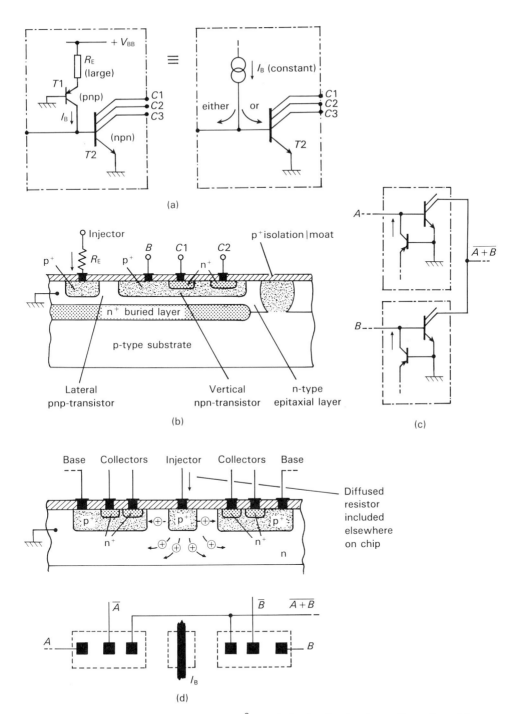

Figure 2.13 Integrated-injection logic (I^2L): (a) the basic circuit configuration; (b) fabrication details; (c) wiring of collector outputs to provide NOR logic; (d) merged fabrication for (c), using one common pnp injector transistor.

current which acts as the base input for $T2$ unless this current is sunk by a low voltage (logic 0) signal applied to the gate input terminal. When acting as the $T2$ base input this current switches $T2$ on, and allows each of the individual collector outputs to act as a sink (low resistance) to 0 V. In the absence of this base input, each collector becomes effectively open-circuit. Thus the circuit configuration inherently has an inverting action, acting on current sinks rather than our more familiar voltage levels. It is also a single-input/multiple-output logic gate, requiring specific logic functions to be made by commoning of outputs, for example as indicated in Figure 2.13(c) and (d).

The I^2L fabrication follows standard planar expitaxial methods. However, the vertical npn transistor is operating in inverse mode compared with the usual mode shown in earlier diagrams, with the n-type epitaxial layer acting as the emitter of $T2$ rather than as the collector. This allows the multiple collectors to be formed by separate n-type diffusions into the p-type diffusion region.

The elimination of all resistors (apart from R_E) means that the packing density of I^2L gates can be very high. Speed performance can be varied by increasing or decreasing the injector current I_B (varying the value of R_B), trading off speed against power dissipation, for example 0.01 μs gate delay at 100 μW dissipation or 10 μs at 0.1 μW dissipation.

However, although the packing density and speed–power product of I^2L is high, its maximum speed performance is not as good as other bipolar technologies. This is due to saturation of the npn transistor plus its low gain due to using the lightly doped epi. layer as the emitter rather than the collector, and also to limitations in the lateral pnp transistor. One improvement is to use oxide-isolation (see Figure 2.6) around each device rather than the pn junction isolation shown in Figure 2.13; the terminology 'isoplanar I^2L' or I^3L may sometimes be used for this development. Additional advantages are the self-aligned base and collector regions, which with the oxide-isolation borders being brought in to meet the contact diffusions means that smaller device areas, and hence higher performance, can be achieved [3]. Nevertheless, the maximum available speed performance remains inferior to Schottky TTL.

Integrated Schottky logic (ISL) is an attempt to combine the best features of both I^2L and Schottky TTL. Figure 2.14(a) shows the basic circuit configuration, with the fabrication details being shown in Figure 2.14(b).

Unlike the original integrated-injection logic, the current to the base of the npn transistor $T2$ is supplied through some external polysilicon resistor rather than through the merged lateral pnp device $T1$. The latter is now acting as a clamp across the collector of $T2$ to prevent $T2$ saturation. Additionally, the vertical transistor $T2$ has been re-inverted compared with the I^2L fabrication, so that the epi. layer is once again the collector of the npn transistor. Finally, Schottky diodes have been added to the multiple-collectors to improve isolation between each collector.

The effect of these differences in fabrication is to maintain the low power consumption of I^2L and I^3L, but to improve the maximum speed performance by a factor of two or three. A trade-off between speed and power consumption is still

Bipolar silicon technologies 35

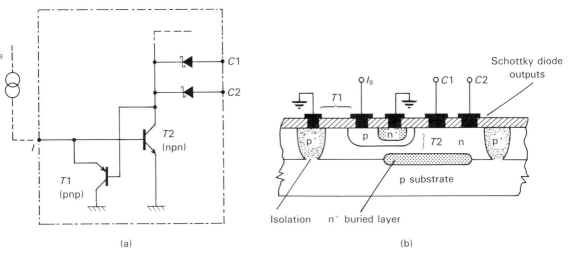

Figure 2.14 Integrated-Schottky logic (ISL): (a) the basic circuit configuration; (b) fabrication details.

available. However, the packing density is not quite as good, although it is still considerably better than Schottky TTL. The Schottky-isolated output swings are also reduced by about 250 mV by the Schottky diodes in comparison with corresponding I^2L and I^3L circuits.

The last of these closely related families of bipolar logic circuits is Schottky-transistor logic (STL), which in some aspects may be viewed as linking back again to Schottky TTL, thus completing a spectrum of bipolar fabrication technologies.

Figure 2.15 gives the basic circuit and fabrication details of STL. The current input comes from a resistor source and the multiple collectors have individual Schottky-diode isolation as in ISL, but the pnp lateral transistor clamp to prevent saturation of the main npn transistor has now been replaced by a Schottky diode as used in Schottky TTL circuits (see Figure 2.10(b)).

The final result is a logic family which maintains the small size of ISL, but offers about twice the gate propagation speed capability. Two extra mask stages are, however, required. Hence in total we have I^2L, I^3L, ISL and STL, all of which have better packing densities than Schottky TTL and are therefore appropriate candidates for large-scale integration, but which in general cannot provide the same speed performance as TTL. At present there is little reason to use these bipolar technologies in custom microelectronics since CMOS (see later) can usually offer cost and performance advantages. However, these injection technologies will be seen to employ the same fabrication steps as Schottky TTL, and it is therefore possible to fabricate both types on the same chip so as to combine the best of both worlds should this be required.

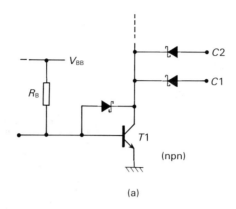

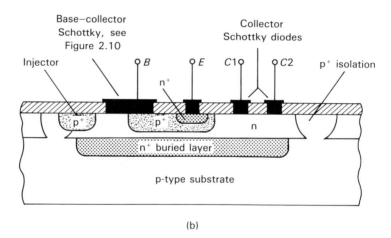

Figure 2.15 Schottky-transistor logic (STL): (a) the basic circuit configuration; (b) fabrication details.

2.1.5 Collector-diffusion isolation

A bipolar fabrication process which is completely dissimilar to the previous n-type epitaxial processes is collector-diffusion isolation technology. This process was originally invented for digital applications by Bell Labs., and has been developed and used extensively in custom microelectronic applications, both digital and analogue, by Ferranti Electronics, now part of GEC–Plessey Semiconductors, UK.

The basic collector-diffusion isolation (CDI) process is shown in Figure 2.16(a). The substrate is p-type silicon upon which is first formed a low-resistance, n^+ buried layer. A thin p-type epitaxial layer is then grown on this buried layer to form the eventual base regions. Device isolation is then made by means of deep n^+ diffusion through the p-type epi. layer to form isolated islands in this epi. layer.

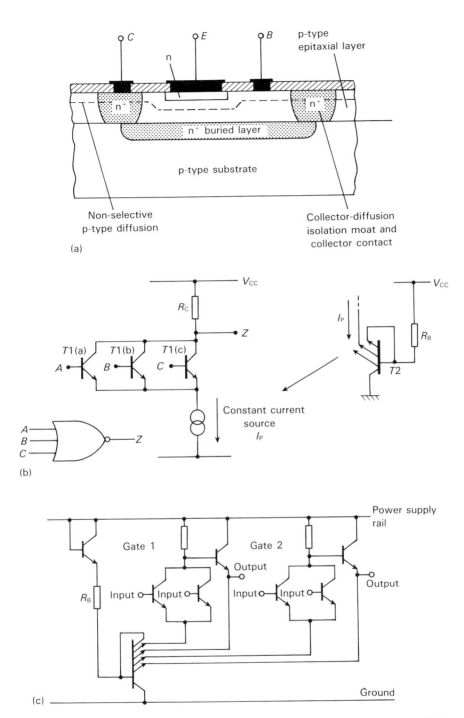

Figure 2.16 Current-mode logic (CML): (a) the collector-diffusion isolation (CDI) fabrication; (b) the basic CML circuit configuration; (c) the multiple-emitter current source used to supply separate gates and also output emitter followers.

This device-isolation moat also forms the collector of the final transistor, thus surrounding the base region on all sides by the collector. A shallow n^+-type diffusion or implantation into the epi. layer to form the emitter is then performed, followed by the final emitter, base and collector contacts to compete the fabrication steps [2].

This process is therefore an extremely simple fabrication process, and can produce high-performance, small-silicon-area devices. The relatively large collector–base and collector–substrate capacitances to some degree limit performance, but this is mitigated by low collector resistance values and very small vertical n–p–n distances.

The CDI process has principally been used in current-mode logic (CML) circuit configurations. Figure 2.16(b) illustrates the basic circuit configuration. Unlike ECL, where a constant current is switched from one transistor path to another (see Figure 2.11), in CML a preset current is allowed to flow or not to flow through the main switching transistors $T1$, the value of this current being fixed so that saturation of the switching transistors does not occur. Thus no Schottky diodes need to be fabricated. A low-output voltage swing, typically 0.8 V, with very fast switching times, is achieved. Level-shifting between these on-chip voltage levels and the outside world is required.

The preset current source in the emitter circuit of the switching transistors is also very efficiently provided by multiple-emitter transistors, each emitter serving a separate logic gate. As shown in Figure 2.16(c), each current source is a current mirror, where the current allowed in each emitter is controlled by the value of R_B. The multiple-emitters are fabricated in the (normal) collector region of Figure 2.16(a), with the emitter used as the collector in this inverted mode of operation; a particular feature of the CDI fabrication is that good inverted transistor performance is available due to the heavy p-type doping of the diffusion isolation region. The gate power consumption can be adjusted over a very wide range by choice of the constant current value, giving a trade-off between power and speed ranging from 10 μW per gate at a speed of a few MHz to 1 mW at 100 MHz. This is comparable to the power dissipation of many CMOS circuits when operating at high speed. Also, the supply voltage tolerance of CML is extremely good since all on-chip voltages track together. Operating voltages as low as 1 V can be provided.

2.1.6 *Bipolar summary*

The basic processes involved in bipolar fabrication are well established, the key steps in most modern processes being the use of a *buried layer* followed by an *epitaxial layer*. Isolation between devices, formerly using pn junctions, has been largely superseded by *silicon dioxide* or *deep-trench* isolation in order to increase packing density and speed. Collector-diffusion isolation (see Figure 2.16) remains an odd man out, not receiving so much industrial support as other bipolar fabrication techniques in spite of certain advantages.

Schottky-diode clamping now features strongly in all high-performance bipolar circuits except ECL and CML in order to prevent transistor saturation. In ECL and CML saturation is inherently avoided by control of the circuit current values, thus saving the fabrication steps necessary to produce the lightly doped n-type silicon/aluminium Schottky interface. The use of *polysilicide interconnect* is increasingly evident.

The several developments of integrated-injection logic have not found widespread use since, in general, higher speeds are available with other bipolar technologies, and higher packing densities with comparative performance are available from CMOS (see Section 2.2.3). The use of lateral pnp transistors with their limiting performance (see Figure 2.13) must now be considered obsolete.

In bipolar technology, the principal fabrication problem limiting yield continues to be collector-to-emitter leakage or short-circuits. This is compounded by the desire to produce very narrow base widths in the fabrication process (see Figure 2.4, for example), but process tolerances or defects can cause this width to be locally reduced or bridged. This critical fabrication feature is not present in MOS technology, which thus has a fundamentally higher yield than bipolar technology.

Further details of bipolar fabrication may be found in more specialized texts [1]–[4], [6]–[10]. A comparison of the performance of bipolar technologies and MOS technologies will be given in Section 2.6.

2.2 Unipolar silicon technologies

Unipolar technology is based upon the use of field-effect transistors (FETs) as the active devices, in which the controlled current flow is by majority carriers only. In n-channel devices the flow of carriers between source and drain electrodes is by electrons, and in p-channel devices the flow is by holes. The potential applied to the third electrode, the gate, controls the number of free carriers flowing in the channel and hence the source-to-drain current. Since electrons have a higher mobility than holes, n-channel devices are inherently faster by a factor of about 2.5 than comparable p-channel devices.

The gate control of an FET may be by the following two distinct methods:

(a) where the gate potential controls the effective conducting cross-sectional area and hence resistance of the source-to-drain channel by creating a reverse-biased pn junction depletion layer in the channel; or
(b) where the gate potential capacitively enhances or depletes the number of majority carriers in the channel.

The former is usually termed a *junction FET* ('JFET'), and can be either an n-channel JFET or a p-channel JFET. The latter may be termed an *insulated-gate FET* ('IGFET') because the gate electrode is separated from the channel by a thin insulating layer of silicon dioxide. Since in the original IGFET constructions the

40 Technologies and fabrication

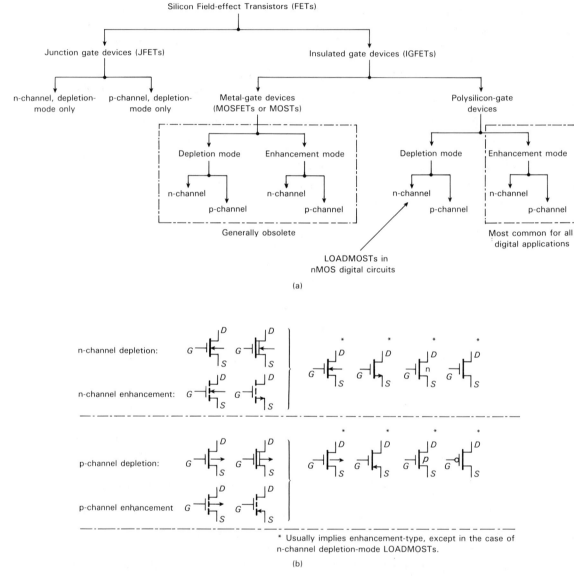

Figure 2.17 Field-effect transistors and their symbols: (a) the FET family tree; (b) some of the insulated-gate FET symbols in use; several are not explicit and require the context in which they are being used to define their exact meaning. The extreme right-hand symbols are commonly used in CMOS digital circuits, representing n-channel and p-channel enhancement devices.

gate was metal, the metal-oxide–silicon cross-section of the gate gave rise to the alternative terminologies 'MOSFET' or 'MOST', terms which have been erroneously maintained even when the metal gate electrode has been superseded by polysilicon, as will be illustrated later. Again, both n-channel and p-channel IGFETs are available. This complete spectrum of unipolar devices is summarized in Figure 2.17(a).

Although the fabrication of silicon JFETs is somewhat simpler than bipolar transistors, not requiring a buried sub-collector layer, their performance is not such as to justify their use except in some applications such as low-power, off-the-shelf operational amplifiers. They do not feature in digital applications since insulated-gate FETs have far greater significance. Hence we will not refer to JFETs any further as a silicon technology, but we will encounter JFET fabrication in connection with gallium-arsenide (GaAs) technology later in Section 2.5.

Confining our interests to the IGFETs, there are four possible types in total:

(a) n-channel enhancement mode, in which the source-to-drain current is zero when the gate-to-source voltage V_{GS} is zero, channel conductivity being increased (enhanced) when V_{GS} is made positive;
(b) p-channel enhancement mode, in which the source-to-drain current is zero when $V_{GS} = 0$, channel conductivity being increased when V_{GS} is made negative;
(c) n-channel depletion mode, in which source-to-drain conduction is possible when $V_{GS} = 0$, V_{GS} having to be made negative to deplete the conductivity and reduce the channel current to zero; and
(d) p-channel depletion mode, in which source-to-drain conduction is possible when $V_{GS} = 0$, V_{GS} having to be made positive to reduce the conductivity to zero.

The various symbols which may be used for these FET variants are given in Figure 2.17(b).

In most digital and other applications enhancement types are of significance, since zero source-to-drain current with $V_{GS} = 0$ is almost always desired. Therefore, we shall only mention depletion mode fabrication briefly in the following discussions.

2.2.1 nMOS insulated-gate devices

Although complementary MOS (CMOS) technology (see Section 2.2.3) is presently dominant in digital custom microelectronics, it is appropriate to consider nMOS as a technology in its own right where it is used extensively in LSI and VLSI memory and similar array-structured circuits, giving extremely high packing densities, as will be considered further in Section 2.3.

The basic structure of an n-channel (nMOS) depletion-mode FET is given in

Technologies and fabrication

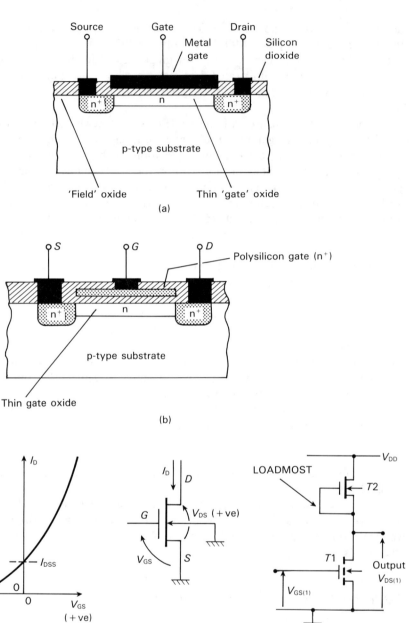

Figure 2.18 Basic depletion-mode nMOS insulated-gate FET fabrication: (a) metal-gate fabrication principles, now generally obsolete; (b) silicon-gate fabrication; (c) typical I_D/V_{GS} characteristics, V_{DS} constant; (d) the use of a depletion-mode FET as a load resistor.

Figure 2.18. In order for source-to-drain electron flow to take place when $V_{GS} = 0$ it is necessary to implant a thin n-type region below the gate to provide majority carriers. With V_{GS} positive the number of free carriers (electrons) is increased, but with V_{GS} negative the number of free carriers in the channel region is decreased and source-to-drain current can be cut off. An alternative method of providing the n-type conducting channel below the gate is to grow an n-type epitaxial layer on the p-type substrate, the n^+ source and drain regions then being formed to complete the source-to-drain regions.

It should be noted that between the conducting n-type channel of an nMOS transistor and the substrate is a pn junction, which if the substrate is connected to the most negative potential in the circuit means that this junction is never forward-biased. The device, including source and drain areas, is therefore isolated from the substrate, and hence is sometimes referred to as 'self-isolated'. The relatively simple controlling equations which relate the device construction and resultant characteristics may be found in many device textbooks [1], [3], [7], [7], [13], [14], and in Appendix D.

Depletion-mode FETs are not employed as active switching devices because of the reverse-bias voltage necessary on the gate to cut off the device. However, it is extensively used as a constant-value load resistor ('loadmost' or 'loadMOST') by connecting together the gate and source electrodes (see Figure 2.18), thus establishing the fixed operating conditions of $V_{GS} = 0$ and giving roughly constant resistance between source and drain [13]. The advantage of forming a resistor in this manner is that it only occupies the area of a transistor, which is far less than resistors formed in silicon by other means.

However enhancement-mode n-channel FETs constitute the best and most widely used FET switching device, either alone in association with loadmosts or with p-channel switching devices in CMOS digital circuits. The basic structures of enhancement-mode devices are given in Figure 2.19. The metal-gate construction is now virtually obsolete except in some early families of off-the-shelf CMOS circuits and possibly where high-voltage operation is required, having been superseded by higher-performance, smaller silicon area silicon-gate devices.

The stages of fabrication of an n-channel silicon-gate FET are shown in more detail in Figure 2.20. The steps are as follows, with re-covering by photoresist being undertaken at every stage where new windows have to be formed:

1. Starting with the lightly-doped p-type substrate, first form a thin layer of silicon dioxide (the 'pad' oxide), followed by a thick protecting silicon nitride layer – cf. the bipolar process given in Section 2.1.1.
2. Open window (mask 1) in the SiO_2 and Si_3N_4, and form a p^+-type isolation border around the perimeter of each device – sometimes termed the 'channel stopper'.
3. Grow a thick SiO_2 layer, the 'field' oxide, over the p^+-channel stopper, which has the effect of driving the p^+-region deeper into the substrate as the silicon dioxide is grown.

44 Technologies and fabrication

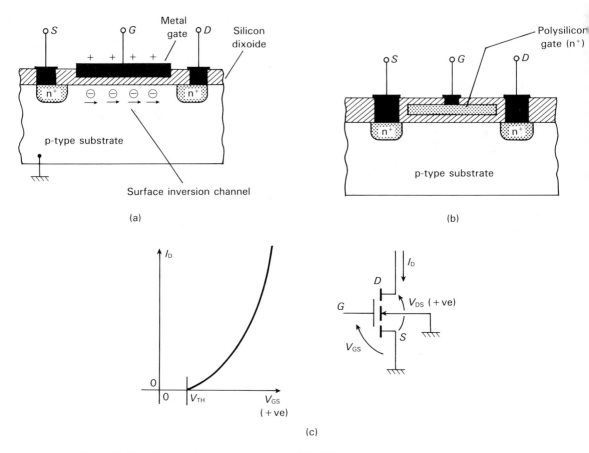

Figure 2.19 Basic enhancement-mode nMOS insulated-gate FET fabrication: (a) metal-gate fabrication, now generally obsolete; (b) silicon-gate fabrication; (c) typical I_D/V_{GS} characteristics, V_{DS} constant.

4. Strip off the Si_3N_4 and pad oxide, and grow a very thin, very precisely-controlled new SiO_2 layer ('gate' oxide) over the area.
5. Adjust the exact p-type doping concentration of the p-type substrate under this gate oxide by boron ion-implantation which passes freely through the thin gate oxide – this provides the threshold voltage adjustment for the final device.
6. Deposit a layer of polycrystalline silicon ('polysilicon') over the whole surface.
7. Open window (new photoresist and mask 2), and remove the polysilicon from everywhere except the gate area of each device – this forms the polysilicon gate for each device.
8. Form the n^+-type source and drain regions of each device by arsenic ion-implantation. The important feature here is that this n^+ implantation can take place through the thin gate oxide regions but is stopped by the thicker-field

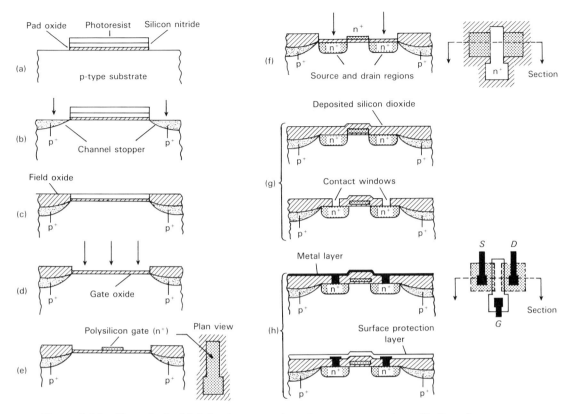

Figure 2.20 The principal fabrication steps in enhancement-mode, self-aligned silicon-gate FET fabrication: (a) formation of pad oxide and Si_3N_4 layers; (b) p^+-type implantation or diffusion; (c) growth of field oxide isolation; (d) removal of pad oxide and Si_3N_4, growth of gate oxide and adjustment of channel doping; (e) deposition and etching of polysilicon to form gate; (f) formation of source and drain regions; (g) silicon oxide covering, with contact windows then cut through; (h) final metalization interconnect patterning and surface protection.

oxide and polysilicon gate areas; hence there is no need for any separate mask to delineate the source or drain areas, this technique being known as 'self-aligning'.

9. Deposit a layer of silicon dioxide over the whole surface, thus burying each polysilicon gate.
10. Open contact windows, known as 'vias', through this SiO_2 layer (new photoresist and mask 3) to meet all source, drain and gate areas, and then cover the whole surface with metal (usually aluminium with 1% silicon).
11. Etch away all unwanted metal (new photoresist and mask 4) so as to leave only the required interconnection pattern between the individual devices on wafer.
12. Apply final surface protection over the whole wafer.

This is the basic fabrication process which gives the very high performance now available from field-effect transistors. The breakthrough to high performance came with the development of the self-aligned polysilicon gate, which eliminated the problems of alignment between source, gate and drain areas inherent in metal-gate devices. With the latter there was always some overlap between the gate and the source and drain areas, which produced much higher gate-to-source and gate-to-drain capacitance values than with the non-overlapping, self-aligned polysilicon gates. These capacitance values, accentuated by the Miller effect when the circuit is in operation, severely limited metal-gate FET performance compared with the silicon-gate FET [1], [14]. Self-alignment and the oxide-isolation techniques shown in Figure 2.20, thus represent key characteristics in many modern fabrication methods.

A further feature of the polysilicon layer of this insulated-gate construction is that the polysilicon is also used as an interconnection level in the final IC to provide underpass connections below the overlying metal interconnections. Although the polysilicon is not as good a conductor as metal, it is adequate for underpasses or other short interconnections. This will be extensively noted in the custom ICs covered in Chapter 4. More recent developments may use silicide rather than polysilicon, which provides a lower resistance for such connections.

The fabrication details of n-channel depletion-mode devices can be basically the same as are detailed above, except that an arsenic ion-implantation to form the thin n-type source-to-drain conducting channel is made through the polysilicon and gate oxide layer. If both depletion- and enhancement-mode n-channel devices are required, as in nMOS memory and similar circuits, then the channel region of each type of device has to be separately implanted to give the two distinct threshold voltages, which therefore requires an additional fabrication mask.

Certain other nMOS fabrication methods have been pursued, but are now applied mainly to discrete devices for high-voltage or high-current applications rather than in monolithic integrated circuits. Figure 2.21 illustrates two such fabrication methods. In the DMOS structure of Figure 2.21(a) the effective channel length is the narrow p-channel dimension between the n^+ source and the intrinsic (π-type) region surrounding the drain. A current-carrying capacity of many amps with a switching time of a few nanoseconds is possible. The p^+ and n^+ diffusions under the source electrode are made through the same window, thus eliminating any problem of alignment.

In the V-groove (VMOS) structure of Figure 2.21(b), which was at one time proposed for memory circuits, the substrate acts as the source, with the channel formed along the four sides of the groove. The channel width is determined by the thickness of the epitaxial layer, which in early days was shorter than could be achieved with conventional MOS fabrication methods using lithography methods — this is now no longer the case. Switching times of a few nanoseconds and a current-carrying capacity of one amp or more are available from current discrete VMOS devices.

It is clear that more difficult fabrication is involved in these devices compared

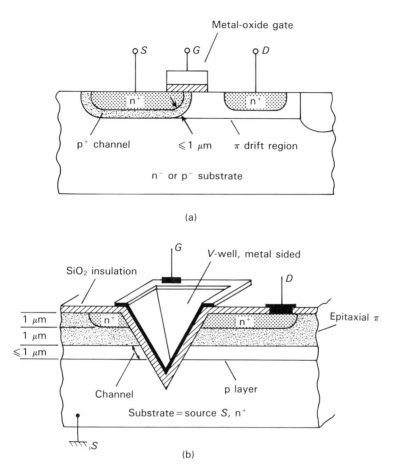

Figure 2.21 Further depletion-mode nMOS fabrication techniques, now generally reserved for discrete devices: (a) the double-diffused DMOS fabrication; (b) the V-grove VMOS fabrication.

with the self-aligned silicon-gate process, and thus they do not now present a challenge to the latter for LSI and VLSI applications. Further details may be found in more specialized publications [2], [14], [15], [16], [17].

2.2.2 pMOS insulated-gate devices

Because of the inherently inferior performance of pMOS compared with nMOS, circuits using pMOS devices only are no longer used. This was not the case in the early days of metal-oxide–silicon devices, since a problem known as 'surface inversion' occurred between the lightly-doped p-type channel region and the

Technologies and fabrication

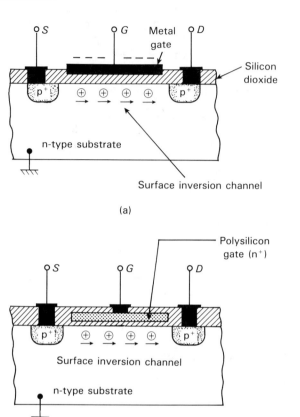

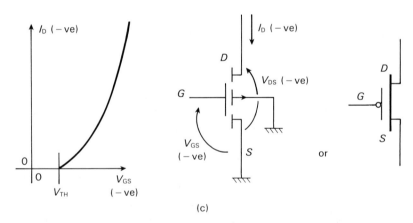

Figure 2.22 Basic enhancement-mode pMOS insulated-gate FET fabrication: (a) metal-gate fabrication, now largely obsolete; (b) silicon-gate fabrication; (c) typical I_D/V_{GS} characteristics, V_{DS} constant.

overlying silicon dioxide of n-channel enhancement-mode MOSTs (see Figure 2.19(a)), giving rise to an unwanted thin conducting channel and hence depletion-mode rather than enhancement-mode operation. This problem was overcome by the introduction of silicon gates with ion-implantation to control the device threshold voltage, thus making nMOS feasible and preferable to pMOS.

The general construction of pMOS devices is similar to nMOS, except that polarities are reversed. Figure 2.22 shows the basic structure of the now-superseded metal-gate and the later silicon-gate FETs.

However, except for some discrete devices the use of pMOS is nowadays exclusive to complementary MOS (CMOS) digital logic circuits, where the enhancement-mode p-channel devices act as a switch to the positive (V_{DD}) supply rail and the complementary enhancement-mode n-channel devices act as a switch to the negative (V_{SS} or 0 V) supply rail. We will therefore continue our discussions on pMOS in the context of CMOS fabrication techniques.

2.2.3 CMOS technologies

The basic CMOS digital logic configurations are shown in Figure 2.23. The p-channel enhancement-mode FETs are configured in an arrangement which is always the dual of the n-channel enhancement-mode FETs, that is, a series configuration of one type is associated with a parallel configuration of the other type; for any x-input logic gate there are always x FETs of each type, the p-channel devices providing a conducting path to V_{DD} and the n-channel devices a conducting path to V_{SS}. Because the resistance of a p-type FET is higher than that of a comparable n-type FET, it is preferable to series the n-type devices rather than p-type, which implies that CMOS NAND configurations are preferable to NOR configurations. However, this principally affects the final interconnection details and not the basic device fabrication methods.

CMOS fabrication requires both pMOS and nMOS devices to be fabricated alongside each other on the same substrate, with appropriate insulation between them. The three basic ways of fabrication are shown in Figure 2.24. However, the method illustrated in Figure 2.24(a) requires that the n-wells (or 'n-tubs') must be rather heavily doped so as to convert the p-type substrate into n-type, and as a result the performance of the p-channel devices is degraded [3]. Similarly, if p-wells (or 'p-tubs') on an n-type substrate are used, the performance of the n-channel devices is impaired. The best method of fabrication is therefore the twin-tub process of Figure 2.24(c), which allows both the n-wells and the p-wells which are formed in a very lightly doped epitaxial layer to be given optimum doping levels. The penalty to be paid for twin-well fabrication is of course additional processing cost.

The methods of isolation between devices are also of particular significance. One early CMOS isolation technique was to diffuse or implant a deep p^+ moat around each n-channel device and a corresponding n^+ moat around each p-channel device, thus giving pn junction isolation. This method, known as 'dual guardband'

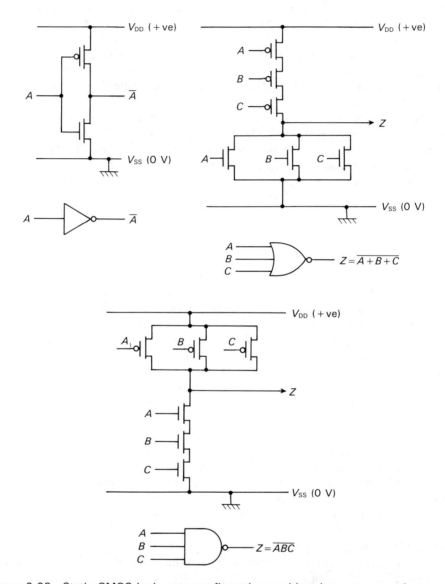

Figure 2.23 Static CMOS logic gate configurations, with enhancement-mode p-type devices connected to the V_{DD} supply rail and corresponding n-type devices connected to the V_{SS} supply rail (cf. n-channel-only logic gates, where the p-channel devices are replaced by a single depletion-mode n-channel 'loadmost').

Unipolar silicon technologies 51

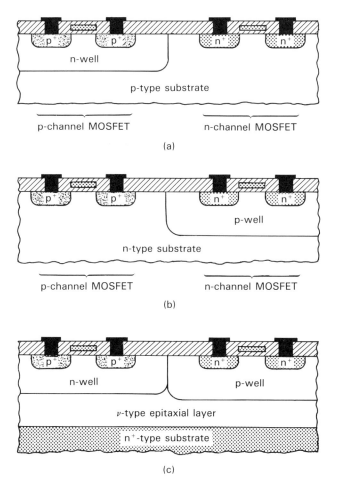

Figure 2.24 The three basic ways of fabricating CMOS: (a) p-channel device in an n-well with p-type substrate; (b) n-channel device in a p-well with n-type substrate; (c) twin-tub epitaxial fabrication on an n-type substrate.

isolation, was largely used with metal-gate CMOS, for example in early off-the-shelf 4000-series CMOS logic circuits, but required a high silicon area for its implementation [14]. It did, however, allow high-voltage working if necessary, but was not appropriate for high-performance LSI or VLSI circuits.

Isolation techniques for CMOS are now entirely by some arrangement of oxide-isolation. The present generation of CMOS circuits is therefore characterized by *oxide-isolated, self-aligned, silicon-gate* fabrication methods. Terminologies such as LOCMOS (local-oxide complementary MOS) and IsoCMOS (isoplanar CMOS) have been used by certain manufacturing companies for their particular fabrication method.

52 Technologies and fabrication

The general fabrication details of an oxide-isolated p-well CMOS process is illustrated in Figure 2.25. This is a planar, not a planar epitaxial process. The various manufacturing steps are as follows, with, of course, re-covering by photoresist being undertaken at each appropriate stage:

1. On the n-type substrate first form the thin pad oxide and thicker silicon nitride layers – see Section 2.2.1.

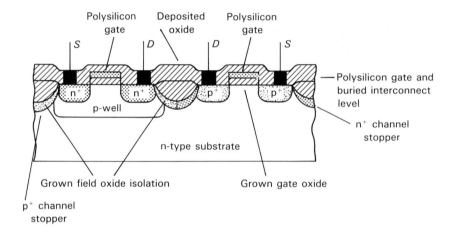

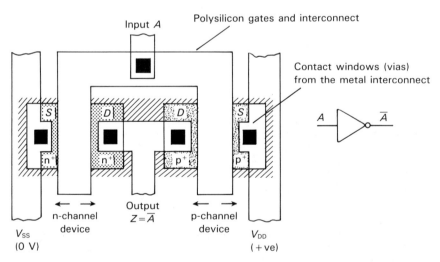

Figure 2.25 The general fabrication details of an oxide-isolated silicon-gate CMOS Inverter gate, with n-channel and p-channel enhancement-mode transistors. Note, for clarity, the transistors are drawn side-by-side as shown here; in practice they are usually located with a straight strip of polysilicon to form both gates, as subsequently shown in Figures 4.6 and 4.18.

2. Mask 1: open window in photoresist and etch away all pad oxide and Si₃N₄ except over the final transistor areas.
3. New photoresist and mask 2: open window to define the area that will become the p-well, and ion-implant the p-well dopant with boron – note this ion-implantation can be made through the pad oxide and Si₃N₄ but the boron is stopped by the photoresist.
4. Mask 3: open further window in the photoresist around the p-well perimeter and ion-implant a p^+-type channel-stopper border.
5. New photoresist and mask 4: open window around the p-channel device and ion-implant an n^+-type channel stopper border with phosphorous.
6. Grow a thick SiO₂ layer, the field oxide, over all the areas not protected by the pad oxide and Si₃N₄ to produce the field oxide isolation around each device; the channel stopper borders and the p-well will be driven in during this high-temperature process.
7. Strip off all remaining Si₃N₄ and pad oxide, and grow the thin SiO₂ gate oxide layer.
8. Adjust the exact doping concentration of the n-channel and p-channel areas by careful ion-implantation through the gate oxide so as to give the required threshold voltage values.
9. Deposit a layer of polysilicon over the whole surface.
10. New photoresist and mask 5: open windows and remove polysilicon from everywhere except the required gate areas.
11. New photoresist and mask 6: open window and form the self-aligned n^+ source and drain regions of the n-channel device by arsenic implantation through the thin field oxide.
12. New photoresist and mask 7: open window and form the self-aligned p^+ source and drain regions of the p-channel device by boron implantation through the field oxide.
13. Deposit a silicon dioxide layer over the whole surface burying the polysilicon gates.
14. New photoresist and mask 8: open windows and cut contact vias in the SiO₂ to all source, drain and gate areas.
15. Cover all surface with metal.
16. New photoresist and mask 9: etch away all unwanted metal to leave the required interconnection pattern.
17. Apply final surface protection.

CMOS fabrication will be seen to be noticeably more complicated than nMOS or indeed bipolar fabrication because of the need to fabricate the two polarity devices on the same substrate. Isolation is particularly important, as also is the necessity to ensure that there is no parasitic pnpn action between the pMOS source electrode and the nMOS source electrode. If such a path with a sufficiently high current gain exists, then thyristor action can occur, with the 'thyristor' switching on and causing a destructive short-circuit between V_{DD} and V_{SS} supply rails [3],

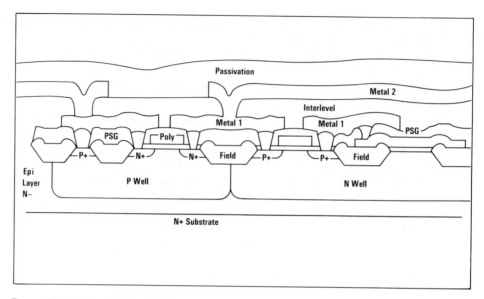

Figure 2.26 The VQ double-layer-metal CMOS process of Plessey Semiconductors, using twin-tub oxide-isolated fabrication. (Courtesy of GEC-Plessey Semiconductors, UK.)

[13]. This is known as 'latch-up', and was particularly troublesome in early CMOS devices. The heavily doped p^+- and n^+-channel-stopper regions are designed to prevent the formation of any effective pnpn thyristor latch.

In practice, CMOS fabrication may involve additional steps to those detailed above. Twin-tub fabrication in particular requires further processing to form the epitaxial layer and the n-well. Double- or even triple-layer metal interconnections add yet further complexity, with the total number of mask stages possibly as high as 15. Figure 2.26 shows the cross-sectional details of a typical present-day, double-layer-metal industrial CMOS process.

The introduction of oxide-isolation was the key feature in producing high-density high-performance CMOS circuits. However, the area requirements of the p^+- and n^+-channel stoppers necessary with oxide-isolation have spurred the development of the deep-trench isolation methods shown previously in Figure 2.7. This isolation technique may therefore be found in recent state-of-the-art CMOS fabrication.

2.2.4 Silicon-on-insulator

Devices where the bulk (substrate) of the device is silicon, sometimes collectively termed 'bulk-silicon' devices, have an inherent problem that the substrate is a semiconductor, which means that the action of all devices must be isolated from the substrate as well as from each other. Also, the substrate must be electrically connected in the final circuit and not left floating, to the 0 V supply rail in the case of MOS, and hence there will be inherent parasitic capacitance between the conducting substrate material and the devices. Therefore, the idea of planar fabrication on the top surface of a non-conducting substrate has fundamental advantages. This possibility is termed 'silicon-on-insulator (SOI) fabrication' to distinguish it from bulk-silicon technologies.

The substrate insulator which has been the object of most research is artificial sapphire, giving rise to silicon-on-sapphire or SOS technology. It has been considered mainly with the view of improving the performance of CMOS, in particular to producing silicon-on-sapphire insulated-substrate CMOS (SOS–CMOS or SOSMOS) [1], [13], [14], [18]. Figure 2.27(a) shows the general fabrication details.

It will be seen that SOSMOS fabrication is a planar epitaxial process, as has been described previously, with oxide-isolation between devices. The advantages

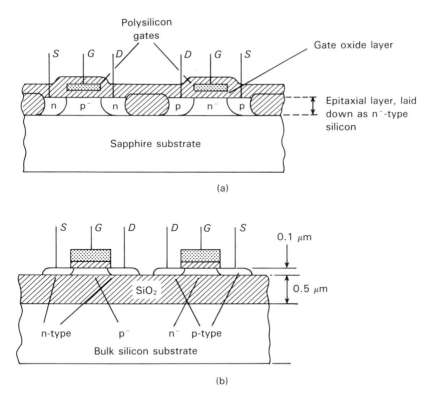

Figure 2.27 Silicon-on-insulator (SOI) technologies: (a) silicon-on-sapphire CMOS; (b) silicon-on-oxide CMOS.

claimed for this fabrication method are increase in speed and packing density by a factor of two compared with comparable bulk-silicon CMOS, and a much greater inherent resistance to radiation damage due to the sapphire rather than silicon substrate. Against these advantages are very much higher cost due to the manufacture of the sapphire substrate, and problems caused by the difficulties of forming the flaw-free silicon on the surface of this substrate reliably. Hence the technology has been largely pursued for military and space applications and is not a commercial competitor to normal CMOS.

An alternative fabrication method retaining bulk silicon as the substrate is shown in Figure 2.27(b). Here, silicon dioxide forms the insulation below all devices, with the devices being built up in small isolated islands on the SiO_2 insulating layer [18]. Little has been heard of this technique recently, and hence it must be regarded as not providing sufficient advantages to be commercially viable or necessary.

2.2.5 *MOS summary*

The basic processes involved in unipolar technologies are all well-proven, with CMOS evolution in particular being the most significant for custom microelectronic applications. The planar fabrication techniques are the same as are used in modern bipolar technologies, with *oxide-isolation* and *self-alignment* being the key elements in providing high-density, high-performance circuits.

The buried (sub-collector) layer necessary in most bipolar technologies is not required in FET fabrication, neither is the formation of any Schottky diodes for antisaturation purposes. However, an *epitaxial layer* and the formation of n-wells or p-wells or twin-wells is desirable for CMOS. In general, CMOS requires more mask levels and processing than does bipolar, the whole thrust of CMOS development being to improve speed so as to be comparable with that available from bipolar. In this respect modern CMOS generally matches the speed of all but the fastest bipolar technologies; see Section 2.6.

The other fundamental and essential difference between bipolar and unipolar technologies is that the in-circuit performance of FETs is influenced by physical device size, in particular by the length L and width W of the source-to-drain channel in any given fabrication process. The smaller the L/W ratio the faster the circuit. However, in bipolar technology the critical factor is the effective width of the base layer, but this is usually a function of processing rather than geometric layout. It is this dependence upon layout rather than processing that allows the layout of MOS or CMOS custom digital circuits to be undertaken by circuit designers outside semiconductor manufacturing companies, using *geometric design rules* supplied by the latter. It is also the reason why the majority of published textbooks and academic courses deal largely with unipolar design rather than bipolar [13], [19], [20], [21], although the whole thrust of custom microelectronics should be to release the circuit designer from having to be involved at all with details at the fundamental transistor design level. Further details of FETs and unipolar fabrication may be found in more specialized texts [1]–[4], [6]–[10], [13].

2.3 Memory circuits

The development of high-capacity, low-cost memory has made possible the evolution of affordable CAD systems for custom IC design and for all other computer purposes. Memory ICs are in the vanguard of VLSI capability and represent the highest commercially available transistor count per chip.

Memory circuits cater for two types of need: (a) where it is necessary to change stored data very frequently during the operation of the equipment, such circuits acting as a temporary memory bank; and (b) where the stored data are fixed and not altered during normal operation. The former are *random-access memories* (RAMs), very fast read and write usually being required, while the latter are *read-only memories* (ROMs) with fast read time being desirable.

With random-access memories the data written into the storage circuits are lost when the power is disconnected and have to be re-written after re-connection of power; such circuits are therefore said to have 'volatile' memory. (The very low standby power consumption of certain CMOS RAMs has made it possible for an internal lithium battery to be included within the IC package, so as to give continuity of memory in the event of a mains supply power failure.) Read-only memories, on the other hand, retain their stored pattern of data under power-off conditions, and are therefore termed 'non-volatile'. The permanent data stored in ROMs may be mask-programmed into the circuit by the supplier, or supplied unprogrammed to the purchaser for his own electrical programming procedure. The latter are 'programmable' or 'field-programmable' ROMs, and may in turn be divided into the following two kinds:

- types which may be programmed once only with required data (PROMs); or
- types where the stored data can be re-written if required by taking the circuit out of service and undertaking an erasing/re-programming procedure (EPROMS).

The latter include several variants such as erasure by ultra-violet light (UV-PROMs) and electrically erasable and re-programmable (EAPROMs or EEPROMs). There is often a limit to the number of times that an EPROM can be re-programmed.

The range of memory ICs is indicated in Figure 2.28. Mask-programmable ROMs are available from many vendors. Bipolar versions provide the lowest access (read) time, but are currently limited to 32 or 64 K-bits storage capacity, but unipolar – particularly nMOS – can provide up to eight times this capacity, although with a slower access time. Note that 'K' in this context means 1024. The same distinction applies to field-programmable types (PROMs); a representative

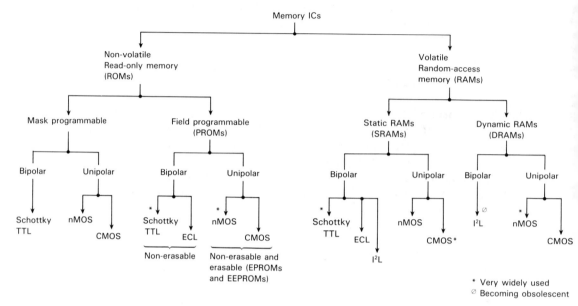

Figure 2.28 The families of commercially available memory circuits. Non-volatile static RAMS using ferroelectric memory, see text, have very recently been added to the above families. However, note that other programmable devices, such as programmable logic arrays (PLAs) and programmable logic sequencers (PLSs), which are closely related to PROMs, are not normally listed under memory devices.

non-erasable 20-pin bipolar IC can provide 16 K of memory organized as 4096 × 4-bit words with an access time of 30 ns, whereas an erasable CMOS EPROM can provide up to 1024 K-bits of memory organized, say, as 64 K × 16-bits or 128 K × 8-bits, but with an access time of 150–200 ns.

The same general distinction of maximum speed vs. capacity is also true in the range of available random-access memories. With static RAMs, sometimes referred to as SRAMs ('ess-rams'), Schottky TTL can, for example, provide 256 × 1-bit memory capacity in a 16-pin IC with a 40 ns access time, whereas CMOS – by far the most widely used RAM technology – can provide 256 K-bits (32 K × 8-bits) capacity in a 28-pin IC with an access time of 120–150 ns. With dynamic RAMs, sometimes referred to as DRAMs ('dee-rams'), unipolar technologies now dominate for manufacturing reasons, see below. Up to 1024 K-bits organized as 256 K × 4-bits or equivalent per IC with access times of 100–150 ns are widely available. A common feature of all these mass-market standard parts is that, in spite of their complexity, prices are usually in the range of $10 or less per IC. It is therefore not often economical to incorporate very large memory capacity within a custom IC.

Full details of the circuit configurations and fabrication methods of RAMs and ROMs may be found elsewhere, including the essential on-chip circuits necessary to refresh continuously the stored information in dynamic RAMs [1], [3], [22]–[24].

However, as will be seen in the following brief overview, the basic fabrication procedures do not introduce any radically new techniques over those already covered.

With the exception of very small memory circuits the basic arrangement of all read-only memories is shown in Figure 2.29(a). The 'memory' consists of either forming the required connections between the horizontal 2^j row address decode lines and the 2^k column lines for each output in the $2^j \times 2^k$ memory array matrix, or destroying those connections which are not required; for m outputs there are m such identical matrices. The size of a ROM is defined by the examples shown in Table 2.1. The heart of all ROMs and PROMs therefore consists of the mechanism used to form the pattern of required connections in each memory array. In non-erasable types, once this connection pattern has been formed it can never be altered, but in erasable types a pattern can be destroyed and an alternative one made.

Figure 2.29 shows the fundamental concepts. In mask-programmable bipolar types the multiple-emitter connections to the decode lines are individually made or omitted by the surface metallization mask pattern. In nMOS types the programming is usually performed by the formation of a thin gate oxide at the locations where an effective n-channel MOS switch (see Figure 2.20) to 0 V is required; where no path to earth is wanted, then a thick field oxide is left over the gate position, which means that an effective gate for this depletion-mode n-channel device is not possible and the source-to-drain conductance remains zero.

Note that depletion-mode and enhancement-mode FETs are present in the circuit of Figure 2.29(d). CMOS is not particularly relevant for programmable array matrices since it would require a much greater silicon area to use a true CMOS logic configuration where both n-channel devices *and* p-channel devices have to be switched on and off. However, CMOS may be relevant in the Enable and other non-programmable parts of the ROM circuit.

Programmable ROMs (PROMs) and erasable-programmable ROMs (EPROMs) employ the same basic fabrication techniques, but with the addition of appropriate means whereby the purchasers of the circuit can select the required connections in the array matrix.

In bipolar PROMs the user-programming is by means of a fusible metallic link in each emitter-connection. This link is either doped silicon or some metal such as nichrome, and the programming operation consists of applying a short-duration pulse of about 20 mA to individual circuits, which is sufficient to blow this 'fuse'

Table 2.1. The relationship between the number of inputs and outputs of a ROM and the specified ROM size

No. of inputs $n = j + k$	No. of outputs m	Size of ROM
5	8	256 bits, or 32 × 8
12	4	16 K, or 4096 × 4
16	8	512 K, or 64 K × 8
18	4	1024 K, or 256 K × 4

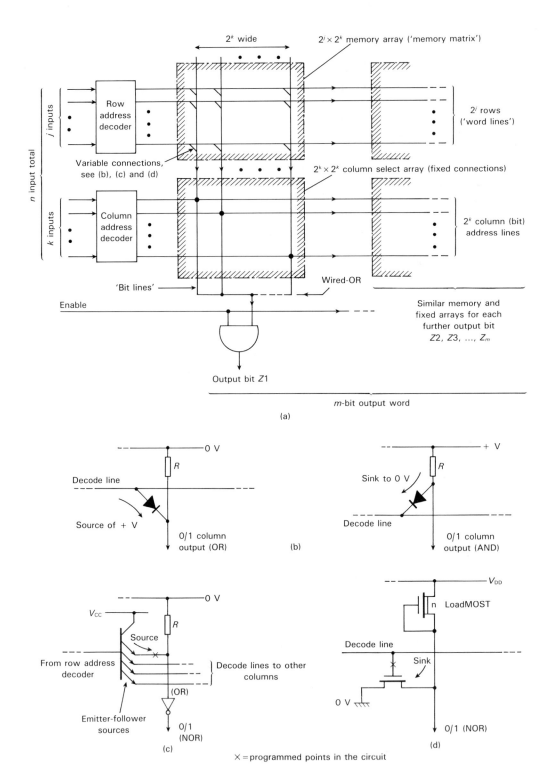

Memory circuits

link but insufficient to impair the rest of the circuit. Figure 2.30(a) shows details of one such product. Clearly, once blown, a fuse path cannot be 'repaired', and hence the device memory is permanent (non-volatile) and non-reprogrammable.

Commercially available unipolar PROMs do not employ fusible links. Other methods are used which effectively activate an FET gate by applying a short electrical voltage pulse when the device is required to be made functional, leaving it inoperative where no switching action is required. The terminology 'electrically programmable' (EPROM) is usually applied to these devices, the generic term PROM often being reserved for bipolar types only. Further, since it is usually possible to erase and subsequently re-write (alter) the programming pattern in the FET switches by opposite polarity electrical pulses or ultra-violet light, the additional designations EEPROM, EAPROM and UV-PROM are found.

Details of the method of EPROM operation may be found in many texts [1], [23]–[26]. Briefly, each FET switch in the memory array is fabricated with an isolated gate buried in the SiO_2 insulation layer, above which is the normal control gate (see Figure 2.30(b)). With no charge on this isolated (floating) gate the control gate is ineffective in controlling the source-to-drain conductance, which is zero. However, when a very short voltage pulse of possibly 50 V is applied between the source and drain electrodes, avalanche multiplication occurs and charge becomes induced and subsequently trapped on this floating gate. The control gate is now effective in switching the source-to-drain current. Because of the very high insulating properties of SiO_2, the trapped charge remains effective for a very long period (possibly a decade or more), but can be removed, for example by exposure of the circuit to intense UV light or X-rays which result in free carriers being induced to cancel the stored charge. The device may then be re-programmed with new data. The memory in EPROMS is still regarded as non-volatile, but re-programmable.

Random-access memories, by definition, require circuits which can always accept new data, with no limitations on the number of times 'written-to' and 'read-from'. The internal circuits must therefore be read/write storage circuits, and unless very high speed is required the lower power consumption of nMOS and CMOS circuits, which allows a greater memory capacity per chip, will be preferable.

All conventional RAM circuits are by definition volatile, since if power to the chip is interrupted the stored data is lost. However, static RAMs retain their data as long as power is on, whereas dynamic RAMs require continuous internal refreshment since the data is held by capacitive means rather than by conventional cross-coupled bistable circuits.

Figure 2.29 Read-only memories (ROMs): (a) the fundamental circuit topology, where only the required row/column connections in the memory array remain present in the final programmed circuit; (b) two concepts for the decoder control of the column outputs, using some form of unidirectional interconnection; (c) bipolar circuit using multiple-emitter transistors, each emitter to one column line only; (d) nMOS circuit, using depletion-mode LOADMOST.

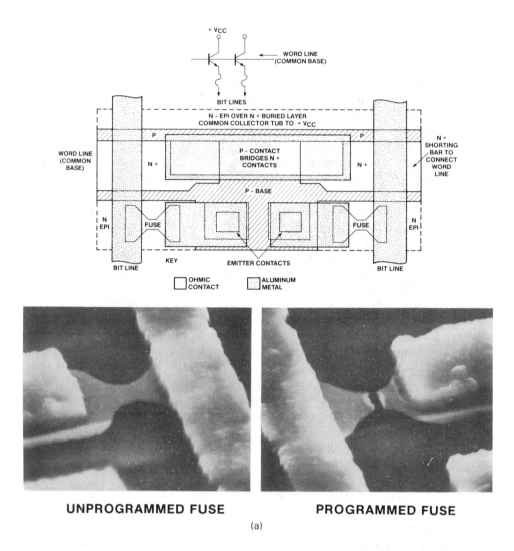

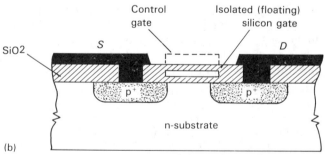

Figure 2.30 Details of programmable read-only memories: (a) the fuse-blowing details of a Schottky TTL bipolar PROM (courtesy of Monolithic Memories, CA.); (b) the floating gate concept used in many unipolar programmable ROMs.

Memory circuits 63

The basic topology of all RAMs whereby individual memory cells in a matrix array can be written to and read from is given in Figure 2.31(a). Typical static memory cells are shown in Figure 2.31(b), (c) and (d). In the case of nMOS the two depletion loadMOSTs may be replaced by two high-value resistors made from lightly doped polysilicon positioned on top of the four active devices, which can decrease the total silicon area of the cell but at the expense of additional processing steps and hence cost. The CMOS memory cell gives the lower power consumption, but speed and packing density is generally not as good as nMOS.

Fabrication of static RAMs does not therefore involve any basic concepts beyond those previously considered, being normal bipolar or MOS planar technologies but with great emphasis upon layout and packing density. Further details may be found elsewhere [9], [23]–[26]. Dynamic RAMs, however, do involve some novel internal concepts which stem from the requirement to pack ever greater memory capacity on a single chip, which means minimizing the number of devices per memory cell. The present status is that the single-FET memory cell shown in Figure 2.32 is now the classic means of producing high-density dynamic memories.

The n-channel enhancement-mode FET shown in Figure 2.32(a) merely acts as an on/off switch to allow the storage capacity of the cell to be charged or discharged (written), and for the state of this charge to be read. However, due to circuit leakage the charge has to be refreshed every two or three milliseconds, which requires additional on-chip circuitry to refresh all the memory cells continuously. Further, the act of reading a cell discharges the capacitor, so that in order to maintain this data bit it must be copied when read into an output buffer register and then re-written back into the same memory cell. Hence, although each memory cell is simple, considerable complexity is necessary in the peripheral circuitry of the DRAM [6], [23].

The fabrication details of a double-level polysilicon memory cell are shown in Figure 2.32(b). There remains intense commercial developments in this field in efforts to provide more memory capability per IC, but as device size is reduced so is the storage capacitance value. This results in more complex fabrication in order to provide adequate capacitance in smaller-dimension cells [27], as illustrated by the details given in Figure 2.32(c).

Whilst the preceding pages survey the most widely used forms of memory, a completely new form of circuit has recently been developed which gives a *non-volatile* static RAM capability. This new form of memory uses a ferroelectric material which maintains a stable polarized state after the application and removal of an externally applied electric field. The market product is known as a 'Ferroelectronic Random Access Memory', or FRAM^{TM}. (FRAM is the registered trade mark of Ramtron Corporation, Colorado Springs, and applies to all their range of non-volatile read and write memories.)

The ferroelectric effect is the ability of the molecular structure of certain crystals to become aligned (polarized) under the influence of an applied field, and to remain polarized after removal of the field. Reversal of the applied field causes

64   Technologies and fabrication

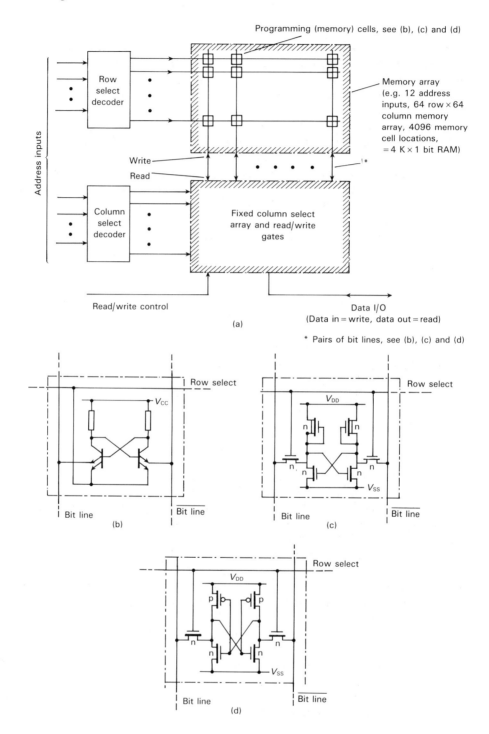

polarization in the opposite direction. Thus ferroelectric material has two stable polarization states, and can be employed as a 'bistable capacitor' with two distinct threshold voltages to switch the device from one capacitive state to the other, no power being necessary to maintain the polarization of the crystal structure in either of the two states. (Note that this ferroelectric effect has nothing to do with any magnetic effect in ferrous compounds, and hence is something of a historical misnomer from the early 1920s. There is, however, an analogy with ferromagnetic materials which maintain a magnetic polarization without applied power.)

Work on ferroelectric compounds has resulted in the development of a complex ceramic material, largely consisting of lead-zirconate-titanate (PZT) which is compatible with, and can be deposited as a thin film over, existing semiconductor processing. Three basic steps are required to add ferroelectric devices to conventional processing, namely the formation of the bottom capacitor electrode, the deposition of the thin film of PZT, and the formation of the upper capacitor electrode [28]. This is illustrated in Figure 2.33(a).

Early circuits disclosed the use of PZT capacitors to convert a conventional volatile six-transistor CMOS memory cell, such as that shown in Figure 2.31(d), into a non-volatile cell, by connecting one PZT capacitor to each bistable circuit output. Under power-off conditions the two capacitors retained opposite states of polarization depending upon which bistable output was at logic 1 or logic 0, and on restoration of power these dissimilar capacitive states would bias the recovery of the bistable circuit so that it re-established its previous stable state. Additional circuit details disconnected the PZT capacitors under normal working conditions [29]. Later disclosures [30] indicate a simpler circuit involving the two-transistor, two-capacitor memory cell shown in Figure 2.33(b). In this memory cell write data to the cell on the two bit lines is used to polarize the two PZT capacitors directly. This information is permanently stored by the PZT capacitors without requiring refreshment or power. Under read conditions the two capacitors are re-connected to the bit lines and a special memory sense amplifier differentially senses the difference in polarization between the two capacitors, and generates a 0 or 1 data output bit as appropriate. The action of reading, however, is destructive, and the data has to be re-entered into memory by re-polarizing the PZT capacitors. Further developments using only one PZT capacitor per cell have also been proposed [28], [29].

Figure 2.33(c) illustrates a currently available product [30]. Performance details in comparison with other alternatives have been quoted in Table 2.2. The forecast areas of use have included gate-array and standard-cell custom circuits, as well as computer core and disk replacements.

Figure 2.31 Random-access memories: (a) the basic topology of all RAMs − one I/O bit only shown; (b) static bipolar single-bit memory cell (simplified) using multiple-emitter transistors; (c) the static 'six-transistor' single-bit nMOS memory cell; (d) the static CMOS single-bit memory cell. (Note, in (b), (c) and (d) additional column-select gating selects the two vertical bit lines to the individual cell.)

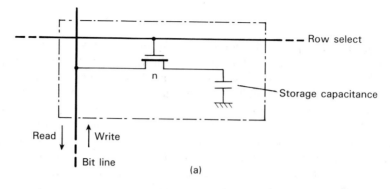
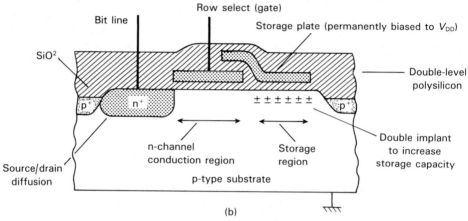
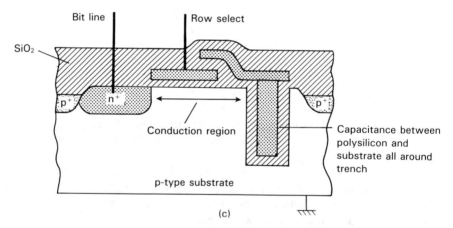

Figure 2.32 Dynamic RAMs: (a) the single-transistor dynamic memory cell; (b) fabrication details involving a double-level-polysilicon structure, polysilicon gate and polysilicon storage plate (not to scale); (c) deep-trench construction for 1 M-bit capability; surface area of cell reduced but capacitance value maintained. Further developments for 4 M-bit or greater capacity involve dropping the polysilicon capacitor electrode even deeper into the substrate and stacking the gate vertically above it (not to scale).

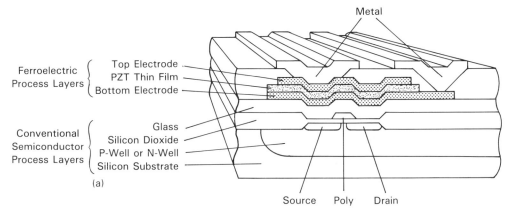

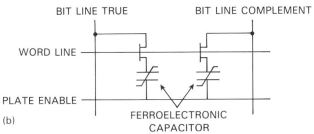

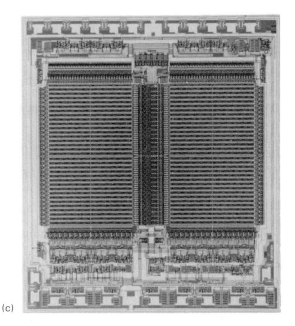

Figure 2.33 Details of the non-volatile ferroelectric static memory cell: (a) the addition of a PZT thin-film capacitor on conventional CMOS processing; (b) the two-transistor, two-capacitor, non-volatile memory cell; (c) a commercial ferroelectric random-access memory chip. (Courtesy of Ramtron Corporation, CO.)

68   Technologies and fabrication

Table 2.2. A comparison of typical FRAM performance with other memory devices

Parameter	SRAM	DRAM	EEPROM	FRAM
Read cycle (ns)	25	150	50 to 200	25 to 100
Write cycle (ns)	25	150	10^7	25 to 100
Operating voltage (V)	+5	+5	+5	+5
Programming voltage (V)	+5	+5	+12 to +21	+5
Operating temperature range (°C)	←		−55 to +125	→
Data retention at zero power (years)	0	0	10 to 100	>10
Relative cell area	3X	X	2X	X

To summarize, all ROM and RAM circuits employ well-established planar fabrication techniques, with appropriate additions or variations from normal digital logic ICs to meet the particular requirements. The OEM design engineer will normally use them as standard off-the-shelf products, except possibly where a vendor has ROM and RAM macros in his standard-cell custom library (see Section 4.2), but the precise design and fabrication details of these memory circuits always remains the province of the semiconductor vendor. Further details may be found in a comprehensive survey paper by Maes *et al.*, which also contains an extensive list of further references [31].

2.4 BiCMOS technology

Because of the strengths and weaknesses of both bipolar and MOS technologies — bipolar consuming more power than MOS, but MOS in general not being as fast or linear and not so capable of driving further loads — it is an attractive proposition to consider merging the two technologies on one chip to provide the best of both worlds. This is BiCMOS technology.

BiCMOS attempts to use the best fabrication method of both bipolar and CMOS. However, there are certain problems, for example bipolar fabrication usually uses an epitaxial layer on an p-type substrate, whereas CMOS may be fabricated without an epi. layer on an n-type substrate. Thus there is a choice between keeping the bipolar fabrication technique the same and 'adding' the CMOS requirements, or vice versa. In either case, however, a more complex fabrication procedure will be involved.

Figure 2.34 shows the general arrangement of two BiCMOS techniques, one with and one without an epitaxial layer. Some of the processing steps can be made common between the bipolar and CMOS devices, for example in Figure 2.34(a) the emitter and collector n^+-regions can be made at the same time as the n-channel FET source and drain regions, and in Figure 2.34(b) the deep n-wells can be simultaneously fabricated. There may, however, have to be compromises between the doping levels in the bipolar and MOS regions if simultaneously fabricated.

The true place for BiCMOS technology in the custom microelectronics area has

BiCMOS technology 69

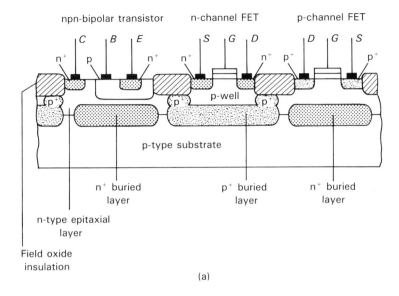

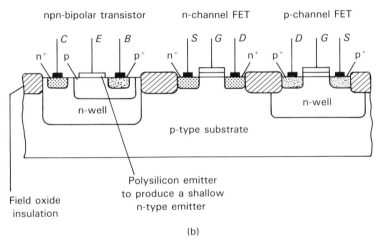

Figure 2.34 BiCMOS fabrication methods: (a) planar epitaxial fabrication on a p-type substrate with vertical npn bipolar transistors; (b) n-well planar fabrication with no epitaxial layer on a p-type substrate.

yet to be assured. At present it is not clear whether the bipolar transistors will be principally used on the output of CMOS digital circuits in order to give the good drive capability which CMOS logic cannot readily provide, or whether they will be used for analogue duties in a mixed analogue/digital environment, or both. Since up to eighteen mask stages are involved in the fabrication it is likely to remain a comparatively expensive and specialized technology [9], [32], [33].

2.5 Gallium-arsenide technology

Gallium-arsenide (GaAs) is a III/V compound which can be made in a perfect crystal form as the basis for semiconductor devices. Its energy gap is higher than that of silicon (1.4 eV compared with 1.1 eV for silicon) and its electron mobility is higher (0.85 m² V⁻¹ s⁻¹ compared with 0.14 m² V⁻¹ s⁻¹ for silicon), thus giving it the potential for high temperature and very-high-speed applications. Its hole mobility, however, is no better than silicon, (0.05 m² V⁻¹ s⁻¹), thus favouring n-channel activity rather than p-channel.

The active GaAs devices used in digital applications are metal-semiconductor junction field-effect transistors (MESFETs), which are normally used in a conventional switching mode to 0 V rather than in any pass-transistor (transmission gate) configuration. The logic circuits are either 'Schottky-diode FET logic' (SDFL), in which additional Schottky diodes perform the logic discrimination with the MESFET providing output invertion/amplification, or 'buffered FET logic' (BFL) in which the MESFETs perform the logic with the Schottky diodes for output level-shifting and signal isolation. Sub-nanosecond switching performance is possible.

The fabrication method in GaAs technology is basically very simple. A region of n-type material is formed by ion-implantation in an almost pure GaAs substrate, and ohmic source and drain connections made directly to it, as shown in Figure 2.35. Between these two contacts a metallic gate (aluminium) is laid down, which forms a Schottky-junction-diode connection with the n-type material. When the gate is made negative with respect to the source, this Schottky junction is reverse-

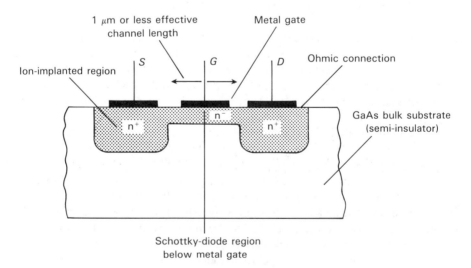

Figure 2.35 GaAs fabrication using ion-implanted n-region and metal-gate control electrode.

biased, and a depletion layer extends into the n-type channel, narrowing the area in which source-to-drain electrons can flow. The drain current can therefore be controlled by the value of V_{GS}. The normal JFET action has depletion-mode characteristics ($I_D \neq 0$ with $V_{GS} = 0$), but enhancement-mode characteristics can be achieved if the implanted n-type channel region is sufficiently thin that the Schottky depletion layer cuts off the channel with $V_{GS} = 0$ [1]. Terminologies D-MESFET and E-MESFET may be found for depletion-mode and enhancement-mode devices, respectively, with the technology sometimes termed 'D-mode GaAs' and 'E-mode GaAs', or, where both depletion- and enhancement-mode devices are used, 'E/D-mode GaAs' [33].

Because the GaAs substrate has a very high resistivity, the problems of isolating individual devices which are present in bulk-silicon technologies do not arise; isolation is inherently present. Also, the negligible device-to-substrate capacitance gives improved high-frequency performance. On the deficit side, however, must be weighed the difficulty and cost of producing the defect-free GaAs substrate, and also the lack of the equivalent of CMOS since a p-channel device in GaAs technology would show no advantage over silicon in terms of speed.

Several variations of the fabrication method shown in Figure 2.35 are possible, including growing an epitaxial n-type layer rather than ion-implantation [1]. For discrete devices a mesa structure built as islands sitting on top of the substrate may be found. However, to date, GaAs has not been widely used except in microwave applications, although some GaAs digital custom products are now available [34]–[36].

2.6 A comparison of available technologies

Having completed this overview of the various semiconductor technologies and their fabrication methods, an overall summary and comparison can now be made. It will, however, be appreciated that all technologies rely upon similar fabrication techniques, and are therefore subject to similar limitations in resolution and accuracy of the manufacturing steps. Any advance in, say, the resolution of optical lithography is applicable to all technologies.

The maximum digital capability of MOS and bipolar has continuously increased, as shown in Figure 2.36(a). MOS has always led in maximum gate count due not only to the generally smaller size of MOS gates but also because of lower power dissipation per gate. New developments in very low-power bipolar of sub-micron geometry may help to close this gap in the near future.

Gallium-arsenide capability is also growing fast, and has the potential to exceed the packing density of bipolar. Figure 2.36(b) gives a forecast of this capability, but cost may limit the take-up of this performance. Silicon-on-insulator technology, however, does not feature in any present general forecasts.

A comparison of the principle properties of the various technologies when used for digital purposes is given in Table 2.3. To some extent speed and power may be

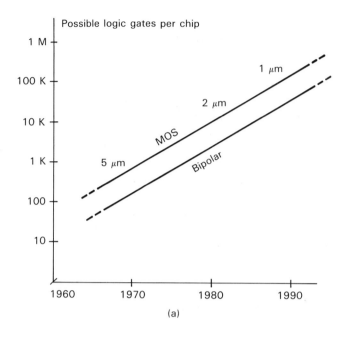

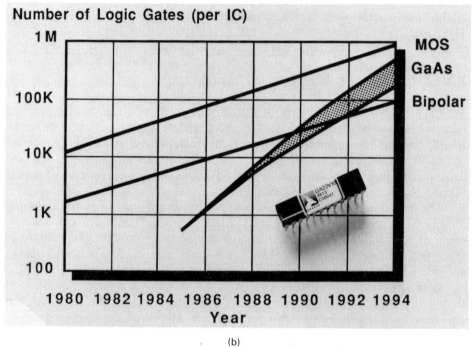

Figure 2.36 The general evolution of maximum digital capability per chip: (a) MOS and bipolar; (b) the forecast for GaAs. (Courtesy of Gazelle Microcircuits [35].)

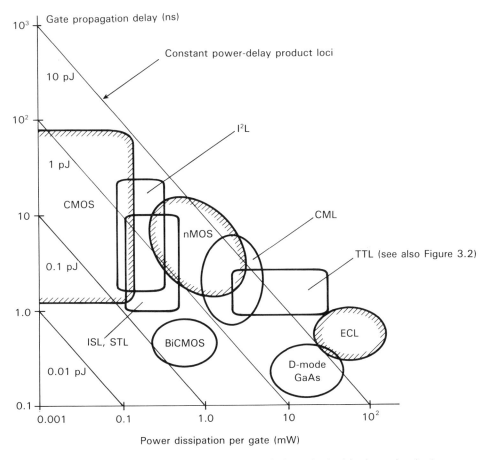

Figure 2.37 The speed–power performance of the principal logic technologies.

traded off against each other, and hence a broad graphical representation of these two parameters (as shown in Figure 2.37) may be more significant.

The continuing reduction in device geometries to sub-micron values is improving the packing density and reducing the power-delay product of all technologies, particularly BiCMOS and gallium-arsenide. GaAs has the potential of exceeding the packing density available in bipolar technology, as well as having a higher speed, but silicon technologies – either bipolar or MOS – will always remain dominant.

From an applications point of view, parameters other than those considered above may be important. For example, in analogue-only applications linearity rather than packing density may be more significant. Table 2.4 contrasts a range of general characteristics which may be of significance for specific applications.

The following two chapters will illustrate where these various technologies are employed in both standard products and custom ICs.

Table 2.3. Typical data for the main logic technologies

Technology	Gate delay (ns)	Gate power dissipation (mW)	Power-delay product (pJ)
Standard TTL	10	10	100
Low-power Schottky TTL	10	2	20
Advanced low-power Schottky TTL	5	1	5
I²L	15	0.5	7.5
ISL	4	0.25	1.0
STL	2	0.5	1.0
ECL	0.5	50	25
High-speed CDI CML	3	0.5	1.5
Silicon-gate nMOS	10	0.5	5
Silicon-gate CMOS	15	0.1[a]	1.5
Silicon-on-sapphire CMOS	10	0.1[a]	1.0
BiCMOS	0.75	1	0.75
GaAs	0.15	5	0.75

[a] Power dissipation proportional to the number of switching operations per second; these figures are typical for 1 MHz switching at 5 V.

Table 2.4 Relative attributes of common bipolar and MOS circuit capabilities: 4 = superior relative performance, 3 = relatively good, 2 = relatively moderate, 1 = relatively poor

	Bipolar npn	Bipolar I²L/ISL	Bipolar CDI	nMOS	CMOS
General					
Supply voltage tolerance	2	3	4	3	4
Power dissipation	1/2	2/3	2/3	3	4
Speed	4	2	3	3	3
Transconductance g_m	4	4	4	1/2	1/2
Drive capability	4	2/3	2/3	1/2	2
Analogue					
Gain per stage	4	[a]	3/4	1	1/2
Bandwidth	4	—	3/4	1	1/2
Input impedance	1/2	—	1/2	4	4
Output swing	4	—	2/3	2	2
Linearity	4	—	4	2	2
Precision passive devices	4	—	4	2	2
Transmission switches	1	—	1	2	4
Digital					
Max. switching speed	4	2	3	2/3	2/3
Logic swing	2	1	1	3	4
Voltage noise margin	2	1	1	3	4
Source and sink drive capability	4	3/4	3/4	1/2	2

[a] Not often employed for analogue duties.

Max. toggle frequency (MHz)	Gate packing density (2-input gates mm^{-2})	No. of fabrication masks required
35	20	7
40	20	8–9
50	50	8–9
15	100	8
40	75	10–12
50	75	10–12
200+	20	8
50	100	8
35	400	7
25	300	10–15
35	300	10–15
100+	200	12–18
300+	50	6–7

2.7 References

1. Goodge, M., *Semiconductor Device Technology*, Macmillan Press, London, 1983.
2. Jaeger, R. C., *Introduction to Microelectronic Fabrication*, Addison-Wesley Modular Series on Solid State Devices, Vol. V, Reading, MA, 1988.
3. Sze, S. M. (ed.), *VLSI Technology*, McGraw-Hill, New York, 1983.
4. Till, W. C. and Luxton, J. T., *Integrated Circuits: Materials, devices, fabrication*, Prentice-Hall, Englewood Cliffs, NJ, 1982.
5. Hodges, D. A. and Jackson, H. G., *Analysis and Design of Digital Integrated Circuits*, McGraw-Hill, NY, 1983.
6. Dillinger, T. E., *VLSI Engineering*, Prentice-Hall, Englewood Cliffs, NJ, 1988.
7. Sparkes, J.J., *Semiconductor Devices: How they work*, Van Nostrand Reinhold, Wokingham, UK, 1987.
8. Pierret, R. F., *Semiconductor Fundamentals*, Addison-Wesley Modular Series on Solid State Devices, Vol. I, Reading, MA, 1988.
9. Maly, W., *Atlas of IC Technologies: An introduction to VLSI processes*, Benjamin/Cummings Publishing, Menlo Park, CA, 1987.
10. Ruska, W. S., *Microelectronics Processing: An introduction to the manufacture of integrated circuits*, McGraw-Hill, NY, 1987.
11. Hart, K. and Slob, A., 'Integrated-injection logic – a new approach to LSI', *Proc. IEEE Int. Solid State Circuits Conf.*, 1972, pp. 92–7.
12. Hewlett, F. W., 'Schottky I^{2}L', *IEEE J. Solid State Circuits*, Vol. SC10, 1975, pp. 343–51.
13. Uyemura, J. P., *Fundamentals of MOS Digital Integrated Circuits*, Addison-Wesley, Reading, MA, 1988.
14. Hurst, S. L., *Custom-Specific Integrated Circuits: Design and fabrication*, Marcel Dekker, NY, 1983.
15. Rogers, T. J. and Meindl, D. J., 'VMOS, high speed TTL compatible MOS logic', *IEEE J. Solid State Circuits*, Vol. SC9, pp. 239–50.

16. Rogers, T. J., Jenne, F. B., Frederick, B., Barnes, J. J., Hiltpold, W. R. and Trotter, J. D., 'VMOS memory techniques', *IEEE ISSCC Digest*, February 1977, p. 74–5.
17. Baliga, B. J. and Chen, D. Y., *Power Transistors: Device design and application*, IEEE Press, NY, 1984.
18. Davis, R. D., 'The case for CMOS', *IEEE Spectrum*, Vol. 20, No. 10, 1983, pp. 26–32.
19. Mead, C. and Conway, L., *Introduction to VLSI Systems*, Addison-Wesley, Reading, MA., 1980.
20. Glasser, L. A. and Dobberpuhl, D. W., *The Design and Analysis of VLSI Circuits*, Addison-Wesley, Reading, MA., 1985.
21. Weste, N. and Eshraghian, K., *Principles of CMOS VLSI Design*, Addison-Wesley, Reading, MA., 1985.
22. Texas Instruments, *MOS Memory Data Book*, Texas Instruments, Houston, 2nd edition, 1986.
23. Schroder, D. K., *Advanced MOS Devices*, Addison-Wesley Modular Series on Solid State Devices, Vol. VI, Reading, MA, 1987.
24. Glazier, A. B. and Subak-Sharpe, G. E., *Integrated Circuit Engineering: Design fabrication and applications*, Addison-Wesley, Reading, MA, 1979.
25. Howes, M. J. and Morgan D. V. (eds.), *Large Scale Integration*, John Wiley, Chichester, UK, 1981.
26. Cirovic, M. M., *Handbook of Semiconductor Memories*, Reston Publishing Company, Reston, VA, 1981.
27. Santo, B., 'Technology '89: Solid State', *IEEE Spectrum*, Vol. 26, January 1989, pp. 47–9.
28. Technical report, *Nonvolatile Ferroelectric Technology and Products*, Ramtron Corporation, Colorado Springs, CO, 1988.
29. S. Weber, 'A new memory technology', *Electronics*, 18 February 1988, pp. 91–4.
30. Ramtron Corporation, *FM1008/1108/1208/1408 FRAM Data Sheets*, Ramtron Corporation, Colorado Springs, CO, 1989.
31. Maes, H. E., Groeseneken, G., Lebon, H. and Witters, J., 'Trends in semiconductor memories', *Microelectronics Journal*, Vol. 20, Spring 1989, pp. 5–58.
32. Dettmer, R., 'BiCMOS – getting the best of both worlds', *IEE Electronics and Power*, Vol. 8, 1987, pp. 499–501.
33. Zimmer, G., Esser, W., Fichtel, J., Hostika, B., Rothermal, A. and Schardein, W., 'BiCMOS: technology and circuit design', *Microelectronics Journal*, Vol. 20, Spring 1989, pp. 59–75.
34. Howes, J. J. and Morgan, D. V. (eds.), *Gallium-Arsenide: Materials, devices and circuits*, John Wiley, Chichester, UK, 1985.
35. Special Report: 'Inside technology: gallium-arsenide', *Electronics*, June 1988, pp. 65–85.
36. Morgan, D. V., 'Gallium-arsenide: a new generation of integrated circuits', *IEE Review*, Vol. 35, September 1988, pp. 315–19.

3 Standard off-the-shelf ICs

Although the principal aim of this text is the coverage of custom microelectronics, it is appropriate to include a reference to standard off-the-shelf integrated circuits for several reasons.

Firstly, the use of standard ICs still heavily outweighs the use of custom circuits both in volume and financial value, and is likely to continue to do so. However, every OEM design engineer has to consider the pros and cons using (a) only standard parts, or (b) custom ICs, or (c) some combination of the two, in every new product design, and thus knowledge of all types of product is necessary. These engineering decisions will form the Subject of Chapter 7.

Secondly, off-the-shelf ICs generally set the standards of capability and performance of the semiconductor industry. The design of such products is invariably state-of-the-art at the time of design, so that each product has a market edge over preceding designs or competitors' products. Custom ICs may be tailored to perform specific duties more efficiently than with an assembly of standard ICs, but the performance of the individual circuits cannot be better than that within a contemporary standard product. It follows that if there is an already-available off-the-shelf LSI product which does everything a design engineer requires, for example a video-display control IC or a keyboard controller, then there is no technical justification for designing the custom equivalent of this item.

The range of standard ICs may be divided into the following:

- technology (bipolar or MOS);
- duty performed (analogue, digital, non-programmable or programmable); and
- size and capability.

This is shown in Figure 3.1. To this classification must also be added the vast range of discrete semiconductor devices such as individual transistors, thyristors, zener diodes, light emitting diodes (LEDs), liquid crystal displays (LCDs), etc., which form an essential part of many systems but are not considered here.

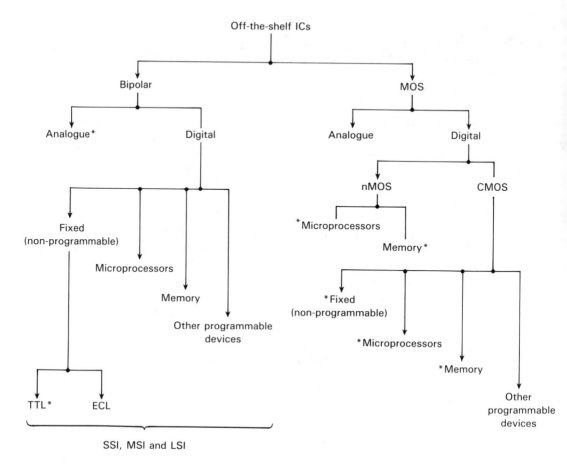

Figure 3.1 The range of off-the-shelf microelectronic components excluding discrete devices. The categories marked with an asterisk are particularly prominent.

3.1 Non-programmable SSI, MSI and LSI digital ICs

The greatest global use of microelectronic circuits remains in standard SSI, MSI and LSI digital ICs. Both bipolar and silicon-gate CMOS digital ICs are widely available, particularly in the 74** series products. The older metal-gate CMOS 4000 series products are becoming obsolescent, but may still be relevant if higher voltage working is desired, giving a 3–15 V operating range compared with the nominal 5 V of the 74** series.

The six principal categories of circuits available in the standard 74** series are as follows:

(a) Standard 74-series TTL;

(b) lower power Schottky 74LS-series TTL;
(c) advanced low-power Schottky 74ALS-series TTL;
(d) isoplanar high-speed FAST™ 74-series TTL;
(e) high speed 74HC-series CMOS; and
(f) high-speed 74HCT-series CMOS.

The four bipolar families are all designed for +5 V working with an operating temperature range of 0 to +70 °C; the two CMOS families allow an operating temperature range of −40 to +85 °C, with +5 V working for the 74HCT series. The general electrical performance is given in Figure 3.2. (Individual manufacturers of 74-series ICs may use slightly different designations from those cited here.) Two new series of CMOS circuits are also becoming more widely available: the advanced high-speed 74AC series and the advanced high-speed 74ACT series, both of which offer higher maximum speeds than the 74HC and 74HCT series.

The 74** range of ICs covers all the standard logic building blocks from the SSI 7400 quad 2-input NAND package to MSI arithmetic functions, counters, encoders, decoders and others [1]–[6]. Fabrication details are as discussed in the preceding chapters. Because of the familiarity of many design engineers with the use of these circuit elements, it is significant that in custom microelectronics most

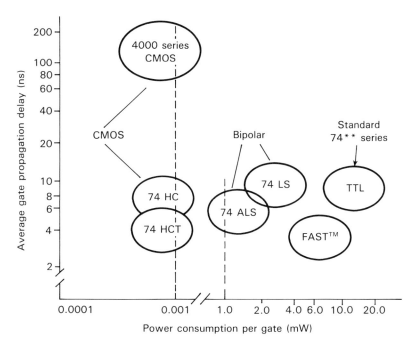

Figure 3.2 The power consumption/propagation delay performance of off-the-shelf 74** series bipolar and CMOS ICs and the older 4000-series CMOS circuits.

vendors have standard circuits for incorporation in custom ICs which mirror these standard products. Indeed, one of the prominent applications of custom microelectronics has been to replace a number of separate 74** series packages with a single functionally equivalent USIC so as to minimize product size and assembly time as well as to enhance overall product reliability.

The use of off-the-shelf emitter-coupled logic is much more specialized, and is only justified when the maximum performance available from 74** series ICs is inadequate. Very great care has to be taken when designing circuits using ECL packages in order that the maximum speed capability is not impaired, including the use of correctly terminated transmission-line interconnections for the high-speed data [9], [10].

3.2 Standard analogue ICs

The design of very-high-performance analogue circuits in microelectronic form largely remains a specialist area, since detailed knowledge of the fabrication process and its parameters is necessary in order to achieve maximum performance. For this reason it is improbable that any custom-designed analogue circuit will have a better performance than a vendor's standard product, although in many applications absolute performance will not be essential. Also, because standard analogue ICs are made in very large volumes, the cost per circuit is usually far less than can be achieved in a custom design.

The range of off-the-shelf analogue circuits is indicated in Figure 3.3. The operational amplifier forms the core of many circuits, and is available in very many versions from many vendors; a main distributor's catalogue may list as many as 200 or more variants in bipolar, mixed bipolar–FET and CMOS technologies [11]–[15]. All these products contain relatively few transistors compared with digital ICs, and thus are not classified by the terminologies SSI, MSI, LSI, ... Strictly speaking, analogue-only circuits lie outside LSI and VLSI discussions. However, since the transistors in analogue circuits are operating in their linear region rather than as on/off switches, the power dissipation per chip is inherently higher than in similar-size digital circuits.

The bipolar 741 operational amplifier was one of the first commercially available unconditionally stable op.amps. Its basic circuit diagram is given in Figure 3.4. There are now a number of variants on this design from different vendors, to improve the slew rate, bandwidth or other parameters [9], [15].

CMOS op.amps. are also widely available, particularly the 7600 series. In general, these do not provide such a good amplifying performance as bipolar, but can have advantages such as higher input impedance, lower cross-over distortion in output stages and lower total power consumption. Further commercial op.amp. families combining the best features of bipolar and MOS technologies are also available [12]–[17]. Representative performance figures are given in Table 3.1.

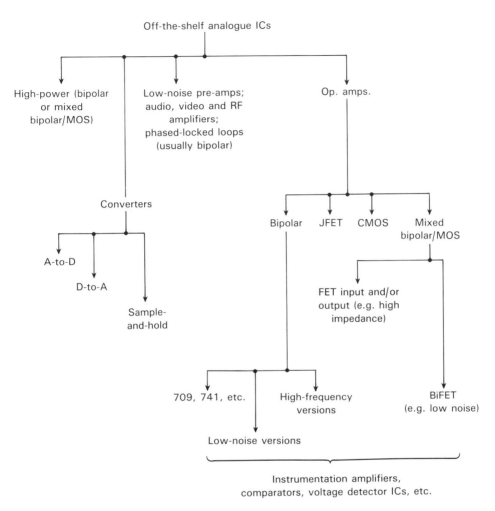

Figure 3.3 The range of off-the-shelf analogue circuits.

3.3 Microprocessors

Although the specific duties which a microprocessor has to perform are dictated by the user-specific software program, the microprocessor IC itself is a standard off-the-shelf product, the purchaser being unable to change the internal circuit configuration.

The microprocessor is basically a single-chip central processing unit (CPU) (see Figure 3.5(a)). To make a complete processing or computing system requires the addition of further circuits, mainly memory to store the software instructions and

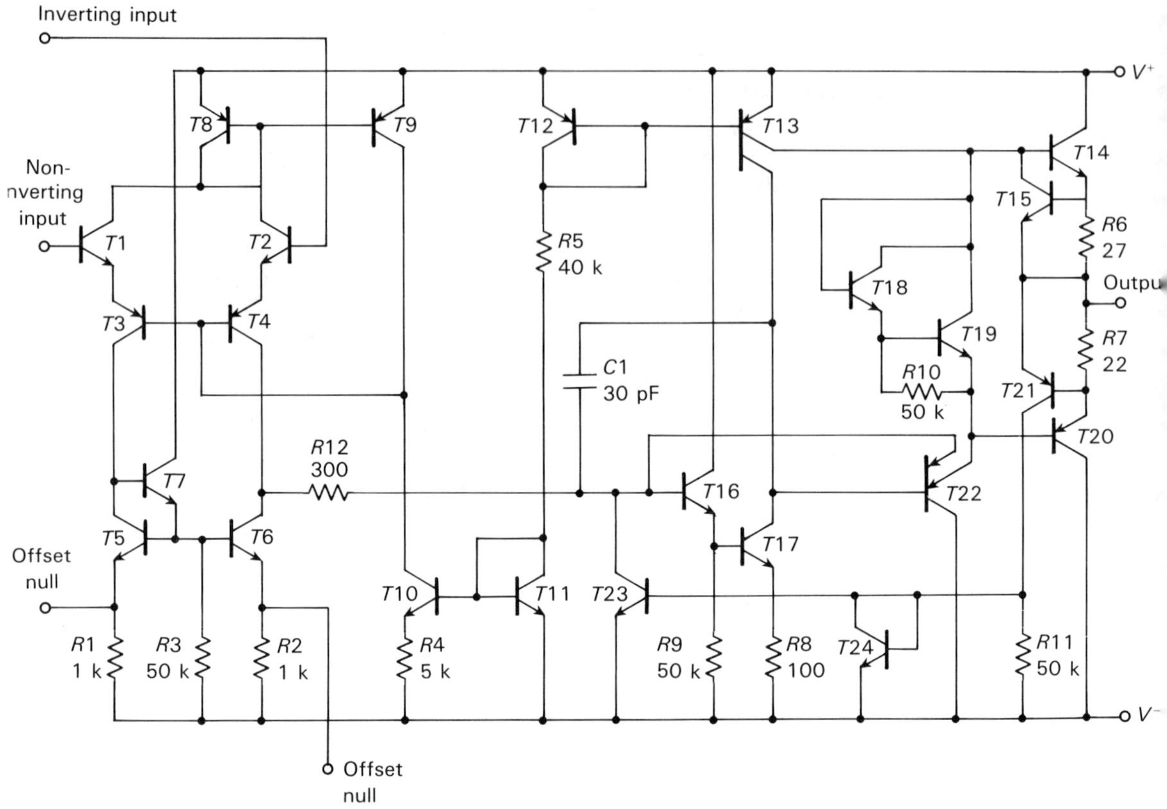

Figure 3.4 The circuit of the basic 741 operational amplifier.

Table 3.1. Performance figures for typical off-the-shelf op.amps.

Op.amp. type	Supply voltage range (V)	Max. differential input voltage (V)	Max. output swing (V)[a]	Operating temperature range (°C)
741N (bipolar)	±5 to ±18	±15	±13	0–70
741S (bipolar)	±5 to ±18	±15	±13	0–70
5539 (bipolar)	±8 to ±12	—	± 2.5	0–70
TL081 (BiFET)	±3 to ±18	±18	±13.5	0–70
LF351N (J-FET)	±5 to ±18	±18	±13.5	0–70
7611 (CMOS)	±1 to ±8	±8	± 4.5	0–70

[a] At max. supply voltage.
[b] Dependent upon operating point.

Microprocessors

other data, and input/output (I/O) circuits to interface the system with outside peripherals, as shown in Figure 3.5(b). Memory ICs will be the subject of Section 3.4.

A wide range of microprocessors and their support ICs are available from many sources. They may be characterized by:

- the technology, which may be bipolar or MOS;
- the width of the data bus which carries the system instructions and other information, and which may be 4-bits, 8-bits, 16-bits, 32-bits or, very recently, 64-bits wide.

The data bus widths used in some products may not be the same for all the system duties, for example for complex computational purposes a microprocessor is available with an 8-bit data bus combined with a 32-bit CPU. The Intel 80386SX family also uses different width busses, namely a 32-bit CPU with a 16-bit external data bus and a 24-bit external address bus. Terminology such as '32-bit architecture' may thus require classification by closer examination of the device specification.

While Schottky TTL, ECL, I^{2}L and CDI bipolar technologies have been used for microprocessors, increasing on-chip complexity and the requirement to reduce power dissipation has meant that bipolar has now very largely been replaced by MOS technologies. Some bipolar products may still be available for very-high-speed applications, or where radiation hardness is required for military or space applications. (Silicon-on-insulator MOS technology has been employed to provide radiation hardness, but this is very specialized and not commercially available.) Hence nMOS and CMOS currently dominates the microprocessor market with an

Max. power dissipation (mW)	Large-signal open-loop gain (dB)	Input resistance	CMR (dB)	Slew rate ($V \mu s^{-1}$)	Band width (kHz)
500	106	2×10^6	90	0.5	10
625	100	1×10^6	90	20	200
550	52	1×10^5	80	600	48 000
680	106	1×10^{12}	76	13	150
500	110	1×10^{12}	100	13	150
250	102	1×10^{12}	90[b]	0.16[b]	500[b]

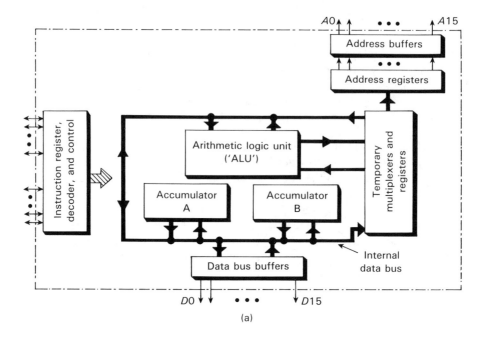
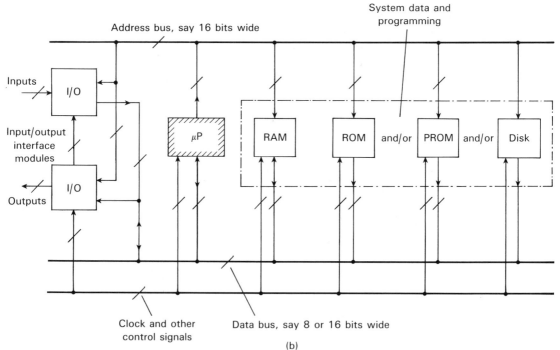

Figure 3.5 The standard microprocessor (µP): (a) the usual µP architecture; (b) the µP with typical associated circuits — in certain products some of this external memory may be on the µP chip itself instead of being additional separate ICs.

increasing availability of compatible CMOS and nMOS products [18]–[23]. The fabrication technologies are as covered in Chapter 2, the microprocessor and its associated chips being at the forefront of all VLSI fabrication developments due to the extremely large mass market for such products.

For many simple domestic and industrial control systems, 4-bit microprocessors such as the Texas 1000, 2000 and 3000 series are appropriate; the global sales of

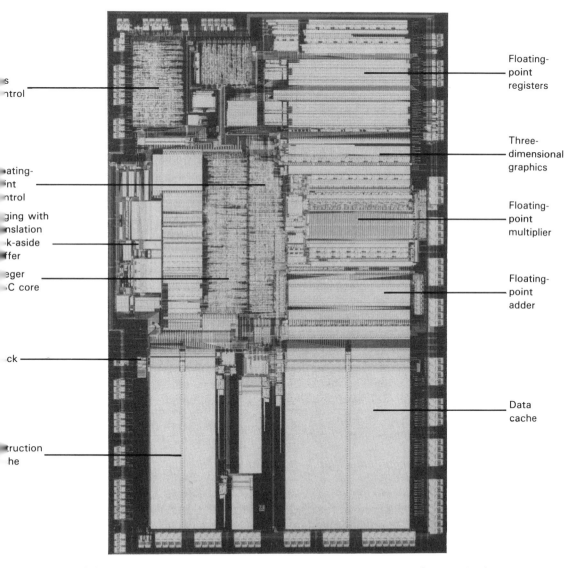

Figure 3.6 The 64-bit Intel i860 microprocessor. (Courtesy of Intel Corporation.)

these and similar 4-bit microprocessors is reported to exceed that of any other size of microprocessor. However, for more general-purpose industrial and scientific applications it is the 8-bit and 16-bit versions which dominate; for very computationally intensive applications the 32-bit, or even 64-bit, versions may be more appropriate. The range of commercially available products with their support ICs includes the following families:

- 8-bit Z80 series, nMOS and CMOS technology
- 8-bit 6500 series, nMOS and CMOS technology
- 8-bit 8600 series, nMOS and CMOS technology
- 8-bit 8080 series, nMOS technology
- 8-bit 8085 series, nMOS and CMOS technology
- 16-bit 8086 series, nMOS and CMOS technology
- 16-bit 68000 series, nMOS and CMOS technology
- 16-bit 9900 series, nMOS technology

In general where both nMOS and CMOS compatible products are present, the CMOS versions are later developments of earlier nMOS products, and are designed to give equal if not better performance with reduced power consumption. For example, the early nMOS 6500 series parts were designed to operate at 1 MHz, whereas the later CMOS parts provide a 2 MHz operating speed. However, the largest microprocessors currently available remain in nMOS technology, with operating speeds of 10 MHz or more. Figure 3.6 shows a state-of-the-art 64-bit product, which is possibly more powerful than is required by most OEM designers.

Microprocessors form an essential element for many digital systems. However, there is no question of any individual OEM designer ever designing a specific version for some custom product, as in no way could the performance and low cost of a commercial product be matched. However, the microprocessor can find a place in custom microelectronics by using a vendor's design in standard-cell ('megacell') form, where the microprocessor forms a part − possibly the major part − of the assembly of standard building-blocks which go to make up the unique custom chip. These considerations will be covered in more detail in Chapter 7.

3.4 Memory

Chapter 3, Section 2.3 gave an overview of the technologies, circuit configurations and fabrication details used in digital memory circuits. As noted there, memory ICs represent the pinnacle of transistor count and device packaging density per IC, aided by the regular characteristics possible with memory circuits.

3.4.1 *Read-only memory (ROM)*

Bipolar and MOS mask-programmable ROMs (see Figure 2.28) are not off-the-shelf ICs since the IC manufacturer has to pattern the required interconnections for a particular application. Original-equipment-manufacturers must therefore liaise with vendors for the supply of such parts and volume must be such as to justify their use.

Field-programmable ROMs, however, are very widely available, particularly in Schottky TTL bipolar technology ranging from 256 bits (32 × 8-bit) in 16-pin dual-in-line packages to 32 K (4 K × 8-bit) in 24-pin DIL packages, with access times of the order of 20 ns being possible. As an alternative to DIL packaging, surface-mount versions are also increasingly available.

In the range of erasable-re-programmable read-only memories (EPROMS), MOS technology employing the floating gate principle shown in Figure 2.30(b) is used in all standard products. In the type of EPROM where bulk erasure of all the chip programming is by exposure to ultra-violet light (the UV-PROM), available standard parts range from 8 K bits (1 K × 8-bit) in a 24-pin package to 4 M (512 K × 8-bit) in a 32-pin package. Both nMOS and CMOS technologies with access times as fast as 150 ns are represented, but with the CMOS versions providing considerably lower power dissipation and with a static current drain of around 1 mA at 5 V, compared with perhaps 50–100 times this value for nMOS equivalents. Figure 3.7 illustrates a typical UV-PROM. It is also possible to purchase single-chip microprocessor ICs complete with UV-PROM, a current example of which contains an 8-bit CMOS microprocessor plus 4 K × 8-bits of UV-PROM, together with RAM and I/O circuitry, all on the one chip.

Electrically erasable ROMs are also widely available. Currently available standard products, known as EE or E^2 PROMS, use either nMOS or CMOS technology, and supersede earlier, much slower, versions which were generally known as electrically-alterable-programmable ROMs (EAPROMs). EEPROMs are not available in such large sizes or with the access time performance of EPROMs: 64 K (8 K × 8-bit) with an access time of 250 ns is typically available, with up to 4 M being reported (1989) at the development stage.

The advantages and disadvantages of use of UV and EE re-programmable ROMs are as follows:

1. The UV-PROM cannot be selectively erased, and the bulk erasure process requires removal of the IC from the circuit and radiation for a considerable time – possibly 45 minutes – to erase the previous dedication pattern.
2. The EEPROM, however, can have selective word erasure, requiring a pulse of 10–20 V for a few milliseconds to effect a selective erasure – this may be done in-circuit if the voltage pulses can be safely applied to the complete circuit.
3. The number of possible erasures/re-programming cycles for EEPROMs is usually higher than for UV-PROMs. However, as previously noted, the

Standard off-the-shelf ICs

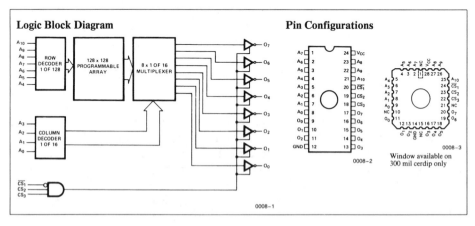

Figure 3.7 A CMOS technology 2048 × 8 bit UV-PROM, with maximum access time of 25 ns. (Courtesy of Cypress Semiconductor Corporation, CA.)

performance of EEPROMs is generally not as good as that available from UV-PROMs.

3.4.2 Random-access memory (RAM)

Random-access memory circuits are available in both static and dynamic form. Static RAMs ranging from 1 K (256 × 4-bit) to 256 K (32 K × 8-bit) are widely marketed. Larger sizes are becoming increasingly available. CMOS technology dominates the larger sizes, with access times approaching 30 ns for high-speed versions. A dominating characteristic of these CMOS RAMs is their competitive cost, currently $10 or less for a 256 K IC, and their low power as low as 5 to 10 μW static dissipation.

Commercially available dynamic RAMs have an even greater capacity than have static RAMs: 1 M × 1-bit configurations in 18-pin packages are widely available, with 100 ns access time and a cost of about $20. Still larger versions, up to 4 M, have been announced. Figure 3.8 illustrates a typical dynamic RAM chip.

Large-capacity memory circuits, therefore, are further building-blocks which are readily available at a price which cannot be approached by any custom design, and which are subject to intense continuous development. However, like the microprocessor, both ROM and RAM designs are available as megacells from many custom microelectronic vendors for incorporation into standard-cell custom chips, although it may still be more satisfactory and economical to keep very large memory requirements as separate ICs which are in large-scale and proven production.

Table 3.2 summarizes some representative information on commercial memory ICs. Further circuit and technical details may be found elsewhere [4], [24]–[29].

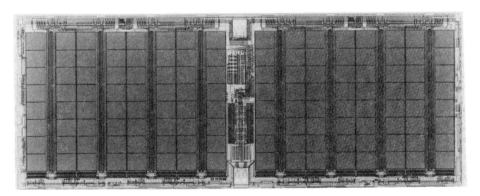

Figure 3.8 Floorplan of a 4 M dynamic RAM, die size 4.92 mm × 13.22 mm. (Courtesy of Fujitsu Microelectronics, UK.)

Table 3.2. Representative data on a small sample of commercially available memory ICs

Type no.	Device	Technology	Organization	Programmable means	Access time (ns)	Packaging (pins)
SN74371	ROM	Bipolar	256 × 8-bit	Mask	30	20
SN74387	PROM	Bipolar	256 × 4-bit	Fuse	30	16
23256	ROM	nMOS	32 K × 8-bit	Mask	200	28
53128	ROM	CMOS	16 K × 8-bit	Mask	250	28
82HS321	PROM	Bipolar	4 K × 8-bit	Fuse	35	24
27512	EPROM	nMOS	64 K × 8-bit	Electrical, UV erasable	200	28
27C1024	EPROM	CMOS	64 K × 16-bit	Electrical, UV erasable	150	40
28C64	EEPROM	CMOS	8 K × 8-bit	Electrically programmable/ re-programmable	200	28
43256	SRAM	CMOS	32 K × 8-bit	—	100	28
50464	DRAM	nMOS	64 K × 4-bit	—	120	18
42456	DRAM	CMOS	256 K × 4-bit	—	100	20
27C4001	EPROM	CMOS	512 K × 8-bit	Electrical, UV erasable	150	32

3.5 Programmable logic devices

The term 'programmable logic device' (PLD) theoretically covers any off-the-shelf IC which can be programmed after purchase by the customer to perform some specific logic duty. Hence PROMs (see Section 3.4, preceding) should come under this heading, although due to their principal use in computer systems they are normally considered as memory devices.

Programmable logic devices are capable of realizing any required logic function within the capacity of the device. It will be recalled that any Boolean logic expression can be written in a sum-of-products form, for example as shown in Figure 3.9. The product (AND) terms may be minterms, that is, containing all the input variables in either true or complemented form (the 'literals'), or minimized if possible so as to contain fewer terms in each AND function [30]–[32].

The direct realization of a Boolean expression or truthtable is therefore an AND level followed by an OR level. All programmable logic devices are based upon this, giving the architectural arrangement shown in Figure 3.10. More than one output is invariably provided.

However, the distinction between the many types of programmable logic device is in the details of the input decoding and in the AND and OR arrays. The OR array, for example, may be fixed rather than programmable. Also, although the two arrays are invariably referred to as 'AND' and 'OR', respectively, they may in practice both be made as either NAND arrays or NOR arrays. This is readily shown by applying De Morgan's theorem [30]–[32] to a sum-of-products expression; for

Programmable logic devices 91

Inputs			Minterm designation	Output function f(x)
x_1	x_2	x_3		
0	0	0	m_0	0
0	0	1	m_1	1
0	1	0	m_2	0
0	1	1	m_3	1
1	0	0	m_4	1
1	0	1	m_5	1
1	1	0	m_6	0
1	1	1	m_7	0

(a)

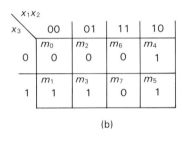

(b)

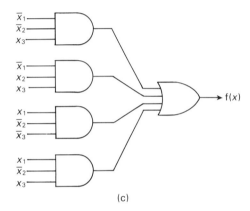

(c)

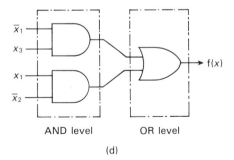

(d)

Figure 3.9 Example realization of a simple 3-variable Boolean function $f(X)$: (a) truthtable for $f(X)$; (b) Karnaugh map representation; (c) sum-of-minterms realization, $f(X) = \bar{x}_1\bar{x}_2 x_3 + \bar{x}_1 x_2 x_3 + x_1 \bar{x}_2 \bar{x}_3 + x_1 \bar{x}_2 x_3$; (d) minimized sum-of-products realization, $f(X) = \bar{x}_1 x_3 + x_1 \bar{x}_2$.

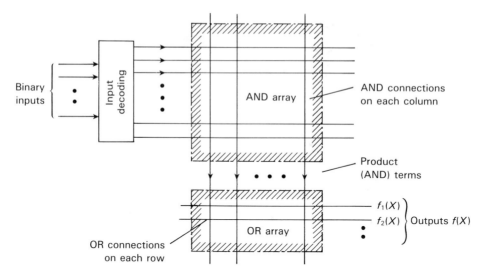

Figure 3.10 The basic architecture of programmable logic devices.

example in Figure 3.9(d) the expression

$$f(X) = \bar{x}_1 x_3 + x_1 \bar{x}_2$$

may be arranged as

$$f(X) = \overline{\overline{\{(\bar{x}_1 x_3) + (x_1 \bar{x}_2)\}}}$$

$$= \overline{\{\overline{(\bar{x}_1 x_3)} \cdot \overline{(x_1 \bar{x}_2)}\}}$$

which is now in all-NAND form. The alternative all-NOR form is obtainable by further manipulation, namely:

$$f(X) = \bar{x}_1 x_3 + x_1 \bar{x}_2$$

$$= \overline{\{\overline{(x_1 + \bar{x}_3)} + \overline{(\bar{x}_1 + x_2)}\}}$$

$$= \overline{\overline{\{\overline{(x_1 + \bar{x}_3)} + \overline{(\bar{x}_1 + x_2)}\}}}$$

Note, the latter equation expresses a sum-of-products realization using three NOR gates followed by a final Inverter gate. This is how some nMOS programmable devices may be structured. Notice also that the opposite polarity of the input signals is present in the NOR case compared with the NAND realization.

Referring back to Figure 2.29(a) it will be appreciated that programmable read-only memories (PROMs) are programmable devices where the n binary input variables are decoded by the input decoders and then passed through the arrays to form the required minterms. These minterms are wire-ORed together to form the output Z, and hence the OR array of Figure 3.10 is fixed and non-programmable in the PROM case. The four or eight or more separate AND arrays of the PROM form the 4-bit, 8-bit or more output word of the PROM, every input combination ('address') being programmed at minterm level to give the required output word. Each output, therefore, is a direct copy of a complete input/output truthtable.

3.5.1 *Programmable logic arrays (PLAs)*

While the PROM can be used as a powerful general-purpose device for combinational logic design, alternative programmable devices specifically designed for random logic applications are generally more appropriate. These can include sequential as well as combinational capability in the same package.

The principal distinction between PROMs and programmable logic arrays is that the PLA input variables $x_1, \ldots, x_n$ are not decoded into all the possible 2^n minterms. Instead, they are first decoded into the $2n$ literals, that is, x_i and $\bar{x}_i$ for each binary input $x_i, i = 1, \ldots, n$. The product (AND) array then can then make product terms containing as few or as many literals as necessary. This is shown in Figure 3.11.

The programming and the AND array thus provides the direct realization of the product terms for any Boolean expression. The OR array combines these terms as required to produce each output function $f(X)$, completing the synthesis of a sum-of-products expression.

The programming means in PLAs and the variants to be described shortly are the same as were discussed in Chapter 2, when considering memory circuits. Fusible links are employed in bipolar circuits and floating gates in MOS circuits (see Figure 2.30(a) and (b) respectively). All bipolar devices are therefore one-time-only programmable, but some CMOS devices are available which are erasable/re-programmable. However, there is generally not such a wide need for re-programmability in PLAs as in PROMs, since the latter are very often used in the software development phase of a microprocessor system design when design details are still fluid.

Conventions which are often used with programmable logic devices include the following:

- The buffered true and complement of each input is shown by a single symbol.
- The effective (programmed) connections in each array are shown by a dot where the vertical and horizontal lines cross.
- The product and sum lines are defined by single input AND and OR symbols respectively.

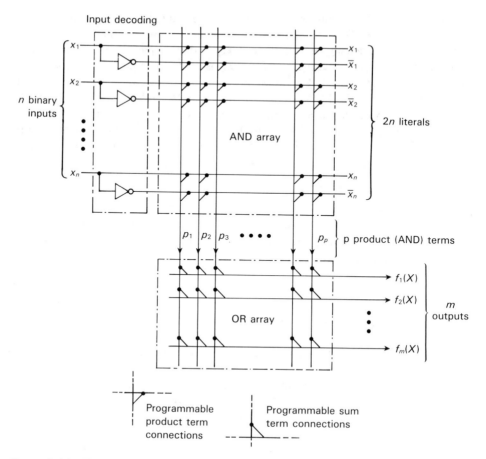

Figure 3.11 The general-purpose PLA with n inputs, p product terms and m outputs, with programmable AND and OR arrays.

These conventions are shown in Figure 3.12(a). (These conventions are used in particular for bipolar PLDs; for MOS devices when the input array may be a NOR structure, then the use of these symbols is perhaps misleading.) Additional features may also be found in commercial products, including the following:

- the ability to feed back some or all of the outputs to the AND array to act as common functions or to enable latches to be constructed; and
- Enable and/or Inversion facilities on the output lines to give additional logic flexibility.

These enhancements are shown in Figure 3.12(b). Additionally, some I/Os may be programmed to act as inputs or as outputs, as required.

Programmable logic devices

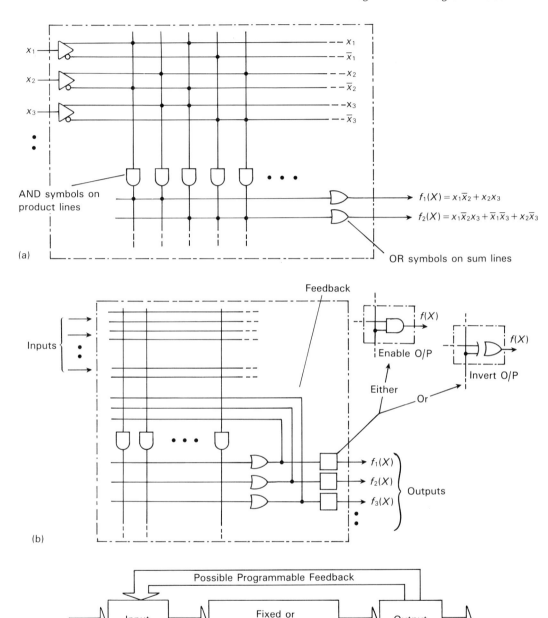

Figure 3.12 Further PLA details: (a) input and other circuit conventions, with two simple output functions illustrated; (b) additional facilities which may be incorporated; (c) the comprehensive block schematic used by some vendors.

The PLA and the following variants thus provide very useful off-the-shelf packages for many small logic design applications. The three principal technical limitations are in capability, specifically:

(a) the number of inputs n available – clearly, one 8-input PLA for example cannot be used if product terms containing more than eight inputs are required;
(b) the number of outputs m available; and
(c) the number of internal product lines p available – clearly if the m outputs collectively involve more than p product terms there will be insufficient internal AND array capacity to synthesize all the required functions.

The use of PLAs thus usually involves multioutput Boolean minimization procedures in order to realize the several outputs with a minimal number of product terms; absolute minimization may not be necessary provided the number of product terms is not greater than the device capacity. Most vendors of programmable devices and programming equipment have appropriate commercial CAD software packages available to undertake this design requirement automatically from, say, truthtables or non-minimum Boolean expressions. What available CAD software will generally not do is to partition a given system requirement which will not fit into one PLA, so that multiple PLAs in some optimum parallel or series/parallel combination can be used. This is a partitioning or decomposition problem for which no simple design algorithm exists, and which in general relies upon human experience.

Further details of the capabilities of, and design techniques for, PLAs may be found in many sources [4], [28], [33]–[38]. The designation FPLA – field-programmable logic array – will also be used by certain manufacturers [37].

3.5.2 Programmable array logic (PALs)

A widely available variant on Figure 3.11 is the programmable array logic or PAL™ device, whose basic architecture is similar to Figure 3.11 except that the programmable OR array is replaced by fixed OR gates. (PAL is a registered trademark of Monolithic Memories, who first introduced this type of PLD to the commercial market.) Additionally, the outputs from some of the OR gates are usually fed back to the AND array (see Figure 3.12(b)) in order to increase the device capabilities. Selective output inversion through Exclusive-OR gates may also be available [33]–[35], [39].

Figure 3.13 illustrates a typical small PAL product. The fixed number of inputs into each internal OR gate is the principal disadvantage of the PAL in comparison with the PLA, although the feedback loops may be used to augment this fan-in at the expense of increased propagation time.

One additional feature with certain vendors' PAL devices may be noted. This is the availability of the equivalent PAL device with mask-programming of the

Programmable logic devices

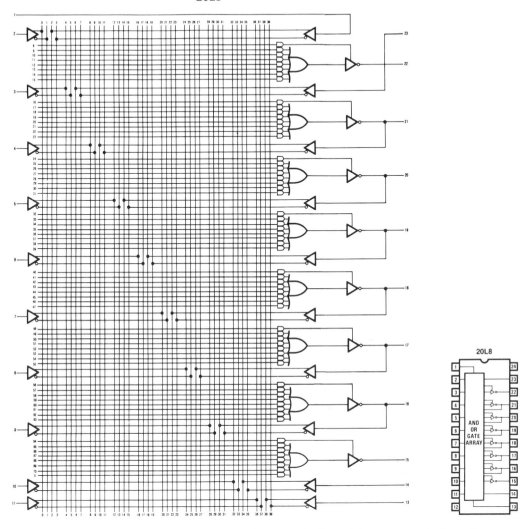

Figure 3.13 A bipolar technology PAL, fourteen dedicated inputs, two dedicated outputs, six programmable input/outputs, 15 ns typical propagation delay. (Copyright © Advanced Micro Devices, Inc., (1988). Reprinted with permission of copyright owner. All rights reserved.)

custom requirements. Such devices are known as HALs – hard array logic devices – and may be more economical where large quantities of a particular PAL circuit are found to be needed. As with ROMs the vendor rather than the OEM has to manufacture these dedicated devices.

3.5.3 Programmable gate arrays (PGAs)

The programmable gate array, frequently called a field-programmable gate array or FPGA, is a yet simpler version of the architecture of Figure 3.11. In PGAs no OR gates are provided, but comprehensive feedback and inversion facilities are available instead. A sum-of-products expression such as $\bar{x}_1 x_2 x_3 + x_1 \bar{x}_3 x_4$ is therefore available in the alternative form of

$$\left\{ \overline{(\bar{x}_1 x_2 x_3) \cdot (x_2 \bar{x}_3 x_4)} \right\}$$

by using the feedback and inversion resources. Figure 3.14 illustrates a typical commercial architecture.

3.5.4 Programmable logic sequencers (PLS)

The programmable devices so far considered do not contain any dedicated storage (memory) circuits. This additional facility will be found in programmable logic sequencers, sometimes termed 'field-programmable logic sequencers' (FPLS).

Normally, both AND and OR programmable arrays are present, with clocked D-type, RS-type or JK-type storage circuits at the outputs. The Q and $\bar{Q}$ outputs of these circuits are usually fed back into the AND array, as illustrated in Figure 3.15.

Some PAL devices with the AND-only architecture of Figure 3.14 are also available with similar dedicated storage circuits, but the designation 'PAL' is retained for these commercial products, rather than PLS.

Table 3.3. Some representative commercial programmable logic devices

Type no.	Device architecture	Technology	Organization $n \times p \times m$[a]	Re-programmable
93459	PLA	TTL	16 × 48 × 8	No
16L8A	PAL	TTL	10 × 64 × 8	No
C16L8	PAL	CMOS	10 × 64 × 8	Yes
C22V10	PLS	CMOS	12 × 120 × 10	Yes
PLS179N	PLS	TTL	8 × 45 × 12	No
10H20EG8	PLS	ECL	12 × 90 × 8	No
405-45	PLS	TTL	16 × 64 × 8	No
EP1810	PLS	CMOS	16 × NA × 48	Yes

[a] Organization $n \times p \times m = n$ dedicated inputs, p internal product terms, m dedicated outputs or reconfigurable I/Os.

3.5.5 Other programmable standard products

Other variants on these programmable logic devices have been marketed, including programmable diode matrix (PDM) devices which are effectively arrays of diodes from which passive diode–AND and diode–OR networks can be made. PDMs are now generally obsolete. Other variants on the PLA, PGA and PLS architectures, however, may be marketed. For very large devices, dynamic rather than static MOS circuits may be encountered, mirroring the developments in ROM technologies. Table 3.3 gives a short representative sample of available off-the-shelf products.

Units for programming these various types of programmable devices are also readily available (see Figure 3.16). The more sophisticated microprocessor-controlled units cater for a very wide range of devices, and include check and verification means to ensure that the device is programmed correctly to the required specification.

There are, however, two further commercial products which are distinctly different from these PLD products. One is the programmable gate array (PGA) and the other is the programmable logic controller (PLC).

PGAs are unique devices falling somewhere between a conventional PLD and a custom gate-array IC. They will be considered separately in the following section. The PLC on the other hand is not a single IC package but, rather, a very specific commercial product designed to form an electronic replacement for electromagnetic control systems. It consists of the following four major parts:

(a) an isolating input section (an input module or modules);
(b) an electronic central processing module (CPU);

On-chip storage circuits	Access time (ns) or max. clock speed (MHz)	Packing (pins)
None	25 ns	28
None	7.5 ns[b]	20
None	25 ns	20
10	25 ns	24
8	45 ns	24
8	3.5 ns	24
8 JK buried, 8 JK brought out	45 MHz	28
48 programmable D, T, RS or JK	35 ns	68

[b] High-speed version.

100 Standard off-the-shelf ICs

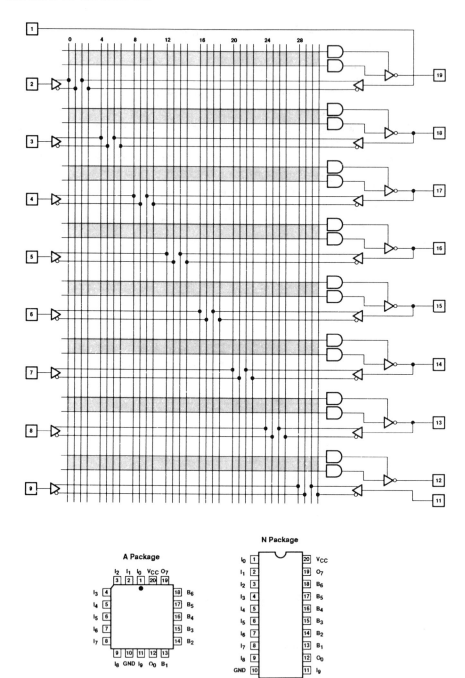

Figure 3.14 The field-programmable gate array (FPGA). (Courtesy of Signetics/Philips Components Ltd.)

Programmable logic devices

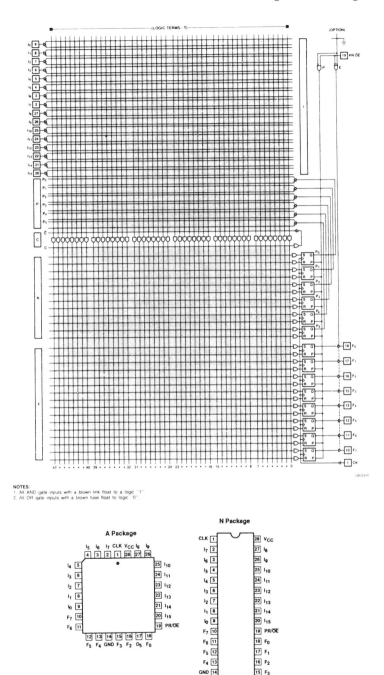

Figure 3.15 The field-programmable logic sequencer (FPLS). (Courtesy of Signetics/Philips Components Ltd.)

102 Standard off-the-shelf ICs

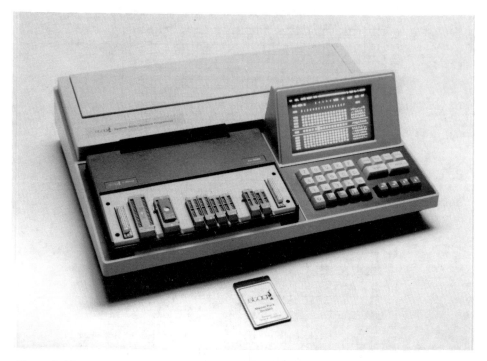

Figure 3.16 A commercial programming unit for dedicating a wide range of programmable logic devices. (Courtesy of Stag Microsystems Ltd.)

(c) an isolating power output section (an output module or modules); and
(d) a power supply unit (PSU module).

Further facilities such as timers, printer interfaces, etc., are also available as part of a boxed PLC system.

The CPU, which contains RAM and PROM or EPROM memory, is programmed by the user to undertake the required control tasks, operating upon system input data supplied via the input modules and providing output control signals via the output modules. The PLC assembly is thus a *power electronics system* with microelectronics providing the central control, rather than an entirely microelectronics system. Further details of these very useful control engineering products may be found elsewhere [40], but will not be referred to again in this text.

3.6 Logic cell arrays (LCAs), sometimes termed programmable gate arrays (PGAs)

Since the logic cell array (LCA^{TM}) is fundamentally different from other PLDs, we will consider it as a further type of off-the-shelf, but user-programmable,

Logic cell arrays (LCAs)   103

component. (LCA is a registered trademark of Xilinx, Inc., who introduced this device to the commercial market.)

In comparison with many other types of PLD, the LCA is a large chip of VLSI complexity. It employs CMOS technology with circuit features as small as 1.2 μm [41], [42]. Its architecture consists of an array of identical cells separated by wiring channels, together with peripheral I/O cells, as shown in Figure 3.17(a). However, unlike a semicustom gate-array IC (see Chapter 4) where the metallic interconnectors in the wiring channels have to be custom-designed and vendor-fabricated, in the LCA there are fixed segments of metal interconnect in all the row

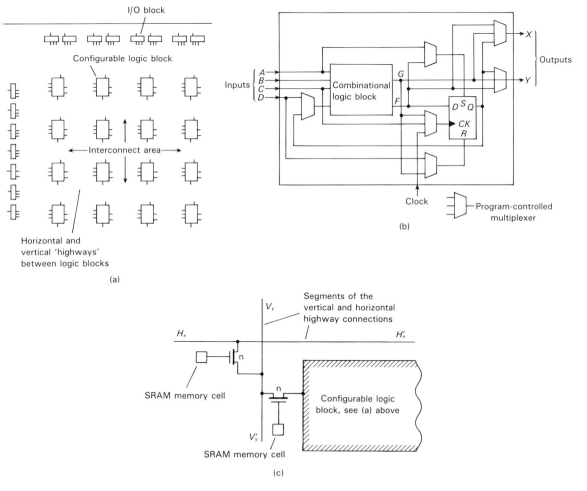

Figure 3.17 The field-programmable logic-cell array (LCA): (a) general architecture of the chip; (b) the standard cell (configurable logic block) details; (c) the FET configuration switches controlled by static RAMs.

and column wiring channels, with programmable means to connect segments together to form any required interconnection pattern between all cells and I/Os.

The cells themselves are also programmable. Each consists of a number of logic elements, as shown in Figure 3.17(b), with multiplexers which can also be programmed to turn the cell into a wide range of combinational and sequential circuit duties.

The actual programmable chip connections are performed by n-channel field-effect transistor switches which are present between all adjacent interconnection segments. With an FET 'off', adjacent segments are effectively isolated from each other; but with the FET 'on', the segments form a connected path. Control of all the FET switches on the chip therefore configures it to a particular custom requirement.

The gates of all these FET switches are, in turn, each controlled by an individual static RAM cell (see Figure 3.17(c)), each of which is programmed to the on or off state by the user's program. Thus, underlying all the functional cells and interconnection segments is a comprehensive array of RAM cells which may be individually addressed and controlled to one state or the other by a separate interconnection (programming) matrix.

The vendors of LCAs provide unique CAD software to enable the OEM designer to design and verify the dedication ('configuration') program. When complete, this program may be stored either on a hard disk or in a separate 8-pin programmable read-only memory (ROM) chip. However, it should be noted that as the LCA chip itself is configured by the internal RAM patterning, the LCA is a *volatile device* which loses its dedication on power-off. When power is switched on or restored, the configuration program has to be re-loaded into the LCA; in this respect the LCA is similar to any microprocessor system where the volatile RAM data has to be re-loaded after any power-off.

It will be appreciated that the LCA chip is a complex device. Nevertheless, due to the absence of area-consuming fuse elements and the use of leading-edge MOS technology, much larger capability is at present available in LCAs compared with most other PLD ICs. Table 3.4 gives a current comparison, although it should be noted that chip capability in all PLD areas is growing continuously.

The disadvantage of the LCA being a volatile device is overcome in another vendor's product which employs 'antifuse' technology to configure the array permanently [43]. The PLICE^{TM} (programmable low impedance circuit element) antifuses are two-terminal devices consisting of an extremely thin dielectric barrier between two electrodes which when an 18 V programming pulse is applied breaks down to form a bidirectional ohmic path of about 1 kΩ compared with its previous resistance of about 100 MΩ. (PLICE is the registered trade mark of Actel, Inc., CA.) A 2000-gate IC contains 186 000 antifuse elements, roughly 100 per logic gate, but only a small percentage have to be blown to configure a chip to a specific duty. Typical product specifications are as shown in Table 3.5. Figure 3.18 illustrates such a product.

Table 3.4. The general comparative characteristics of five ways of making random digital logic networks. Details of vendor-customized gate-array and standard-cell ICs follow in Chapter 4

	Off-the-shelf devices			Vendor-customized devices	
	SSI and MSI ICs	Programmable logic devices (PLDs)	Logic cell arrays (LCAS)	Gate-array custom ICs	Standard-cell custom ICs
Capacity in no. of gates	Tens to few hundred	Hundreds to few thousand	Thousands to tens of thousands	Thousands to tens of thousands	Tens of thousands to hundreds of thousands
Speed	Medium to fast	Slow to medium	Slow to medium	Slow to fast	Medium to fast
Functionality defined by user	No	Yes	Yes	Yes	Yes
Time to customize	—	Seconds	Seconds	Depends upon many factors – possibly months' turn-around	
User-programmable	No	Yes	Yes	No	No

Source: based upon [42].

Table 3.5. The product specification for a range of PLICE™ devices

Device type	ACT 1010	ACT 1020	ACT 1230	ACT 1260
CMOS process	2 μm	2 μm	1.2 μm	1.2 μm
No. of logic cells	295	546	720	1404
Equiv. number of gates	1200	2000	3000	6000
Max. no. of latches	295	546	700	1400
Max. no. of D-type circuits	147	273	540	1052
No. of PLICE antifuses	112 000	186 000	333 000	666 000
No. of I/Os	57	69	108	150

It will be appreciated that the volatile LCA is fully re-programmable without any restrictions, but the non-volatile antifuse device is one-time-only programmable.

At present it is not established what the market for these products will become, since they strongly overlap the capabilities of vendor-dedicated semicustom gate arrays. This will be referred to again in subsequent chapters. And finally there is a problem with terminology: these volatile and non-volatile cell array devices are sometimes referred to as 'programmable' or 'field-programmable gate arrays' [42]–[44]. This can cause confusion with other products such as the simpler PGA products covered in Section 3.5.3, and is another example where care has to be taken to identify the internal architecture of available products correctly.

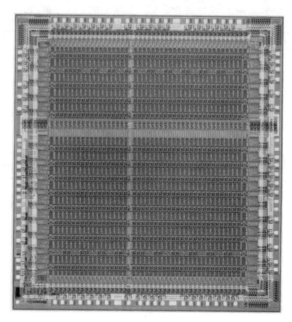

Figure 3.18 The one-time-only ACT ('application configurable technology') cell array which employs antifuse programming. (Courtesy of Actel, Inc.)

3.7 Specialized off-the-shelf ICs (ASICs)

(See p. 18 at the beginning of Chapter 2 on the use of the terms ASIC, USIC and CSIC. In this section we are talking about ICs for very specific duties, which are therefore increasingly referred to as ASICs. A further term ASSP ('application-specific standard part') has also recently been seen. It is a pity that there is not yet any consensus of opinion or standardization on all these terms.)

There is an increasing number of standard ICs, usually of LSI or VLSI complexity, which are becoming available for very specific purposes, having been designed and optimized for the one duty. It is an interesting evolution in that an IC for a special purpose may start its life as a user-specific custom IC (USIC), but when a supplier appreciates its global market potential then its design may be fine-tuned to improve its performance or capability and marketed as a catalogue item. The product then ceases to be the property of one customer. Some vendors who were initially suppliers of custom ICs only, for example LSI Logic Corporation, now list standard parts in their catalogue; all this makes it increasingly important to try to distinguish between USICs – specific to one customer – and ASICs, which are specific to an application but available to anyone.

Among the products which may be found in the vendors' and distributors' catalogues are the following:

- digital clock and clock driver circuits,
- LED and LCD drivers,
- keyboard, printer, video-display unit, disk drive and communications controllers,
- speech synthesis circuits,
- smoke detection circuits,
- temperature, strain gauge and other instrumentation circuits,
- signal processing circuits,

and others. If such a product were available to match a particular OEM requirement, then in purely financial terms it would rarely be appropriate for an OEM to consider designing a custom IC (USIC) for this duty.

3.8 Summary

This overview of off-the-shelf parts has been considered necessary since custom microelectronics will never supersede the use of standard parts in all circumstances. The choice between the use of standard parts and custom ICs is therefore a crucial engineering decision which every OEM designer must face when considering a new product design.

Having covered this review we may now move on to the principal theme of this text, namely *custom microelectronics*, but in Chapter 7 we will return to consider the choice of design style for a particular product and the pros and cons of custom ICs vs. standard parts.

1.6 References

(Note, the many very comprehensive handbooks and data books produced by suppliers are regularly updated, and therefore no date of publication has been cited below in such references.)

1. Williams, A. B., *Designer's Handbook of Integrated Circuits*, McGraw-Hill, NY, 1984.
2. Parr, A. E., *Logic Designer's Handbook*, Granada Press, 1984.
3. Marston, R. M., *CMOS Circuits Manual*, Heinemann, London, 1987.
4. Texas Instruments, *The TTL Data Book*, Vol. 1, Vol. 2 and Vol. 3, Texas Instruments Inc., Dallas, TX.
5. National Semiconductor, *Logic Data Book*, Vol. 1, Vol. 2 and Vol. 3, National Semiconductor Corporation, Santa Clara, CA.
6. Texas Instruments, *High-speed CMOS Logic Data Manual*, Texas Instruments Inc., Dallas, TX.
7. GE–RCA, *High-speed CMOS Logic Data Book*, General Electric–RCA, Somerville, NJ.

8. Mullard-Signetics, *Technical Data Book 4*, Parts 2, 4, 5, 8 and 8(a), Mullard Ltd, London.
9. Sedra, A. S. and Smith, K. C., *Microelectronic Circuits*, Holt, Reinhart and Winston, NY, 1987.
10. Motorola, *MECL High Speed Integrated Circuits*, Motorola Semiconductor Products, Inc., Pheonix, AZ.
11. Texas Instruments, *Linear Circuits Data Book*, Texas Instruments, Inc., Dallas, TX.
12. National Semiconductors, *Linear Data Book*, National Semiconductor Corporation, Santa Clara, CA.
13. Analog Devices, *Data Acquisition Book*, Analog Devices, Inc., Norwood, MA.
14. Linear Technology, *Linear Databook*, Linear Technology, Inc., Burlington, Canada.
15. Gayakwad, R. A., *Op-Amps and Linear Integrated Circuits*, Prentice-Hall, Englewood Cliffs, NJ, 1988.
16. Texas Instruments, *LinCMOS Design Manual*, Texas Instruments, Inc., Dallas, TX.
17. Texas Instruments, *BiFET Design Manual*, ibid.
18. Comer, D. J., *Microprocessor-based System Design*, Holt, Rinehart and Winston, NY, 1986.
19. Cassell, D. A., *Microcomputers and Modern Control Engineering*, Prentice-Hall, NJ, 1973.
20. McGrindle, J. A., *Microcomputer Handbook*, Collins, London, 1985.
21. Whitworth, I. R., *16-bit Microprocessors*, Collins, London, 1987.
22. Mitchell, H. J. (ed.), *32-bit Microprocessors*, Collins, London, 1986.
23. Ciminiera, L. and Valenzano, A., *Advanced Microprocessor Architectures*, Addison-Wesley, NY, 1987.
24. Goodge, M., *Semiconductor Device Technology*, Macmillan Press, London, 1983.
25. Dillinger, T. E., *VLSI Engineering*, Prentice-Hall, Englewood Cliffs, NJ, 1988.
26. Texas Instruments, *MOS Memory Data Book*, Texas Instruments, Inc., Dallas, TX.
27. Intel, *Microprocessors Handbook*, Vol. 1 and Vol. 2, Intel Corporation, Santa Clara, CA.
28. Cypress Semiconductor, *CMOS Data Book*, Cypress Semiconductor, San Jose, CA.
29. *IEEE J. Solid State Circuits*: special issue on semiconductor memory annually each October.
30. Muroga, S., *Logic Design and Switching Theory*, John Wiley, NY, 1979.
31. Lewin, D., *Design of Logic Systems*, Van Nostrand Reinholt, NY, 1985.
32. Barna, A. and Porat, D. I., *Integrated Circuits in Digital Electronics*, Wiley, NY, 1987.
33. Bolton, M. J. P., *Digital System Design with Programmable Logic*, Addison-Wesley, Wokingham, UK, 1990.
34. Bostock, G., *Programmable Logic Handbook*, Blackwell, Oxford, 1987.
35. Hurst, S. L., *Custom-Specific Integrated Circuits: Design and fabrication*, Marcel Dekker, NY, 1985.
36. Advanced Micro Devices, *Programmable Logic Handbook*, Advanced Micro Devices Inc., Sunnyvale, CA.
37. Signetics, *PLD Data Manual*, Signetics Corporation, Sunnyvale, CA.
38. Intel, *Programmable Logic Handbook*, Intel Corporation, Santa Clara, CA.
39. Monolithic Memories, *PAL Programmable Logic Handbook*, Monolithic Memories, Inc., Santa Clara, CA.
40. Kissell, T. E., *Understanding and Using Programmable Logic Controllers*, Prentice-Hall, Englewood Cliffs, NJ, 1986.
41. Xilinx, *Programmable Gate Array Design Handbook*, Xilinx, Inc., Santa Clara, CA.

42. Freeman, R., 'User-programmable gate arrays', *IEEE Spectrum*, Vol. 25, December 1988, pp. 32–5.
43. Actel Inc., *The Beginner's Guide to Programmable ASICs*, Actel Inc., Sunnyvale, CA.
44. Harding, W., 'New design tools revive in-circuit design verification', *Computer Design*, Vol. 28, 1 February 1989, pp. 28–32.

4 Custom microelectronic techniques

The custom microelectronic techniques to be discussed in this chapter, and which will constitute the subject area of all the remaining chapters of this text, now form a highly significant sector of the complete microelectronics spectrum. The technical reasons for the emergence of custom microelectronics have been covered in Chapter 1, namely the increasing maturity of both silicon fabrication and computer-aided-design resources; the practical reasons for considering the use of custom circuits in preference to that of standard off-the-shelf ICs is to attempt to achieve a final product which is better in some way than can be obtained using existing standard parts.

Among the advantages which can or may be achieved in adopting custom microelectronics are the following:

- The final market product is unique, so potential competitors cannot readily copy the design using standard parts.
- The performance of the product may be improved above that possible with standard parts.
- The product designer may have a greater freedom to innovate if necessary, not being constrained by standard parts.
- The final product size may be reduced considerably.
- Power consumption may also be reduced, allowing smaller power supplies.
- Product assembly, test and documentation may be reduced in comparison with an assembly of standard parts.
- Product reliability may be improved due to the reduction in the number of individual parts and their interconnections.
- Possibly a reduced design and development time due to appropriate computer-aided-design resources.
- Possibly a reduced final product cost due to the saving in component cost and product size.

These are the principal advantages claimed for custom circuits. Whilst there are products where the adoption of a custom IC is obviously advantageous or indeed, necessary, for example in hand-held battery-operated products where size is at a premium, not all the above claims are necessarily realized or indeed are significant for a particular product design. Company decisions when or when not to use custom ICs for a new product design will form the major part of the penultimate chapter.

The principal sub-divisions of custom microelectronic circuits are shown in Figure 4.1. As briefly introduced in the first chapter, full custom is where every detail of the custom IC is specifically designed to achieve the best end-product. Standard-cell and gate-array techniques, however, rely upon pre-designed circuits or chip layouts, respectively, in order to minimize expensive design time in completing the final IC fabrication details. These categories and their sub-divisions will be considered in the following sections.

Programmable logic devices (PLDs) have also been included in Figure 4.1, since they are a form of custom-dedicated device. In particular, the larger sizes are becoming increasingly competitive with gate arrays. However, in this chapter we shall have no need to add to the technical data on PLDs which has already been

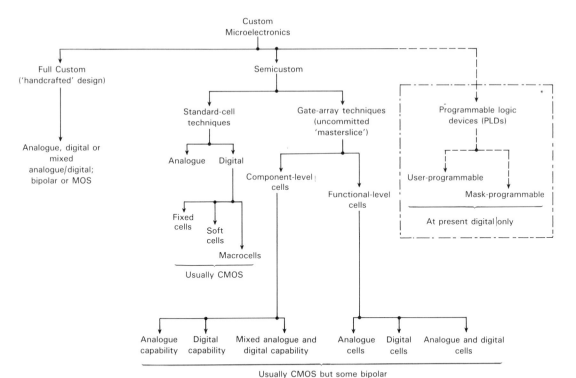

*See Chapter 3, Sections 3.5 and 3.6

Figure 4.1 The hierarchy of custom microelectronic ICs (USICs); at present digital-only products dominate the market. The terminology 'gate array' is perhaps inexact for products used for analogue purposes.

given, but they will be referred to again in subsequent pages dealing with CAD and the choice of design style.

It should also be recalled that custom ICs do not involve any novel fabrication techniques. Indeed, proven stable fabrication processes are essential for the reliable and economic production of custom circuits. The fabrication details considered in Chapter 2 thus cover all the techniques that may be involved in the custom field.

Finally, the term USIC (user-specific IC) rather than ASIC (application-specific IC) will be used in the following pages, for the reasons given on page 18.

4.1 Full hand-crafted custom design

Full-custom USICs are designed to meet the required product specification in every detail, the emphasis being on maximum possible circuit performance and minimum possible die area. A complete set of masks (a 'full mask set') is required for wafer fabrication. Minimum area means the lowest eventual fabrication cost since more ICs can be made on a single wafer, but against this must be weighed the considerably higher initial optimum design costs.

The accepted definition of full custom is that the chip is designed at the silicon level, with complete flexibility to design the individual transistors and the floorplan layout without any pre-designed restrictions such as are present in semicustom design techniques (see Sections 4.2 and 4.3, following). The designer of a full-custom chip must therefore be familiar with the following:

- complex circuit design techniques in order to be able to make full use of the freedom of design available;
- how monolithic circuits make new types of circuit configurations possible, for example by using the inherent good matching of component parameters but possibly with a wide tolerance on absolute parameter values;
- full details of the fabrication technology, in order to be able to size all transistors and other components correctly; and
- detailed simulation procedures in order to verify the performance of the whole circuit as far as possible before any manufacturing costs are incurred.

Thus full-custom design will be appreciated as being a highly skilled specialist area if maximum benefits are to be achieved. It is a task for all IC manufacturing companies when producing a new LSI or VLSI standard circuit or range of circuits, but it is not one which a circuit designer in a non-IC manufacturing company can easily accomplish. Indeed, the latest details of a fabrication process and its critical dimensions may not be available for use outside a restricted circle, and therefore it is not possible for outside designers to become involved in any unique device design activity at the silicon level.

One exception to this generalization has been the Mead–Conway structured approach to full-custom design. This was based upon a set of nMOS geometric design rules, which were simplifications of the various mask layouts used in nMOS chip fabrication, and which in their simplified form were applicable to any conventional nMOS production line [1]–[4]. These design rules were thus generalized or 'portable', and as such did not represent the exact geometric design rules of any particular vendor or involve the ultimate state-of-the-art fabrication (and hence performance) limits.

The fundamental rationalization introduced by Mead and Conway was the normalization of the critical geometric dimensions involved in the nMOS silicon layout. This involved the introduction of a basic dimensional parameter lambda (λ) by which all the critical dimensions of an nMOS layout, such as line widths, line spacings, mask alignment tolerance, overlaps between polysilicon and diffusion and between polysilicon and metal, and so on, could be expressed. This concept is illustrated in Figure 4.2. Typically, λ may be, say, 1.5 μm and all feature details are then expressed as 2λ, 4λ, etc., for a particular design and subsequent mask-making.

Lambda is therefore a conservative and possibly coarse parameter which allows independent designers to design and lay out nMOS integrated circuits, the appropriate value for lambda being incorporated immediately prior to mask-making, in agreement with the subsequent producer of the IC. The whole viability of this approach clearly depends upon the fundamental feature that MOS device parameters are very largely controlled by the device geometry for a given fabrication process (see Section 2.2.5 and Appendix D) and, within limits, a given MOS layout can be scaled up or down to meet a required specification. Such an approach is completely unavailable with bipolar technology.

Extensions of these generalized lambda-based MOS design and layout rules to CMOS technology have also been pursued [2], [5]–[7]. There has also been some consideration of analogue requirements [8]. However, because of the rationalization and generosity of the lambda-based design rules the following three disadvantages have to be accepted:

1. The silicon area required for a given design is inevitably larger than can be realized with conventional full-custom design methods, possibly as much as 50–60% greater.
2. Performance will be inferior because of its very conservative design rules.
3. It is not possible to scale down all the layout dimensions linearly once actual dimensions have reached about 2 μm, since non-linear layout requirements then begin to dominate.

As a result of these practical disadvantages lambda-based MOS design has received little commercial support, and must be considered inappropriate for state-of-the-art geometries. Where this design philosophy has received greatest support, and still does to some extent, is in academic institutions, since it is an approach which gives students some practical work in the fundamentals of MOS design at the silicon level, and is both academically demanding and respectable.

To revert, therefore, to full-custom design techniques which are commercially

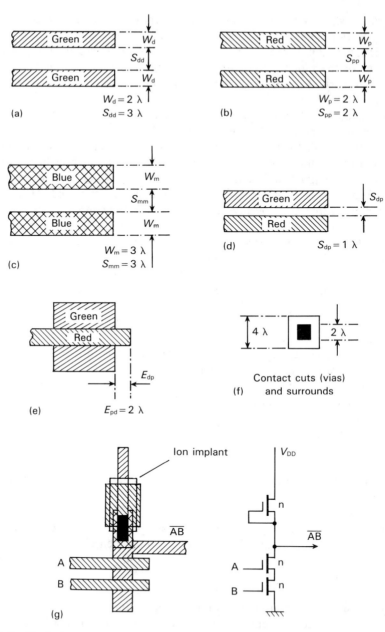

Figure 4.2 Typical geometric layout rules in terms of λ in the Mead–Conway nMOS full-custom design approach. Green, red, blue are conventionally used in colour plotters for MOS layouts: (a) diffusion widths and spacing; (b) polysilicon widths and spacing; (c) metal widths and spacing; (d) diffusion to poly. spacing; (e) poly./gate overlap; (f) contact cut overlaps; (g) an example 2-input NAND gate layout.

viable, detailed design must, in practice, be undertaken by either the manufacturer of the final IC, or by a specialized design house which has expert in-house knowledge both on design and on the process technologies of chosen manufacturers. In general, the market for full-custom ICs will either be where very large volume requirements are anticipated, which therefore justifies the expense of the design, or where very specialized performance is required which cannot be

Figure 4.3 A full-custom IC which executes all the line level functions and check sums for a 10 Mbit local area network (LAN) system. (Courtesy of Swindon Silicon Systems, UK.)

116   *Custom microelectronic techniques*

achieved by using off-the-shelf standard parts. Complex signal processing circuits such as that shown in Figure 4.3 are particularly appropriate for full-custom consideration.

4.2 Standard-cell techniques

Standard-cell techniques rely on the existence of previously designed and fully characterized standard circuits ('building blocks'), layout and performance details of which are held in a CAD database. The custom IC design procedure is thus the

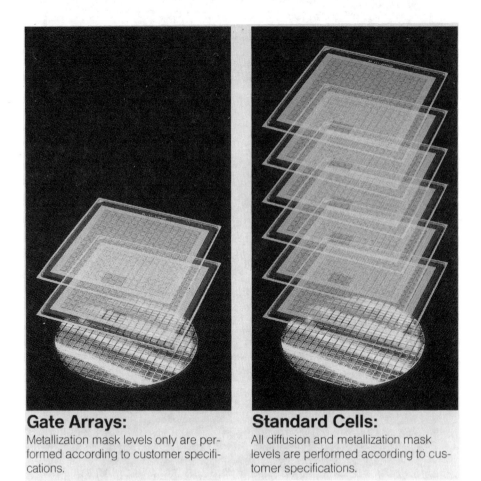

Gate Arrays:
Metallization mask levels only are performed according to customer specifications.

Standard Cells:
All diffusion and metallization mask levels are performed according to customer specifications.

Figure 4.4 The fundamental distinction between a standard-cell USIC and a gate-array USIC. Note that the chip size and routing areas are fully flexible in the former but not in the latter. (Courtesy of NEC Electronics, UK.)

assembly of the required cell designs to form a chip layout, together with simulation to ensure acceptable performance before any manufacturing costs are incurred. A full mask set is required for manufacturing purposes since the chip layout, number of I/Os and chip size are fully flexible. This is in contrast to the fixed chip architecture of the gate-array design style (see Section 4.3) where only the final interconnection details are unique, thus permitting a reduced mask set to be used. This distinction is illustrated in Figure 4.4.

However, the standard-cell approach may be divided into sub-divisions depending upon the precise nature and capabilities of the 'standard' cells. These sub-divisions include the following:

- digital fixed-cell architectures, where there is no flexibility in the CAD system to alter the pre-designed standard cells in any way;
- digital soft-cell architectures, where there is some flexibility in the CAD system to alter the cell designs but not their function in some restricted way;
- digital macro cells, which are large rectangular building-blocks such as RAMs, ROMs and PLAs, and which are usually flexible in exact size and capability; and
- analogue cells, which are usually fixed and not flexible.

These distinctions will be considered further in the following sections. Section 4.5 will give a final overall summary and comparison of both standard-cell and gate-array design techniques.

4.2.1 Fixed-cell architectures

Fixed-cell architectures are custom ICs compiled from pre-designed functions available in a vendor's library which cannot be altered by the CAD system in any way. These fixed-cell designs (sometimes known as 'hard' or 'hard-coded' cells) cover the range of standard logic building-blocks with which all digital designers are familiar, such as logic gates, latches, adders, counters, shift registers, and so on. The range of many vendors' designs often deliberately mirrors the off-the-shelf standard 74^{**} series SSI and MSI parts, which promotes user confidence when transferring from standard design activities to USICs. Indeed, the analogy has been made between a product using standard 74^{**} series ICs, which involves the design of a printed-circuit board (PCB) to carry and interconnect the individual ICs, and a standard-cell USIC design, which involves the placing and interconnect routing of standard-cell layouts on silicon instead of on a PCB. Functionally, this analogy is reasonably good, but the standard-cell chip design activity is more complex. As well as standard logic cells, vendors also have a number of I/O cell designs available, covering a range of input and output system requirements.

In order that the CAD system can assemble (place) the individual cell designs in a compact manner without appreciable waste of silicon area, and then begin to route all the required interconnections from cells and I/Os, it is desirable that the original design of each cell shall conform to some dimensional standardization and not have a random size layout. This is invariably achieved by standardizing upon a fixed height of cell but varying width, thus enabling cells to be placed alongside each other in uniform rows across the chip layout. Cells which conform to this standard-height topology are sometimes referred to as 'polycells'. The routing between cells and to the I/Os is then done in wiring channel areas between the rows of cells, as shown in Figure 4.5.

Whilst this standardization on cell height becomes impractical to maintain when the capability of the cells reaches LSI or VLSI proportions (see Section 4.2.3 below), it is appropriate for all SSI/MSI functions. It is particularly appropriate for CMOS technology where all the logic functions are composed of p-channel and n-channel transistor-pairs, thus allowing the height of all cells to be dictated by the dimension required for a p-channel and n-channel transistor-pair plus V_{DD} and V_{SS} supply rails. This is shown in Figure 4.6. Note that a conventional circuit diagram may not mirror the physical layout arrangement. The width of a cell clearly depends upon the number of transistor-pairs involved in the logic function. Partly because of this layout convenience and partly because CMOS technology is particularly appropriate for the vast majority of USIC applications, it is CMOS that currently dominates the standard-cell market. (Recent surveys have indicated that there were

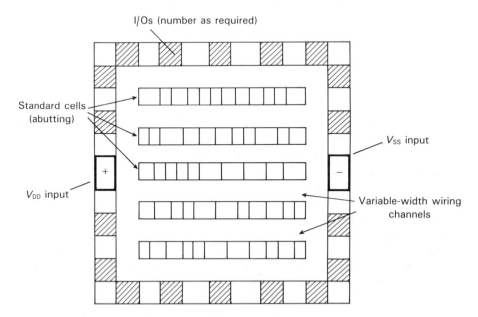

Figure 4.5 The usual topology of a standard-cell chip design, with the required number of constant-height cells assembled in rows and appropriate width wiring channels between rows (cf. Figure 4.11 later).

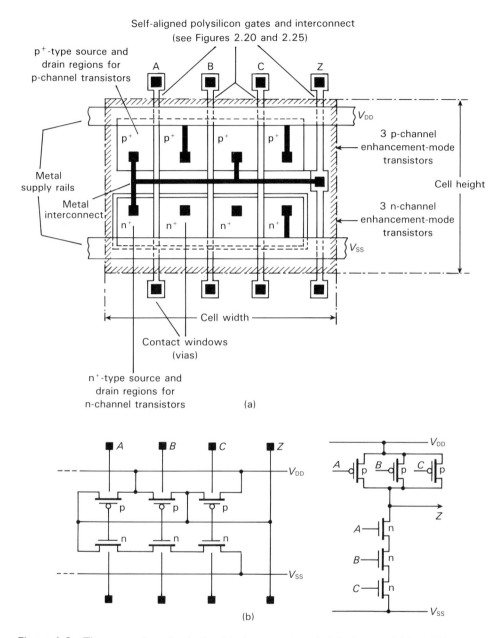

Figure 4.6 The general method of achieving constant height but variable width standard-cell layouts in CMOS technology; a 3-input NAND gate layout is illustrated here with inputs and output available at both top and bottom edges ('both-sides-access'). The number and position of the contact windows ('vias') may vary considerably between different vendors' designs: (a) the basic layout; (b) the equivalent circuit diagram.

fewer than ten vendors worldwide who market other than CMOS standard cells for digital applications [9], [10]. Exceptions include ECL and GaAs for extremely high-speed specialist applications.)

Because of the rows of fixed-height cells in a standard-cell USIC, the fixed chip layout superficially resembles the layout of a gate array (see Section 4.3 and Figure 4.11). However, with the standard-cell design it is important to appreciate the following:

- The final standard-cell chip contains only the cells required, with no unwanted or unused cells.
- The width of the wiring channels between the rows of cells is only that necessary to contain the final required interconnections, not fixed as in a gate-array layout.
- Only the required number of I/Os are present.

The result is that a standard-cell USIC is generally smaller and more efficient than a gate array configured to perform the same duties (see Section 4.5), but the design is constrained by having to use the cells which are available in the vendor's library and which may not always be exactly what the chip designer requires.

The standard-cell design process is initially very similar to system design using standard parts. This consists of system partitioning down to the level of the available logic building-blocks, but because the standard cells in a vendor's library are not available as functional entities it is not possible to make a breadboard prototype circuit for evaluation purposes, as may be done when using off-the-shelf standard parts. Instead, complete reliance upon the CAD simulation is necessary to ensure that the initial design is functionally correct. Standard-cell design procedures are therefore inherently much more heavily CAD-dependent than are design procedures using standard parts, which may be advantageous if the CAD capabilities are comprehensive, quick and accurate.

The complete standard-cell design procedure is as shown in Figure 4.7. Details of simulation, placement and routing, test vector generation and other software activities will be covered in Chapter 5, together with considerations of the possible interface levels between the original-equipment-manufacturer (OEM) and the vendor. The vendor clearly has to be involved in the final manufacturing processes of mask-making, silicon fabrication, packaging and test, but the OEM may wish to do as much of the chip design work as possible before handing over the design to a vendor for manufacture. CAD costs and OEM capability enter into these interface decisions, factors which will also be raised in Chapter 7.

Precise details of the design of standard cells at transistor level are not usually available to OEMs, being the commercial property of the vendor who has hand-crafted them so as to achieve minimum silicon area and optimum performance. Available details are sometimes referred to as 'limited technical data' (LTD), and will include the dimensions and performance of each cell, for example as shown in Figure 4.8, but not its exact internal geometric details. General data for the whole

Standard-cell techniques 121

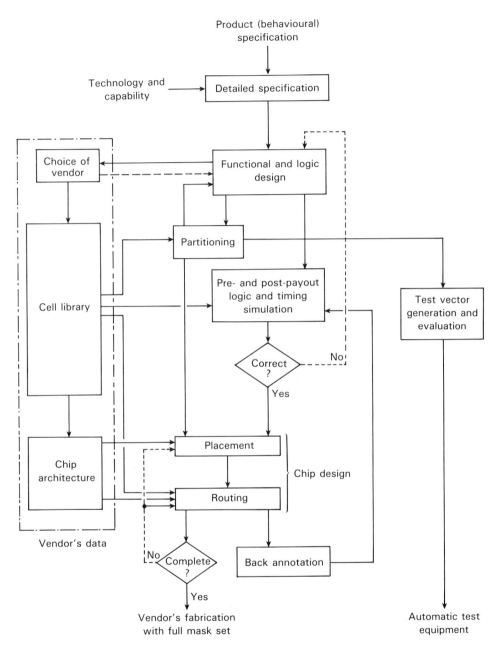

Figure 4.7 The hierarchy of the standard-cell USIC design process, from the original product specification at the top to final fabrication at the bottom. The dotted paths are possible iterations if problems or constraints are found.

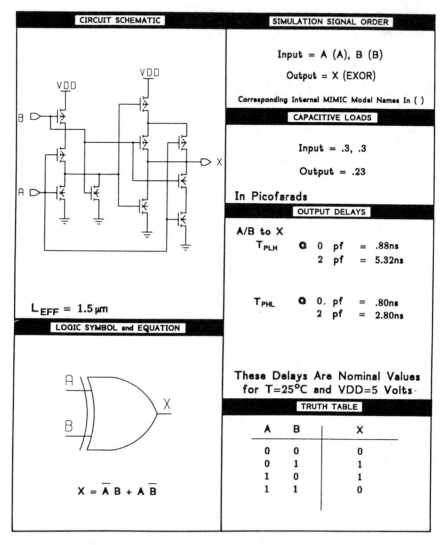

Figure 4.8 Example data sheet for one cell in a vendor's standard-cell library, with cell dimensions of 48 × 90 μm. T_{PLH} and T_{PHL} are the cell propagation times when the output changes from low to high and from high to low, respectively, under given loading conditions. (Courtesy of Harris Semiconductor, Inc., FA.)

cell library will give supply voltages, whether single-layer metal or two-layer metal, and other information. Full technical data (FTD) which includes the geometric details of cells may be available to certain large OEMs under confidentiality agreements and who may wish to design their own special-purpose cells which are compatible with the vendor's standard cells, but this is not usual.

4.2.2 *Soft-cell architecture*

One of the constraints with fixed standard cells is that some compromise must be present in the original cell designs with regard to performance, particularly the fan-out capability of each cell output. Unlike off-the-shelf standard ICs, all of which have on-chip buffered outputs to give good fan-out capability, the provision of high fan-out capabilities on each standard cell would be wasteful in silicon area, since the majority of cells used on-chip would not require this. Nevertheless, such capability may be required in some circumstances.

One way of providing different output drive capabilities is to have two or more versions of a hard-cell design in the cell library; for example one vendor offers three variants of NAND, NOR and other gates, with cell widths in the ratio of 1:1.33:2.33 for increasing drive strength. A more sophisticated method is to allow the CAD system to modify the detailed silicon layout so as to alter the cell specification (but not function) within limits by varying p-channel and n-channel device areas. Cells which may be fine-tuned in this way may be known as 'soft-' or 'soft-coded' cells.

It is important to note that the rules for any change in transistor details are built into the CAD software, so that the designer using the system does not have to be familiar with detailed transistor design or the silicon layout rules. However, vendors of standard-cell USICs may still prefer to offer several variants of hard-coded SSI/MSI cells rather than provide the facility for the customer to fine-tune them for particular applications.

4.2.3 *Macrocells*

Whilst MSI standard cells in a vendor's library may be built up from a compacted assembly of fixed-height SSI cells, it is unrealistic to try to maintain this constant cell-height feature for cells of VLSI complexity. Instead, large standard cells are invariably designed by the vendor with a more efficient layout, ideally square, which means that they cannot be placed in a row layout with the smaller-size standard cells. As will be seen in Chapter 5, this involves a different concept of automatic routing in a CAD system, requiring a block router rather than a channel router.

A vendor's library may include large building-blocks such as the following:

- RAM and ROM;
- PLAs;

- 16-bit multipliers;
- microprocessors.

and others. An increasing range of digital signal processing (DSP) blocks is also becoming available. However, with increasing size and complexity goes the problem of the increasing difficulty of meeting a customer's requirement exactly in terms of size and capability, plus difficulties of placement and routing if only one layout of each large block is available. Hence it is common practice for very large cells to be flexible in the CAD system, rather than hard cells. (When referring to macrocell design programs in a CAD system some authorities use the adjectives 'hard' and 'soft' in a particular way. A 'hard macrocell program' is defined as one which produces all the layout layers for a particular fabrication process, whereas a 'soft macrocell program' produces a more general output such as a netlist which can be subsequently placed and routed on any regular layout structure. Here we will not use the terms 'hard' and 'soft' in this sense.) In particular, the regularity of RAM, ROM and PLA structures affords several degrees of freedom without altering basic transistor and other detailed dimensions, for example:

- freedom to alter the number inputs and outputs;
- freedom to scramble the inputs and outputs in any order to ease placement and routing problems;
- freedom to mirror-image or orientate these layouts in any way.

Similarly, in bus-structured architectures, arithmetic units and the like, the freedom to increase or decrease bus widths is frequently provided.

Macrocells are therefore inherently soft-coded rather than hard-coded cells in order to give the necessary flexibility for a range of custom requirements. Indeed, the increasing capabilities of available CAD software is having further effects such as:

- giving built-in flexibility to accommodate improvements in fabrication processes, rather than having to undertake a costly re-design of the cell library by hand;
- blurring the distinction between full-custom design and cell-based design, since the increasing flexibility of the latter allows optimized RAM, ROM and other blocks to be used in a full-custom design environment;
- possibly allowing special modules to be readily compiled so as to augment existing cell designs to meet new customer requirements.

Figure 4.9 illustrates the layout of a standard-cell custom IC designed for graphics imaging processing which contains both large macros and standard logic cells, the total chip design involving the equivalent of about 75 000 logic gates [11]. This typifies the complexity which can be handled by standard-cell design methods.

In summary, therefore, the libraries offered by vendors are becoming

Standard-cell techniques 125

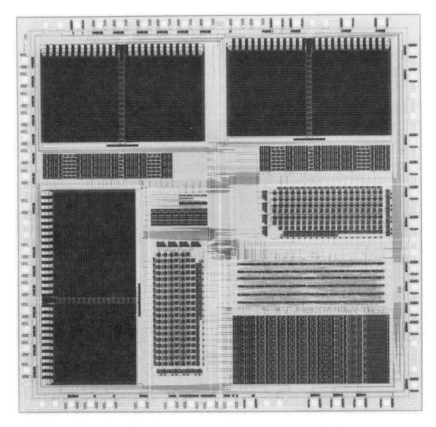

Figure 4.9 A standard-cell chip design involving three large RAM blocks, other smaller datapath and arithmetic blocks and about 750 standard cells for random logic. (Courtesy of VLSI Technology, CA., and Evans and Sutherland, Inc., UT.)

increasingly comprehensive and flexible, with only the smaller SSI and MSI logic cells having a fixed layout design. The use of silicon compilers (see Chapter 5) is a part of this still continuing CAD evolution [12]–[16]. Perhaps, regrettably, the most non-standard part of this design methodology is the terminology. In particular, the nouns used to identify the size of standard cells in a vendor's library is a source of possible confusion. In generally increasing size and capability the following are among the terms which may be encountered for a hierarchy of increasing-complexity digital building blocks:

- leaf cells;
- cells, basic cells, primitives, polycells, standard cells;
- microcells;
- macros, macrocells, macro blocks, macro functions;

- composition cells, compiled cells;
- paracells, parameterized cells;
- megacells, megamacros, megafunctions;
- supracells, gigacells.

All the preceding discussions relate primarily to CMOS products, bipolar technology not being a strong commercial contender in this area. However, recent developments in very-high-speed standard cells using gallium-arsenide technology have been reported, although the number of vendors marketing this technology is limited [9]. The standard cells at present available are generally of SSI or MSI complexity only, but with a speed performance considerably higher than are available from CMOS libraries. Comparative figures are given in Table 4.1.

Figure 4.10 gives the circuit of a 3-input NOR standard cell as used in the Gigabit Logic, Inc. range of GaAs cells, the total cell area being 40 μm × 120 μm,

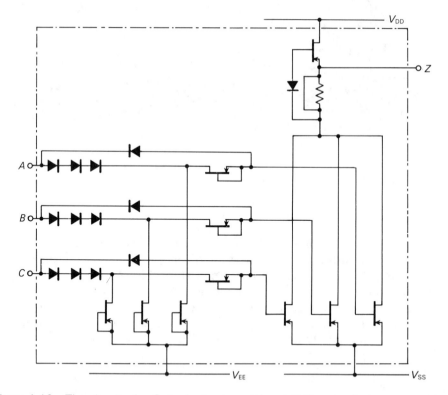

Figure 4.10 The circuit of a GaAs technology 3-input NOR gate design as used in a vendor's standard-cell library, featuring low static drive conditions to reduce the static power dissipation but enhanced dynamic drive to increase switching times [17].

Table 4.1. Comparative performance figures for CMOS and GaAs

	Typical CMOS performance 1.5 μm geometry	Typical GaAs performance
Min. gate propagation time	0.6 ns	0.1 ns
Max. clock frequency	200 MHz	1 GHz

exclusive of power bus and wiring channels, and with a power dissipation of 2 mW per gate [17].

4.2.4 *Analogue cells*

While it is entirely straightforward to define and market a very wide range of self-contained digital building-blocks suitable for most users, typified by the 74[**] series circuits and their equivalents, with analogue the problem is much more difficult. The following are among the circuit parameters which users may specifically need to meet for their product requirements:

- voltage, current or power gain,
- input and output voltage levels,
- input sensitivity,
- output power,
- bandwidth,
- slew rate,
- filter characteristics,

and others [18]–[21]. Indeed, the exact analogue performance required by each new system design may be unique.

Further distinctions between digital and analogue circuits and systems include the following:

- Analogue circuits require the presence of resistor and capacitor elements to provide good signal amplification and frequency performance – no wide-band linear amplification can be achieved without the presence of fixed resistance.
- Analogue performance needs to be accurate and stable, unlike digital circuits where variations in exact electrical performance are not critical.
- The preferred silicon technology for analogue circuits is bipolar (see Section 2.1.1 and Table 2.2), whereas MOS is generally more appropriate for digital applications and standard cells.
- Analogue systems usually involve relatively few analogue circuits in comparison with the very large numbers of digital circuits which can build up in digital systems; hundreds rather than tens of thousands of transistors may therefore be involved in an analogue system, but as each transistor is usually operating in a linear mode power dissipation per circuit may be high.

A further difficulty facing a vendor wishing to market analogue standard cells is that, ideally, analogue cells should be capable of working alongside digital cells in order to provide a comprehensive mixed analogue/digital capability, but the vendor may encounter the following problems:

- The analogue cells may then need to be MOS rather than bipolar so that they can be fabricated using the same processing stages as the digital cells.
- The necessary operating voltages of the on-chip analogue and digital cells may be different.
- The MOS silicon processing may require modification in order to achieve acceptable analogue performance, which may reflect on the performance of the digital cells.
- Great care will need to be taken in floorplan layout in order that analogue cells working at low signal levels do not pick up interference from the digital switching parts on the same substrate.
- Simulation of a large mixed analogue/digital IC design is not possible with currently available CAD resources.

Because of all these potential difficulties, uncommitted arrays (see Section 4.3.4) are at present most often used for analogue and mixed analogue/digital applications, most commonly using bipolar technology rather than MOS. Nevertheless, an increasing number of vendors are making MOS analogue cells available in their standard-cell libraries. The variety of different types of analogue cells listed is, however, small in comparison with the list of possibly hundreds of digital standard cells, and the analogue performance may not be as high as that available from bipolar arrays. Library types include general-purpose operational amplifiers, analogue-to-digital and digital-to-analogue converters, voltage comparators and timer-oscillators [22]–[24]. Exact circuit details are not usually revealed.

This area is currently the most unsatisfactory area of standard-cell technology, and considerable development work is being undertaken to improve the capabilities of the CAD for mixed analogue/digital USICs. BiCMOS technology has been proposed as a possible candidate for this field, but recent developments in very-low-power bipolar technology may prove to be a better solution for the majority of mixed analogue/digital applications.

4.3 Gate-array techniques

Semicustom gate arrays rely upon the availability of wafers which have been fully processed up to, but not including, the final chip interconnections, and which are fabricated as stock items. Each die on an uncommitted wafer contains an array of identical general-purpose cells, the subsequent interconnection design of which is the customization. The uncommitted wafers may be held in stock with a complete first-layer metal covering, the polished surface of which gives rise to the term 'mirror blanks' for such stock items. This is only possible if all via holes (see p. 45), from the first-layer metal through the field oxide to the underlying circuit are made before the first-layer metal is applied. If the via holes need to be located as part of the custom interconnection design, then the first-layer metal cannot be applied before the required vias are subsequently fabricated, and the wafers are therefore held in stock without any metallization.

As has been noted in Chapter 1, the prime advantage of the gate-array approach is to minimize the amount of custom design and fabrication activity necessary to complete a particular circuit. Since only the interconnection design, manufacture of the interconnection masks (a reduced mask set) and fabrication of the final wafer interconnection levels is customer-specific, the time and cost of the final USIC should be less than that of a standard-cell design. However, performance penalties may have to be paid, as will be detailed later.

The very wide range of semicustom arrays which are commercially available may be classified by the following:

- the specification of the general-purpose cell replicated on the uncommitted chip;
- the number of such cell per chip;
- the number of I/Os per chip;
- the fabrication technology, whether bipolar, MOS or otherwise, and
- the number of interconnection levels which have to be customized.

Simple arrays have a single layer of metal for their customization (single-layer-metal arrays), but more complex arrays have two layers (two-layer-metal arrays) such as are illustrated in Figure 2.26, or, in some recent very complex products, more than two customized layers. (Some complex products, particularly high-speed ECL arrays, have one layer of metal for fixed interconnections such as power supply rails, together with two layers for customer-designed interconnection duties. Hence a total of three layers of metal are present, but only two are customer-dedicated.)

The number of customization masks necessary in the reduced mask set therefore increases as illustrated in Table 4.2. If the first-layer via holes to the underlying circuit also have to be fabricated, then one further mask is required in addition to the totals below.

Table 4.2. The total mask requirements for different construction uncommitted arrays

	Via (contact window) masks	Metal patterning masks	Total
Single-layer-metal	–	1	1
Two-layer-metal	1	2	3
Three-layer-metal	2	3	5

Whilst single-layer-metal arrays are clearly less costly to customize than those with two (or more) layers, routing the required interconnections is frequently difficult using just one level. The increased flexibility provided by more than one level and smaller interconnect area are the principal reasons for multilayer metallization.

Figure 4.11 shows the basic chip floorplan which is adopted for virtually all gate-array designs, except for channel-less architectures, which will be considered separately in Section 4.3.3. As will be seen, the layout consists of rows of identical cells separated by wiring channels, with I/Os around the perimeter of this active area. This resembles the standard-cell topology shown in Figure 4.5, but with no flexibility to alter the number and type of cells in each row, or the width of the wiring channels, or the number of I/Os to the exact requirements of a customer. Hence a major disadvantage compared with a standard-cell design is that silicon area utilization will not be as good since, inevitably, there will be unused parts of the fixed-array structure, and performance will generally be inferior due to looser on-chip packing densities. These and other comparisons will be considered further in Section 4.5.

The vendor's choice of the general-purpose cell used in a semicustom array is a fundamental decision which has to be made when developing a gate-array chip. This depends, to a large extent, upon whether the array is targeted for digital-only

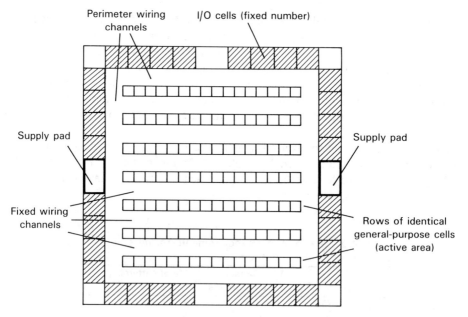

Figure 4.11 The floorplan layout of a typical gate-array chip, sometimes termed a 'channel-array' architecture (cf. Figure 4.27).

Gate-array techniques 131

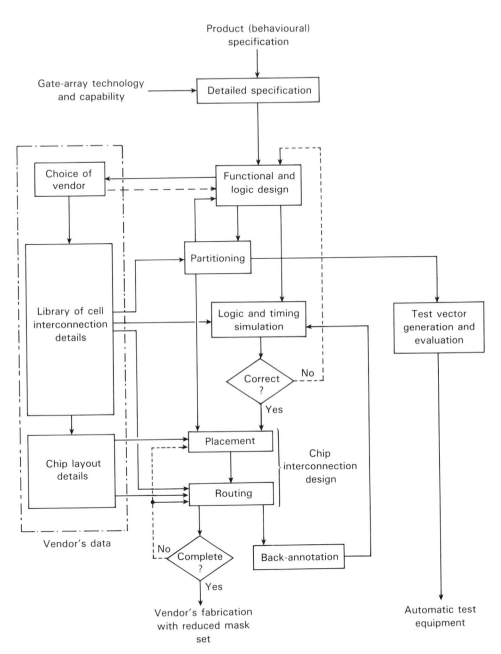

Figure 4.12 The hierarchy of the gate-array design process (cf. Figure 4.7 for a standard-cell design procedure).

applications, or mixed analogue/digital, or solely analogue purposes. The latter categories will be considered separately in Section 4.3.4. For digital-only applications the choice of silicon technology will also influence the choice of the general-purpose cell; most bipolar technologies, including ECL, I^{2}L, ISL, STL, CDI and GaAs (see Chapter 2), have been used, but CMOS is currently the majority choice. The following section will examine these considerations, together with the meaning of 'cell libraries' within the context of gate-array architectures.

In considering this area the terminology 'gate array' may imply that a chip contains an array of digital logic gates only and is a misnomer if the array is designed for analogue purposes or contains separate components rather than functional gates. Terminologies such as masterslice array, uncommitted component array (UCA), uncommitted gate array (UGA) and uncommitted logic array (ULA^{TM}) have been used, (ULA^{TM} is the trademark of GEC–Plessey Semiconductors UK, and applies to all their range of semicustom products), but unfortunately the term 'gate array' is commonly employed for all forms of uncommitted array, irrespective of design and application. We will continue to use this term herewith where generally appropriate.

The design process for a gate-array IC is very similar to that for a standard-cell design, as may be seen by comparing Figure 4.12 with Figure 4.7. However, the exact design activities are dissimilar in detail, since with a gate-array the chip floorplan is pre-designed with no flexibility to modify the uncommitted layout in any way. Also, the placement activity is different: with a gate-array "placement" means the allocation of the fixed cells on the array to chosen duties so as to ease the subsequent routing procedure, but with a standard-cell design "placement" means the physical placement of cells from a library in a suitable layout arrangement for subsequent routing. Further details of these actual design activities will be covered later.

4.3.1 Choice of general-purpose cells for digital applications

The diversity of commercially available gate arrays arises in part from the choice of the general-purpose cell replicated on the chip. Whatever is chosen must be capable of being interconnected so as to form all possible types of digital logic functions. Figure 4.13 shows the principal choices, with component-level cells or functional-level cells as an initial distinction. With the former the custom interconnection design requires the internal cell interconnections which connect together the components to be specified, as well as the interconnections between cells and I/Os; with the latter each cell is a functional SSI entity, sometimes termed a 'hard-wired' cell, requiring only the intercell connections to be considered.

Component-level arrays
The bipolar silicon technologies which have at some stage been considered for component-level digital arrays include Schottky TTL, ECL, I^{2}L and CDI (see

Gate-array techniques 133

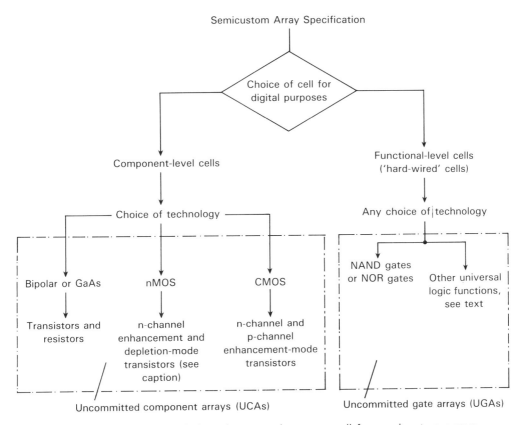

Figure 4.13 The possible choice of a general-purpose cell for semicustom arrays. nMOS technology is now no longer employed by any vendor for this product area, having been superseded by CMOS.

Chapter 2), but only ECL and CDI remain commercially available [9], [10]. ECL caters for very-high-speed specialized applications, whilst CDI has particular advantages for non-standard d.c. supplies and for combining with analogue requirements.

As an example of a high-speed ECL component-level array, consider the product illustrated in Figure 4.14. Its specification is as follows:

- 416 cells arranged in twenty-one rows;
- 76 transistors and 60 resistors per cell, arranged in four equal '$\frac{1}{4}$' cells;
- 1.5 μm minimum feature size with two layers of custom metallization;
- speed/power adjustable from 175 ps on-chip gate delay at 3 mW/gate to 300 ps at 1 mW gate;
- 256 signal I/O pins;
- outputs can drive 25, 50 and 60 Ω transmission lines;

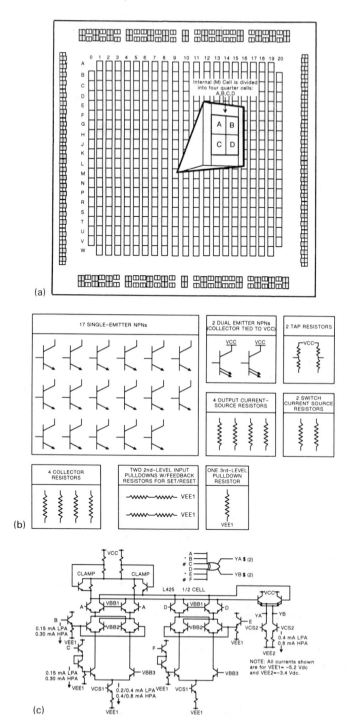

- maximum dissipation 30 W;
- packaging 289 pin pin-grid array or other alternatives.

It is evident that this is a very sophisticated product and is not intended for general-purpose use. Like other ECL products its applications are primarily in computers and in high-speed communications and signal processing. The vendor has a library of proven cell interconnection designs available to cover a range of digital functions, together with CAD resources for the complete custom-design activity. It may be noted that great care has to be taken with the interconnection of such circuits on a printed circuit board in order not to degrade the very-high-speed capabilities; matched transmission line PCB interconnections as shown in Figure 4.15 are required.

A recent alternative to ECL for very-high-speed work has been the marketing of GaAs uncommitted arrays. A number of different choices of cell components have been produced or forecast, but the simplest cell for general-purpose applications consists of one depletion-mode and two enhancement-mode GaAs MESFET transistors from which a 2-input NOR gate can be constructed. The specification of one commercially available product is as follows:

- 4000 cells arranged in ten rows;
- 1 depletion- and two enhancement-mode MESFETS per cell;
- 0.8 μm minimum feature size with two layers of custom metallization;
- typical 2-input NOR gate delay 120 ps at 340 μW;
- 120 signal I/O pins;

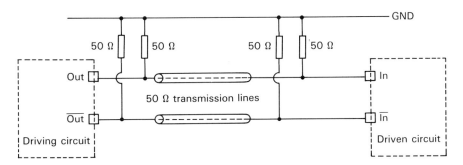

Figure 4.15 The use of terminated balanced transmission-line techniques which may be mandatory for very-high-speed interconnection duties between ECL and other very-high-speed digital circuits.

Figure 4.14 The Motorola MCA III ECL uncommitted component array: (a) the channel architecture floorplan; (b) the component devices available in one-quarter of a cell; (c) example 6-input Exclusive-OR cell commitment, using one-half the components of a cell (two '$\frac{1}{4}$' cells). (Courtesy of Motorola, Inc., AZ.)

- inputs and outputs can be TTL or ECL compatible;
- typical dissipation 2.5 W;
- 149 pin pin-grid-array packaging or other alternatives.

Figure 4.16 illustrates the general arrangements of this GaAs product. Larger and slightly faster arrays are also available.

Like ECL, gallium-arsenide will be seen to provide extremely high performance. In comparison with ECL the speed is generally higher, although newer ECL technologies are giving increasing speeds, but the power consumption is less than with ECL. In general, the speed/power product of GaAs is an order of

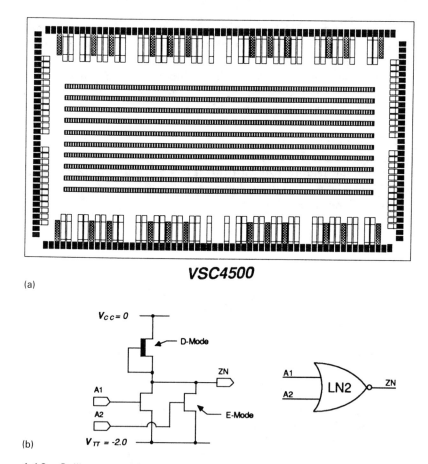

Figure 4.16 Gallium-arsenide uncommitted component array with 4000 uncommitted cells and programmable I/O cells for different interfacing requirements: (a) the chip floorplan; (b) the commitment of a cell to a 2-input NOR gate. Multiple interconnected cells provide a standard range of digital logic functions. (Courtesy of Vitesse Semiconductor Corporation, CA.)

magnitude better, but ECL has a longer history of development and hence remains the major force in the very-high-speed logic area [25]–[28].

Collector-diffusion isolation (CDI) technology also has a long history, being pioneered by Ferranti Electronics, UK (now GEC–Plessey Semiconductors), in their range of uncommitted logic arrays (ULAs). These also occupy a somewhat special niche in the array market. As covered in Section 2.1.5, CDI is a bipolar process which, like ECL, does not drive transistors into a saturated mode, and which also allows a trade-off between speed and power dissipation. The latter parameters are, however, lower than with ECL. The other unique feature of CDI is that it allows efficient analogue circuits to be made with the same basic components as are used in the digital circuits, and hence most ULA products have both analogue and digital capability. We will, therefore, mention ULAs again in Section 4.3.4.

Early versions of the ULA series of custom circuits had component-level cells which were roughly square in layout, as shown in Figure 4.17(a) and (b). These cells were positioned on the chip with wiring access to all four sides, rather than in abutting rows as shown in Figure 4.11. This grid layout, sometimes referred to as a 'block-cell' architecture, is in contrast to the more usual channelled architecture of Figure 4.11. However, current versions have a different cell composition, (as shown in Figure 4.17(c) and (d)), with cells assembled in continuous rows on the chip to give the more familiar array layout. A range of chip sizes up to the equivalent of 10 000 2-input NAND or NOR gates available, with a choice of speed/power options for system speeds up to 100 MHz. The average gate power is 210 μW at 100 MHz, less at lower speed/power options.

Turning from these bipolar products to CMOS technology uncommitted component arrays, there is now no fundamental difficulty in defining a basic cell. The simplest is clearly a single p-channel and n-channel transistor-pair from which all CMOS digital logic circuits may be constructed. It is also clearly relevant to arrange such transistor-pairs in rows, giving a floorplan as shown in Figure 4.11.

However, there is a choice of either one transistor-pair per cell or more than one, as Figure 4.18 shows. Commercially available arrays may have one, two or three transistor-pairs arranged within common p^+-type and n^+-type regions. The polysilicon crossunders, which must be available somewhere in the layout to allow the gate outputs to be brought out from the cells under the common V_{DD} and V_{SS} supply rails, may be considered as part of a cell, or as a necessary addition between cells.

It will be obvious that these uncommitted CMOS cells resemble very closely the designs available in a vendor's standard-cell library, see Figure 4.6. Indeed, exactly the same physical layout of the individual CMOS transistors may be present in a vendor's standard-cell and uncommitted-array products, but each standard-cell design may possibly occupy a smaller width than uncommitted transistor-pairs due to it being specifically optimized for the one functional duty.

The commitment of the cells in a CMOS uncommitted array involves internal-cell metallization to interconnect the p-channel and n-channel transistors in series and in parallel as required, plus the remaining interconnections between cells and

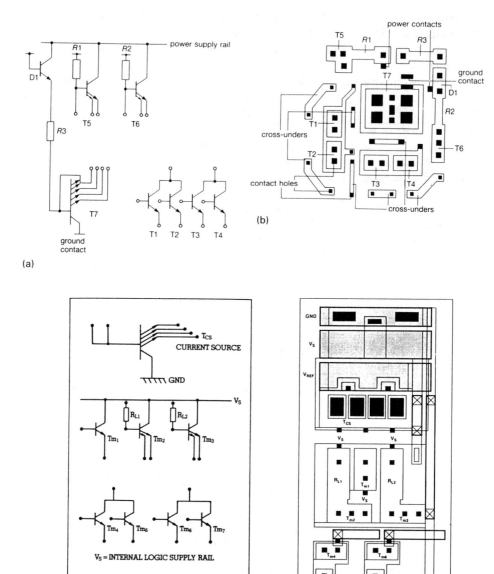

Figure 4.17 The Plessey ULA series of uncommitted component arrays using bipolar CDI technology: (a) the component devices available per cell in early-series ULAs, from which the gate shown in Figure 2.16(c) or other logic functions may be constructed; (b) the physical layout of (a), with wiring access made available in the chip floorplan to all four sides; (c) the component devices available in later series; (d) the physical layout of (c), with a chip floorplan arranged in rows of abutting cells. (Courtesy of GEC–Plessey Semiconductors, UK.)

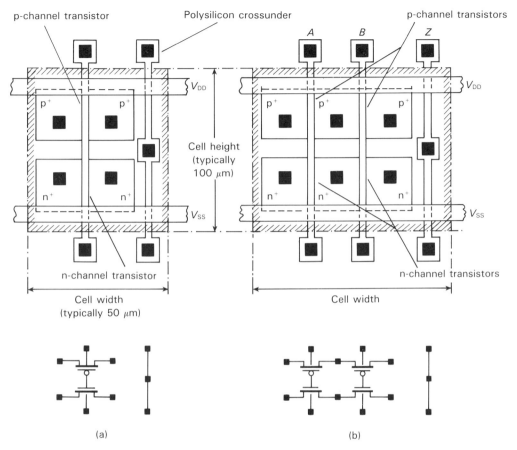

Figure 4.18 The choice of component cells for CMOS technology uncommitted arrays (cf. the CMOS standard-cell layout of Figure 4.6): (a) a single pair of transistors per cell, with an additional polysilicon 'crossunder' for output connections; (b) two pairs of transistors per cell, also with a polysilicon crossunder for output connections.

I/Os. However, all vendors now supply the interconnection details necessary to commit their cells to a range of given logic duties, as well as details of how many cells are required to form given functions [29]–[38]. The terminology 'cell library' used in this connection should be distinguished from the library which vendors provide for standard-cell USICs; with the latter the complete cell fabrication details are contained in the CAD library, whereas in the uncommitted array context the cell library is the *final interconnection design details*.

As an illustration of a commercial product, consider the Texas Instruments TAHC product family [29]. The chip floorplan is a normal channelled architecture, with rows of identical cells separated by wiring channels. The standard cell in each row, however, illustrates the quandary facing a vendor choosing how many transistor-pairs to provide in each p- and n-region. One pair, (as shown in Figure 4.18(a)) may be too restrictive, but three or more pairs may be wasteful. The solution adopted in the TAHC arrays is to repeat three pairs, two pairs, two pairs, three pairs in each row, with polysilicon crossunders for output connections between each. Hence a 'cell' is, in this series, defined as ten transistor-pairs, and has a total logic capacity of two 2-input and two 3-input NAND or NOR gates. (This may be referred to as 'five 2-input NAND gate equivalents', but it should be noticed that it is not possible to make five 2-input NAND gates with the ten transistor-pairs.) Figure 4.19 illustrates this product, together with the interconnection details of a 2-input Exclusive-OR gate.

However, most commercially available uncommitted CMOS arrays standardize on three transistor-pairs per cell rather than the mixed choice shown in Figure 4.19. The number of cells required to build up a range of digital logic building-blocks is then as indicated in Table 4.3. The vendor's CAD system contains all the necessary interconnect information together with simulation data for the customization of the array.

Functional-level arrays

Because of the flexibility of the above type of CMOS array there is little necessity to market alternative forms of uncommitted CMOS arrays containing hard-wired gates rather than component transistor-pairs. However, hard-wired cells may be found in other technologies, including ECL, ISL, LSTTL, BiCMOS and GaAs.

The logic function (often termed a 'primitive' in this context) chosen by a vendor as the functional cell must be capable of being used as a basic building-block from which all digital circuits can be assembled. It must therefore be a *universal logic primitive*.

The simplest choice of logic primitive is either a NAND or a NOR gate. Both are universal primitives, as was seen in Chapter 3 when discussing programmable logic devices, and are widely chosen for uncommitted gate arrays. A 2-input NAND or NOR gate is the smallest possible primitive, but three or more inputs are sometimes chosen in preference to 2-inputs – again, this is a compromise since the actual required fan-in of the logic circuits in a final commitment may vary widely.

The choice of NAND or NOR gates may be inefficient due to the limited logical discrimination of such Boolean gates; by themselves each n-input gate can only distinguish one of the 2^n different possible binary input combinations uniquely, the remaining $2^n - 1$ input combinations producing an identical gate output. It is this property that leads to a large number of simple Boolean gates having to be used in cases where more complex logic functions are required. However, in addition to simple NAND and NOR gates a number of other more comprehensive universal logic configurations can be proposed. These are characterized by the feature that their inputs are not symmetrical, that is, they realize different duties depending upon the pattern of their input connections, unlike simple Boolean gates, which always have the same logic relationships irrespective of any input connection

Gate-array techniques   141

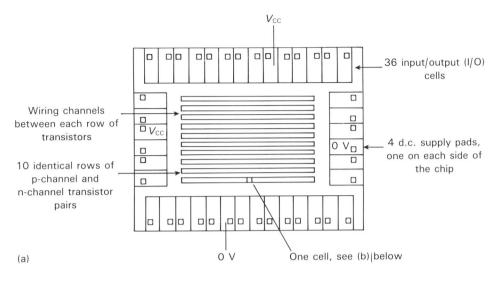

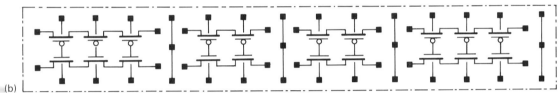

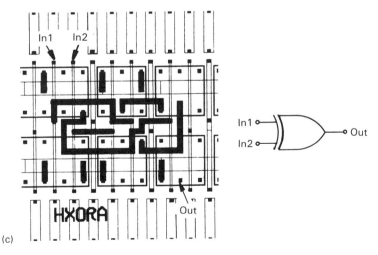

Figure 4.19 The Texas Instruments TAHC 06 uncommitted CMOS array: (a) the chip floorplan with ten rows of uncommitted cells; (b) one cell with a total of ten transistor-pairs divided into four groups; (c) the cell metallization to make an Exclusive-OR function taken from the vendor's design manual, five transistor-pairs being used. (Courtesy of Texas Instruments Inc. [25].)

Table 4.3. A short extract from a vendor's library of functions for a CMOS gate array giving the number of cells required per function, each cell being two transistor-pairs, as shown in Figure 4.18(b)

Function block	Block type	Function	Cells
NOR	F202	2-input NOR gate	1
	F203	3-input NOR gate	2
	F204	4-input NOR gate	2
	F208	8-input NOR gate	7
OR	F212	2-input OR gate	2
	F213	3-input OR gate	2
	F214	4-input OR gate	3
NAND	F302	2-input NAND-gate	1
	F303	3-input NAND gate	2
	F304	4-input NAND gate	2
	F305	5-input NAND gate	3
	F306	6-input NAND gate	3
	F308	8-input NAND gate	7
AND	F312	2-input AND gate	2
	F313	3-input AND gate	2
	F314	4-input AND gate	3
EX-OR	F511	2-input Exclusive-OR gate	3
EX-NOR	F512	2-input Exclusive-NOR gate	3
Full adder	F521	1-bit full adder	7
	F523	4-bit full adder	30
Multiplexer	F569	8-1 multiplexer	17
	F570	4-1 multiplexer	8
	F571	2-1 multiplexer	4
	F572	Quad 2-1 multiplexer	11
Latch	F595	R-S latch	4
	F601	D-latch	3
	F602	D-latch (with reset)	4
J–K flip-flop	F771	J-K F/F	9
	F774	J-K F/F with set–reset	11
Counter	F961	4-bit sync. binary counter with $\overline{\text{reset}}$	50
	F962	4-bit sync. binary counter with $\overline{\text{reset}}$	34
Comparator	F985	4-bit magnitude comparator	46

Source: NEC Electronics Inc., [31].

permutations. The specification of such universal primitives can be derived from function classification theory, and they have the capability of being able to realize any Boolean function of n variables, depending upon their input connections, where n is 2, 3, ... [33]–[37]. Some of the possibilities for $n = 2$ are shown in Figure 4.20, from which it will be evident that each primitive is more complex than a simple NAND or NOR gate. (These complex primitives need not be made exactly as shown in Figure 4.20 from an assembly of individual Boolean gates, but can be hand-crafted in a more compact way to give the same functional relationships. This may be noticed in the design of many commercially available multiplexer circuits, for example.)

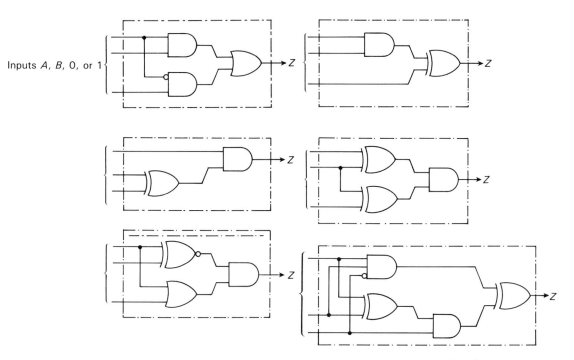

Figure 4.20 Some single-output universal logic primitives which are capable of realizing any function of two variables A, B, depending upon the input connections, the inputs being A, B, logic 0 and logic 1 arranged as necessary. The first circuit will be recognized as a multiplexer. More complex universal primitives capable of realizing any function of three (or more) variables have also been proposed.

In comparing the use of NAND and NOR gates and these more comprehensive primitives, the conflicting factors that arise are as follows:

NAND and NOR gates
- simple cell design, with small silicon area per cell;
- large number of cells required to realize most logic macros, with cumulative propagation delays;
- high local wiring density between cells for cell interconnections, thus requiring wide wiring channels in the gate array;
- the use of NAND or NOR gates is familiar to most logic designers.

Universal logic primitives
- possibly complex cell design with a higher silicon area than a simple NAND or NOR cell;
- fewer cells required than with NAND or NOR cells, giving potentially higher speed;

- lower wiring density between cells than with NAND or NOR cells, thus allowing narrower wiring channels in the gate array;
- use of such complex primitives is initially unfamiliar to most logic designers.

Although studies have shown that there is an overall economy in silicon area in adopting a more comprehensive primitive for gate-array use than NAND or NOR primitives, the more familiar NAND and NOR gates remain the usual choice of a vendor. The multiplexer cell (see Figure 4.20) has, however, been used by some vendors, particularly for high-speed applications. Difficulties in using compre-

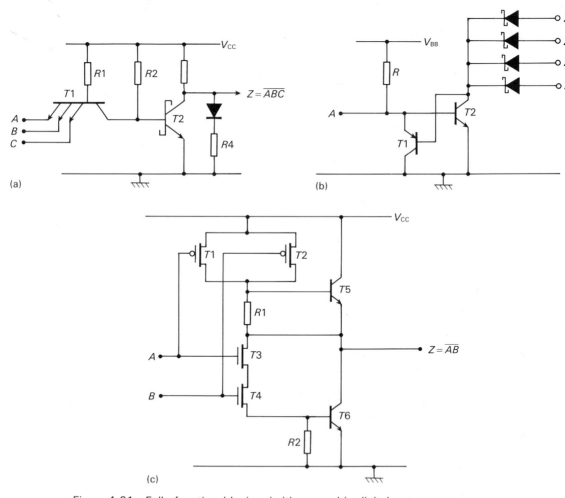

Figure 4.21 Fully functional logic primitives used in digital gate arrays: (a) lowpower Schottky TTL 3-input NAND; (b) ISL four-output NOR; (c) high-speed BiCMOS NAND.

hensive primitives on a gate array should not arise if the vendor provides a comprehensive library of interconnections, as is done for CMOS component-level arrays and other uncommitted products.

Figure 4.21 illustrates some of the functional cells which have been chosen by vendors for their gate-array products. The low-power Schottky TTL circuit of Figure 4.21(a) is intended to be compatible with standard off-the-shelf STTL circuits, and is therefore relevant for use in an all-TTL product environment. The ISL technology circuit of Figure 4.21(b) is similar to the circuit shown in Figure 2.14, but with a fixed resistor to provide the constant bias current I_B. Its low power dissipation lends itself to large, reasonably fast bipolar arrays, possibly as high as 5000 gates per die, but there is the inherent designer unfamiliarity in using a logic gate with a fixed fan-out and a fan-in of one.

Finally, BiCMOS gate arrays with a logic cell (as shown in Figure 4.21(c)) have been announced recently. The basic cell is a 2-input NAND with bipolar output drive, but the particular distinction of these products is that a CMOS static RAM block is included on the same uncommitted die as the BiCMOS array. Sub-nanosecond gate propagation delays are claimed, which is faster than CMOS alone can provide when driving output loads, and the word × bit length of the memory is configurable by the custom metallization [36], [37].

The above and all other functional-cell USICs usually rely heavily upon the vendor's CAD to provide library interconnection details and commitment software.

4.3.2 Routing considerations

The principal design activity required with all types of custom ICs which use some pre-designed basis — cell designs in the case of standard-cell USICs and pre-fabricated base wafers in the case of gate-array USICs — is the routing required to complete the final custom chip. Placement of the required functional parts on the chip layout is a necessary pre-requisite whose sole function is to facilitate the subsequent routing so as to provide a compact and acceptable interconnection topology. Bad placement can cause impossible-to-route or unacceptable routing conditions, or failure to meet the performance specification due to excessively long interconnection runs in critical paths of the design.

With gate arrays, the problems of subsequent routing have to be considered by the vendor at the initial chip design stage, in particular how much silicon area shall be left in the layout for eventual interconnections. With the usual floorplan consisting of rows of cells interspersed with wiring channels, this means: how many interconnections are likely to run along each channel, and therefore what shall be the width of the wiring channels in the chip design? Clearly, too narrow a choice will make the subsequent routing difficult if not impossible, whilst too wide a choice will waste silicon area and be uneconomic. Standard-cell USICs do not of course pose this problem, but it may still be desirable to have guidelines as to the expected or average number of interconnections in a typical design.

The number of connections required between building-blocks in digital systems has been studied. For macros with a very repetitive functionality, for example the connections required between cells to make a shift register assembly, the problem is straightforward, but for random or 'average' logic requirements the problem is more difficult.

Consider the partitioning of a digital system into D sub-divisions, each of which consists of a number of individual functions or cells. With B cells per division the complete system S consists of $B \times D$ cells. Further, let each digital cell have an average of K input/output terminals. This is illustrated in Figure 4.22. An analysis of this situation was first made by Landman and Russo [38], who considered a number of real-life digital systems varying the number of sub-divisions from $D = 1$ (all cells contained within one boundary) to $D = S$ (only one cell per sub-division). It was empirically shown that, provided the number of sub-divisions of the system was greater than about $D = 5$, then

$$P = K \times B^r$$

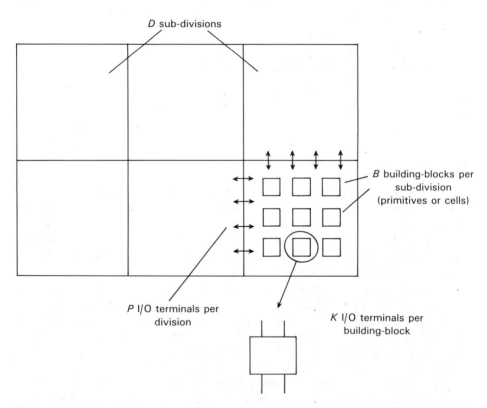

Figure 4.22 The division of a digital system consisting of S cells for building-blocks into D sub-divisions, with B cells per sub-division.

where

P = number of interconnections around the perimeter of the sub-division
K = number of terminals per cell
B = cells per sub-division

and where the index r had a value in the range

$$0.57 \leqslant r \leqslant 0.75$$

depending upon the structure of the system. This relationship has been termed Rent's rule, with the index r termed Rent's index. (Rent was involved in this subject area at IBM in the 1960s, but does not personally appear to have published his involvement.) Note for the extreme case of only one cell per sub-division, $P = K(1.0)^r = K$, which is correct.

The above result by Landman and Russo was found not to hold as D became very small. For $D < 5$ some alternative more complex relationships were suggested involving two indices r_b and r_s [33], [38], but difficulties in establishing values for these indices are evident. The general equation given above is thus usually cited. Further statistical studies by Heller, Mikhail and Donath [33], [39], and others [40], may also be found, but the work involves a number of parameters whose values are also not easy to determine or estimate.

Rent's rule may be applied to channelled architectures as follows. The system partitioning of Figure 4.22 may be applied to a single chip, and the blocks re-drawn as a string of cells, as shown in Figure 4.23(a) and (b). The connections of the B cells now become the number of interconnections in the wiring channel to and from the block. However, if we assume that there is no local peak interconnection density, which is generally true in gate-array circuits or can be made true by appropriate placement, then this number of connections will be the same wherever we window our boundary of cells in Figure 4.23(b). If we also assume that half this number of connections crosses the left-hand boundary and the other half crosses the right-hand boundary, then the wiring channel interconnection density will be one-half of the value of P.

As an example, consider a 400-cell gate array arranged in twenty rows of 20 cells per row, each cell having four I/O terminals. Then the total number of connections associated with a row of twenty cells is given by

$$P = KB^r$$
$$= 4(20)^r$$
$$= 22 \text{ taking } r = 0.57$$
$$= 39 \text{ taking } r = 0.75.$$

Hence the estimated wiring channel capacity is eleven tracks wide for $r = 0.57$, or nineteen tracks wide for $r = 0.75$.

148   Custom microelectronic techniques

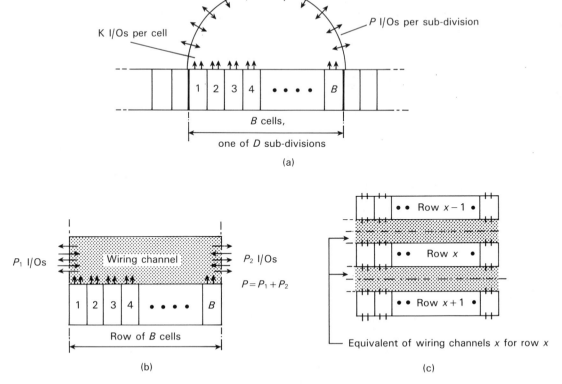

Figure 4.23 The development from Figure 4.22 into a row and wiring channel topology: (a) B cells in each of D sub-divisions; (b) re-drawing of (a) into a wiring channel form; (c) alternative possibilities to (b).

Should the cells in an array have both-sides access, as most commercial products do, the connections from a row of B cells may be considered to be shared equally between two wiring channels. This is shown in Figure 4.23(c). However, it will be appreciated that all the wiring channels, except those at the perimeter of the array, will still have the same total number of interconnections, and thus there is no change in the required wiring channel capacity.

The problem with Rent's rule is obviously the indeterminacy of the value of r. The more functionality that is contained within a cell, the lower the value of r, but with, say, a 2-input NAND cell, r tends to be higher since a greater number of intercell connections is now required. Most USIC vendors by now have sufficient company experience with previous gate-array layouts to be able to judge what capacity wiring channels to provide in new products; most gate-array layouts will be found to have a maximum channel wiring capacity of between ten and twenty interconnections.

The required capacity of the wiring channel around the perimeter of the array is very problematic, since it can vary widely depending upon whether the chip I/Os have to be allocated in some fixed order to suit a customer's requirements or whether they can be freely allocated to the I/O duties. In general, the perimeter wiring channels usually have a capacity similar to that of the wiring channels between the rows of cells.

Floorplan layouts other than continuous rows of cells separated by wiring channels have been used for both gate-array and standard-cell USICs. Some of the arrangements are shown in Figure 4.24. Arrangement (b), with wiring access on all four sides of each cell, was the floorplan used for the early bipolar cells shown in Figure 4.17(a). A theoretical estimation of wiring channel capacities with these alternative layouts, however, is not very useful since there are too many degrees of freedom in the possible routing paths. These alternative floorplans are now generally obsolete except for analogue cells (see Section 4.3.4), but 'channel-less' floorplans have recently been introduced as a completely new concept for digital arrays. The following section will cover this new alternative.

As well as providing sufficient width to run the required number of interconnections, the basic gate-array design also has to cater for the connections to the cells and means to cross interconnections under each other to achieve the final routing. With single-layer-metal gate arrays this is achieved by using the polysilicon level of the fabrication to provide connection paths at right-angles to the longitudinal metal tracks in the wiring channels. Even with two-layer-metal arrays some polysilicon connections are still required at this lower level.

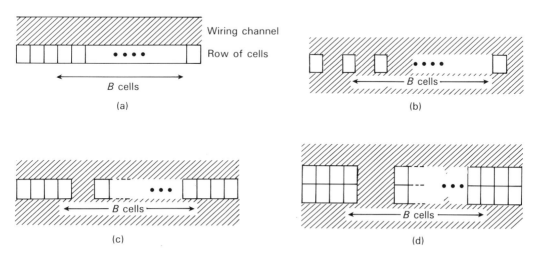

Figure 4.24 Some alternative floorplan concepts: (a) the channelled array with abutting cells as previously considered; (b) block-cell architecture with four-sides access to each cell; (c) quadruple-cell architecture with both-sides access per cell; (d) back-to-back eight-cell architecture with single-sided access only per cell.

Previous illustrations (see Figures 4.18 and 4.19, for example) have shown how the polysilicon gate connections of cells, together with additional crossunders, are provided to the edges of the cells. These connections give both-sides access and also a means of making connections from one wiring channel to an adjacent one through a cell. However, this does not solve the problem of how to cross over connections within a wiring channel so as to connect to the cell I/Os.

The basic solution is to provide a number of polysilicon crossunders or 'fingers', sometimes also known as 'tunnels', across the wiring channels. These may be extensions of the polysilicon connections from the cells themselves, or completely separate crossunders spaced regularly along each wiring channel.

The choice is complicated by the following four factors:

1. Excessive use of polysilicon rather than metal in interconnections will degrade the chip performance due to the higher resistivity of polysilicon compared with metal.
2. If the contact windows (vias) to the polysilicon level have fixed grid positions it may not always be easy to route to some via locations as they may be blocked by previously routed connections.
3. On the other hand, if variable position vias are allowed to the polysilicon, then an additional customization mask (see page 129) is required, but completion of routing will be considerably easier.

There is no best solution to this problem, and as a result very many different arrangements will be found in practice. Figure 4.25 indicates some of the concepts which have been used, with many variations on these arrangements.

Routing considerations for the peripheral wiring channels around the array also dictate the use of polysilicon crossunders, particularly where the logic signal interconnections cross d.c. supply tracks. Again, many and varied arrangements may be found in practice.

The final ease, or otherwise, of routing a channelled gate array thus depends upon the original floorplan decisions made by the vendor. The only flexibility available at the customization stage is in the placement (allocation of the functional duties) of the cells in the array, which must be done so as to avoid bottlenecks in the channel routing. If an originally chosen placement proves to be impossible to route due to too many tracks required in a wiring channel or the inability to cross over to a required via, then a re-placement procedure is necessary to solve such wiring constraints. Human intervention is often the best if not the only way of solving such problems if the CAD system fails to 100% complete the required routing. Two-layer-metal customization clearly gives a greater routing freedom, but at the expense of the additional costs and fabrication time.

Routing of small, single-layer-metal gate arrays may be done by hand using large sheets which represent the floorplan of the chosen chip. This procedure is still available for customer's use from some vendors, since it allows the customer to do all the routing part of the custom design work without the need to purchase any

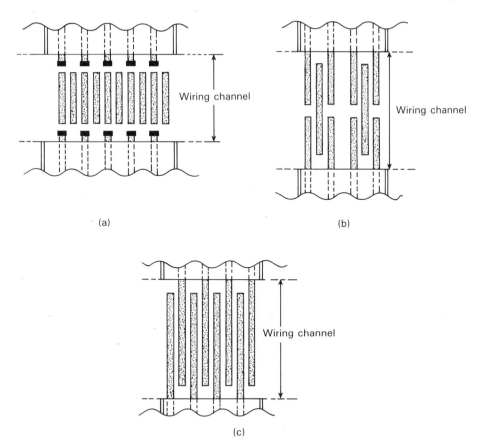

Figure 4.25 The use of polysilicon crossunders across wiring channels: (a) fixed via positions on each cell I/O with separate polysilicon crossunders in the channel, possibly with two or more fixed vias (not shown) per crossunder; (b) extended crossunders from each cell I/O, with fixed or variable via positions per crossunder (not shown); (c) interleaved crossunders from each cell I/O, ideally with variable via positions per crossunder (not shown).

CAD resources [29]. Indeed, it may be preferable to hand-route some gate arrays which do not have a channelled architecture because the human designer is very skilful at solving difficult or flexible routing problems, but in general larger and more complex chips necessitate the use of CAD resources such as will be considered in Chapter 5.

Figure 4.26 illustrates the principle of using large vendor's sheets for manual placement and routing of channelled CMOS arrays. In Figure 4.26(a) transparent decals are used which represent the internal cell metallization patterns for different functional commitments (see also Figure 4.19(c)). The routing is subsequently drawn in using the fixed via positions which occur on a rigid grid pattern. In Figure

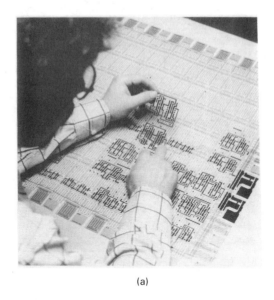

(a)

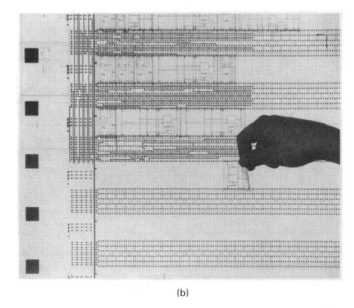

(b)

Figure 4.26 Manual placement and routing of early-generations CMOS channelled gate arrays: (a) decals to give the p-channel and n-channel transistor interconnection details prior to channel routing; (b) decals to act as functional placement prior to channel routing. (Courtesy of (a) Interdesign Inc., CA; (b) Micro Circuit Engineering, UK.)

4.26(b) the designer uses logic symbols rather than the cell metallization patterns for different functions; in this case the vendor subsequently fills in the internal cell metallization details from his CAD database. The positions of a large number of fixed via positions in the wiring channels will be noted.

The great disadvantage of manual routing, apart from the time taken, is that the final hand design has subsequently to be entered into the vendor's CAD system for pre-manufacture checking and simulation and subsequent mask-making, and this transfer of design detail is very prone to human error. Nevertheless, the pre-manufacture simulation should pick up errors of this kind. Manual commitment of digital arrays therefore has drawbacks and is largely obsolete, but it is still very relevant for analogue arrays (see Section 4.3.4).

4.3.3 Channel-less architectures

One of the difficulties of using gate arrays with specifically provided wiring channels is that this floorplan is exceedingly inefficient in making regular structures such as ROM and RAM. For such structures transistors can be packed very closely together with two axes of interconnection (see Chapter 2), but a gate-array floorplan with its wide wiring channels does not lend itself to this requirement.

A channel-less architecture was introduced in 1985 in order to provide a more flexible gate-array topology, and is now available from many vendors. The technology of these arrays is CMOS, and the target market is the very-high-speed sophisticated product area. The terminology used for these products, however, is still rather chaotic, and includes the following:

- Channel-free™
- Compacted Array™
- CHANNELLESS™
- CHANNELLESS ARRAY
- sea-of-gates
- sea-of-cells
- gate forest

and possibly other variants. (Channel-Free and Compacted Array are registered trademarks of LSI Logic Corporation, and CHANNELLESS the trademark of California Devices, Inc. Also, the terms 'sea-of-gates' and 'sea-of-cells' may be used to refer to a floorplan with non-abutting individual cells such as that illustrated in Figure 4.24(b).)

As the name suggests, channel-less architectures contain no pre-defined wiring channels between cells for the custom metallization, see Figure 4.27. Instead, the whole active area of the die is filled with p-channel and n-channel transistors, with the metal levels of interconnect being routed over unused transistors. This gives

154   Custom microelectronic techniques

Figure 4.27 The floorplan of a small CMOS 1.5 μm channel-less array, containing 3808 cells in 56 rows of 68 cells per row, each cell providing three uncommitted transistor-pairs, with 84 I/O pads, V_{DD} supply +3 to +6 V. (Courtesy of Silicon Technology SpA, Italy.)

much greater flexibility in the dedication, including the ability to make compact RAM and ROM macros, but at the expense of greater complexity in the CAD routing software compared with channel routing. The percentage of those transistors on a die which are finally used may be as low as 25%, but this is compensated for by the very large number which are packed on an uncommitted die. However, for regular RAM and ROM macros up to 90% transistor utilization may be possible within these areas [41]–[44].

Like channelled arrays, the channel-less arrays are built up from a chosen basic cell which is replicated on the chip, but unlike the former the cells abut both horizontally and vertically, leaving no dedicated wiring channel spacings. Two-layer metallization is employed, and adjacent cells are often mirrored along both the x- and y-axes so that metal supply and clock lines may be shared between adjacent cells.

The exact arrangement of the p-channel and n-channel transistors provided per cell varies considerably between vendors. However, it is not unusual to have different sizes of transistors in a cell, possibly larger-size p-channel transistors than n-channel so as to provide equal drive capabilities, or 'small' and 'large' transistors for different logic duties. More n-channel transistors than p-channel may be provided, which is relevant for RAM applications, for example. A further innovation may be the inclusion of a very small p-channel and n-channel transistor pair whose gates are permanently connected to V_{DD} and V_{SS}, respectively, and which in their permanently cut-off state provide both isolation for the active devices and also tie-down paths to the well and silicon substrate [43].

Figure 4.28(a) shows one possible basic cell structure consisting of 'small'-size transistor pairs in the centre, and additional 'larger' pairs on the outside. If the middle pairs are not used in a cell, up to ten wiring runs may pass over them, while if the outer transistors are unused, up to fourteen wiring runs may pass over them [45]. With two-layer metallization – sometimes three-layer with certain products – very compact cell libraries can be provided, giving the dedication of such cells for a range of duties up to LSI or VLSI complexity. Figure 4.28(b) shows the dedication interconnection pattern for a nest of eight cells to form a scan-path flip-flop (see Chapter 6), this particular cell structure being mirror-imaged in the floorplan layout in both axes [43]. Notice that interconnections are not confined to run in straight lines but are free to 'dog-leg' to complete the cell connections.

It is clear that channel-less gate arrays cater for a higher level of complexity than does the general range of channelled arrays. Typical specifications are currently as follows:

- 1.5 μm CMOS technology;
- 0.75 ns or faster propagation delay for a 2-input NAND gate driving a fan-out of two;
- total gate count per die ranging from thousands to over one hundred thousand, taking one gate as two transistor pairs;
- estimated useable gates 40%;
- up to 350 or more I/Os per chip;
- TTL/CMOS I/O compatibility.

However, the total market potential for channel-less arrays is still debatable. It is claimed that they will provide the same performance and system complexity as standard-cell USICs (better performance at any given time is perhaps obtained, since it may be possible to bring improved fabrication capabilities faster to the market by re-designing channel-less arrays than by re-designing a complete standard-cell library to cater for technology changes), but with a quicker and cheaper design time and fabrication cost due to the reduced mask set needed for their customization. But as possibly five or more masks are required for customization (via masks plus metal interconnection masks), simple channelled arrays will remain more appropriate for the less sophisticated custom requirements [46]. The more complex routing requirements will be discussed in the following chapter.

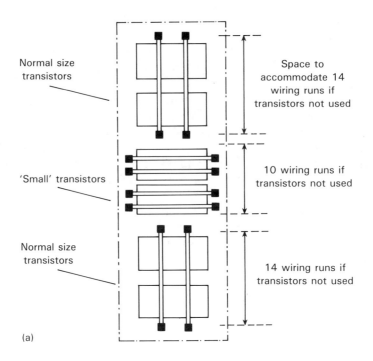

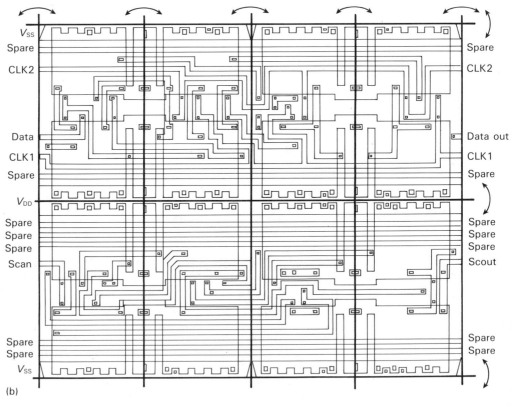

4.3.4 *Analogue and mixed analogue/digital arrays*

In contrast to the very large number of transistors present in channel-less arrays, uncommitted arrays for analogue (linear) purposes usually contain a much less impressive number of devices. They may, however, include high-voltage devices, and will certainly dissipate more power in their linear mode of operation.

Early analogue arrays consisted of a number of individual resistors, capacitors and bipolar transistors on the uncommitted die, with no specific floorplan topology, possibly tens of npn and pnp transistors, a few additional higher-power transistors, and a few capacitors and perhaps a hundred or more resistors of various values. Since these components were scattered all over the floorplan layout, custom routing was invariably done by hand. This lack of a formal grid layout meant that every custom design began from scratch, no vendor's library of interconnection designs being available. However, one accessory was introduced, namely off-the-shelf SSI packages, sometimes termed 'kit parts', each containing three or four of the components provided on the die, from which a breadboard prototype could be physically assembled and verified. Because the analogue frequencies originally involved were not high, usually under 1 MHz, the performance of the discrete component breadboard was very close to that of the committed IC.

However, the new generation of analogue arrays employs a much more structural floorplan. This consists of 'tiles', sometimes termed a 'mosaic', each tile being the analogue equivalent of the uncommitted digital cell but consisting of a fixed structure of uncommitted transistors, resistors and, possibly, capacitors per cell. These tiles are replicated on a fixed grid pattern across the chip layout [21], [47]. Wiring space is available between components and between cells, but wide wiring channels are not usually provided since the interconnection density of an analogue array is much more local and less distributed than in digital arrays [48]–[53].

The problem which is still present when utilizing a tile topology is the choice of components to provide per tile so that each may be of maximum use. This dilemma facing vendors is illustrated in Figure 4.29. Among the solutions which have been chosen are the following:

- the use of two or more different standard tile specifications per die to cover general-purpose duties;

Figure 4.28 CMOS channel-less gate array architectures: (a) a possible basic cell containing normal size and 'small' size transistor-pairs, omitting any supply rails; (b) an alternative cell design, showing the commitment of a floorplan of eight cells to realize a scan-path flip-flop macro. The mirror image floorplan has been indicated. Note that wiring runs marked 'spare' are unused through-space only and not interconnection metal, and that the actual interconnections can take random routing paths. (Sources: (a) based upon [45], (b) based upon [43].)

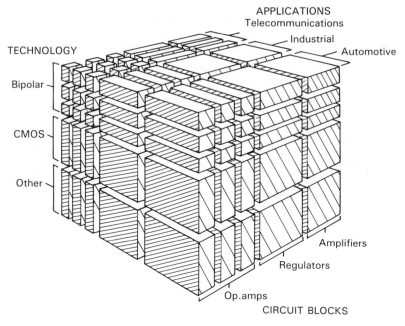

Figure 4.29 A graphical representation of the dilemma in finding a truly universal solution to the specification of an uncommitted analogue IC. (Source: based on Bray and Irissou [47].)

- the inclusion of additional separate resistors and capacitors and sometimes other components such as zener diodes on the die; and
- the inclusion of additional separate transistors for high-power duties.

Figure 4.30 shows one product which contains two types of tile, a 'standard' tile and a 'power' tile. Component details are given in Table 4.4. The small npn transistors have a typical f_T of 350 MHz and the corresponding pnp transistors an f_T of 300 MHz, each with a collector-to-emitter reverse breakdown voltage of 33 V minimum. All internal cell connections are on a rigidly defined grid, which allows the customer to carry out hand-routing on a master layout sheet if required, aided by macro interconnection library details supplied by the vendor [52].

An alternative design, in several ways, is shown in Figure 4.31. Here there are two types of tile again, but now each type consists of small-geometry npn and pnp transistors only, varying between the two types in their placement detail. Additional large transistors, capacitors and zener diodes are provided in columns and around the perimeter of the chip. Another different feature is that the space between all tiles and other components is given a thin-film nichrome cover from which required resistor values can be customized by etching. The resistor values are given by $R = \rho(l/w)$ where ρ is the sheet resistivity, l the effective resistor length and w the

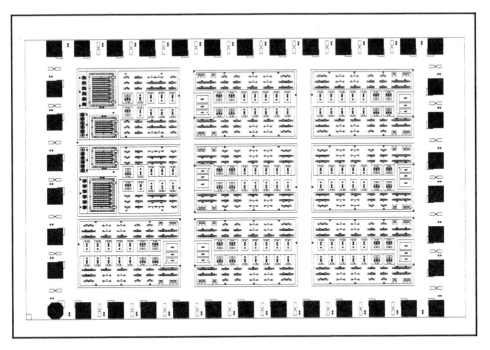

Figure 4.30 A tile-based bipolar uncommitted analogue (linear) array, containing seven standard tiles and two power tiles, with thirty-six I/Os. (Courtesy of AT&T Technologies, PA.)

effective resistor width. This is claimed to give good matching ability, very low parasitic capacitance, and the ability if required to laser-trim resistor values to high absolute accuracy. As with the product considered previously, this linear array may if necessary be routed by hand, with the help of the vendor's macro function library.

Both these two products are typical in using bipolar technology, with silicon dioxide isolation between components so that they are effectively in SiO_2 islands. This gives extremely good isolation properties with high voltage breakdown. Very-high-frequency performance is available from many products, but because the transistor sizes are fixed, system performance cannot match that of a full-custom

Table 4.4. The component details of the analogue array shown in Figure 4.30

	Each standard tile	Each power tile
npn transistors (various)	8	6
pnp transistors (various)	8	6
resistors (various)	50	30
capacitors	1	—

160   *Custom microelectronic techniques*

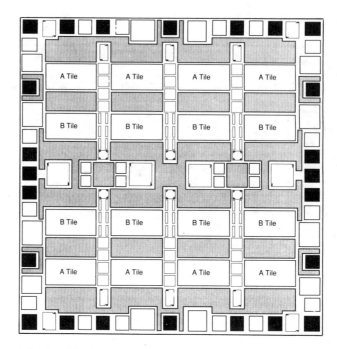

Figure 4.31 A further bipolar tile-based linear array; the plain areas between the cells and other fixed components are nichrome areas which may be etched to form precision resistors. (Courtesy of DataLinear, CA.)

design where full freedom to size the transistors and resistors to achieve optimum amplifier bandwidth is available [21], [48].

One serious drawback with uncommitted linear arrays, such as that illustrated in Figure 4.31, is that the components provided in each tile cannot be conveniently used should any digital logic also be required, for example analogue-to-digital and digital-to-analogue conversion. A component utilization of possibly less than 20% is achieved if analogue tiles are used for logic functions. Earlier generations of uncommitted logic arrays (ULAs^{TM}) using CDI bipolar technology and marketed by (then) Ferranti Ltd and Interdesign offered more satisfactory mixed capabilities, the digital circuits being current-mode logic, as detailed in Figures 2.16 and 4.17, the CDI technology also providing good linear performance. I^{2}L has also been successfully used for combined analogue/digital duties by Cherry Semiconductor Corporation in particular, but present trends are to have separate analogue and digital parts on the same chip in order to optimize each duty.

One such ULA^{TM} arrangement provides a centre area of cells identical to that shown in Figure 4.17(c) and (d) for the digital duties, and a border of analogue tiles around this centre core for linear duties. This combination gives up to 100 MHz analogue capability with 1 ns gate propagation delays in the digital core. An alternative floorplan architecture uses a channel-less CDI sea-of-gates configuration

Maskless fabrication techniques 161

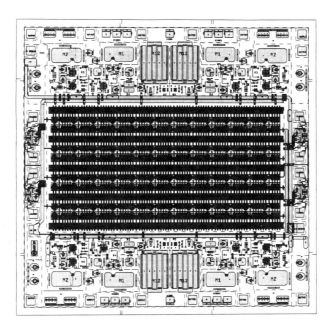

Figure 4.32 A mixed bipolar/MOS technology array, with CMOS transistor-pairs as the central digital core, and analogue components around the periphery. Chip size 5000 μm × 4830 μm, 24 chip I/Os. (Courtesy of AT&T Technologies, PA.)

as the digital core. Both an analogue and a digital macro library are available for custom dedication.

Other recent products, however, combine bipolar and MOS technology for mixed duty arrays, such as that illustrated in Figure 4.32. This particular product features a CMOS channelled array of forty-eight cells, each cell consisting of five transistor-pairs, surrounded by eight main analogue tiles plus other discrete components. This product is aimed at robotic and similar duties, and provides high voltage ratings (>250 V) in some of the analogue components.

It is apparent that the many and varied analogue requirements which may arise necessitate a wide range of different array specifications. Hence the OEM designer must always carry out a wide vendor search when considering the use of such USICs in company products.

4.4 Maskless fabrication techniques

The normal means of producing the required geometric areas on a wafer during fabrication is by photolithography, using individual masks to define each separate level of fabrication. At each mask stage the wafer is first given a silicon dioxide protective layer followed by a thin layer of photoresist which is subsequently illuminated through a mask with ultra-violet light. If positive photoresist is used, then the areas receiving the UV illumination can be etched away, and the areas which have not been illuminated will remain; if negative photoresist is used, then the areas receiving no UV illumination are removed. (For VLSI fabrication, positive photoresist is preferable because it does not swell as much during development as negative photoresist, and hence reproduces very small dimensions more accurately.) Silicon processing then continues through the windows in the photoresist.

Masks are made of special glass with a chromium surface film which is patterned by E-beam lithography. This involves covering the chromium with a resist which is sensitive to an electron beam, rather than photoresist, which is sensitive to UV light, and 'writing' the required pattern on the resist by means of an E-beam machine which focuses, deflects and switches the beam of electrons on and off. Where illuminated by the E-beam the chemical bonds of the resist are weakened and can subsequently be washed away, leaving windows through which the underlying chromium can then be chemically removed, thus producing the optically transparent pattern on the mask. However, because the E-beam can only be deflected by one or two millimetres without unacceptable distortion, the table holding the mask must be moved ('stepped') very accurately under computer control in order that a complete wafer-sized mask may be written.

There are, however, three distinct methods of using masks in the photolithographic process, as follows:

1. Contact exposure: when the mask physically touches the photoresist surface on the wafer during exposure.

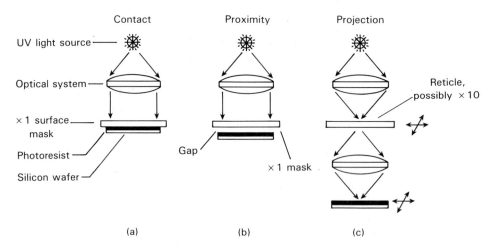

Figure 4.33 Optical lithographic methods: (a) contact; (b) proximity; (c) projection with computer-controlled movement of reticle and wafer.

2. Proximity exposure: when the mask is held a few microns above the surface during exposure.
3. Projection printing: where the mask is held some distance above the wafer and an optical system is used to project the UV illumination on to the surface.

This is illustrated in Figure 4.33 [54]. With the latter method the mask no longer defines the whole wafer patterning but only a small area, possibly one die only, in order to maintain high resolution. The mask is now called a 'reticle' and may be ten times the actual die size, the optical system projecting this enlarged mask pattern down to the required dimensions. A complete wafer is now exposed by a step-and-repeat process using a precision machine called a 'wafer stepper'. (A $\times 5$ or $\times 10$ reticle may also be used to make a full-size $\times 1$ mask by an optical step-and-repeat procedure, the reticle being held as the pattern data.) The resolution obtainable with these methods is about 1 μm or slightly better.

4.4.1 E-beam direct-write-on-wafer

It is evident that the production of masks for wafer fabrication is an expensive and frequently time-consuming process. One mask for a 150 mm diameter wafer may cost several thousand $US, and be manufactured by a specialist company separate from the actual IC vendor. For custom microelectronics where small-quantity production runs may be involved, means to eliminate this cost and the potential bottleneck are attractive.

E-beam direct-write-on-wafer (E-beam lithography) is one commercially available possibility. Here an E-beam machine such as that shown in Figure 4.34 is driven by the same design information on magnetic tape as would be used to make the glass/chromium optical masks, but the E-beam is used to pattern resist directly on the wafers. With careful control of the E-beam energy the resist can be patterned without causing radiation damage to the underlying doped silicon areas. Greater care to avoid damage may have to be taken with MOS technologies than with bipolar.

The advantages of direct-write-on-wafer include the following:

- elimination of the cost in making optical masks;
- possibly improved delivery time for small quantities due to not having to wait for optical masks;
- compatibility of the CAD software output which defines the required geometry with the E-beam machine computer control;
- the ease with which design errors can be corrected;
- the ability to mix different customers' designs on one wafer ('multiproject' wafers) in order to reduce individual costs (see Figure 4.35).

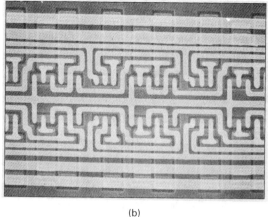

Figure 4.34 E-beam lithography for direct-write-on-wafer: (a) E-beam machine with linewidth capability down to 0.125 μm and workpiece position control by laser interferometry; (b) typical metallization results. (Courtesy of Cambridge Instruments Ltd., UK.)

Maskless fabrication techniques 165

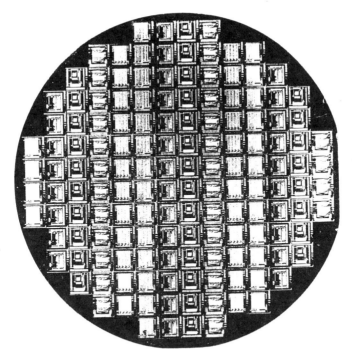

Figure 4.35 A multi-project wafer showing several different designs per chip. (Courtesy of University of Edinburgh, UK.)

Against these real or potential advantages are the following disadvantages:

- very high capital costs of E-beam machines, possibly $4 million per machine, which must be covered in the IC selling prices;
- slow writing speed, possibly more than ten minutes to write just one wafer pattern.

Because of the latter factor it is not relevant to consider E-beam direct-write for volume production purposes, particularly where ten or more mask levels per wafer may be involved. Direct-write becomes significant in the custom IC area to provide ten or twenty prototype circuits for verification purposes, or for a small initial production run to test the market before full-scale production begins [9], [10].

Although at least one company, European Silicon Structures, uses E-beam lithography to direct-write all twelve layers of standard-cell custom designs [9], it is more usual to consider direct-write for the customization of uncommitted arrays, where a reduced mask set would conventionally be used. The ideal candidate for direct-write is a single-layer-metal array with fixed via positions, since this necessitates only one direct-write procedure; multilayer arrays clearly have to be written more than once.

There are certain problems with E-beam array commitment, particularly if this is done by persons and machinery other than those employed by the wafer manufacturer [55]. These include the following:

- alignment of the E-beam with the underlying silicon fabrication;
- possible dimensional tolerances and optical distortions across the full wafer area, sometimes termed 'run-out';
- wafer surfaces not being perfectly flat, particularly with multilayer metallization.

Nevertheless, successful direct-write is now commercially established, and should provide a faster route than mask fabrication to prototype circuits for users of uncommitted arrays.

4.4.2 Other technologies

Laser technology is another technology which has been applied in the manufacture of microelectronic circuits. Lasers are extensively used in sophisticated production equipment for accurate positioning and stepping using helium–neon inferometry techniques, and in inspection and marking [54], but they can also be used for part of the on-chip processing [56], [57]. This latter area with which we are principally concerned here includes the following:

- laser photolithography;
- laser cutting and trimming;
- laser welding; and
- laser deposition.

The use of a laser source to pattern photoresist is receiving increasing attention as geometries shrink to sub-micro values. An excimer-type laser is normally used as the light source, giving high-intensity UV radiation exposures as short as 20 ns with 0.1 μm resolution. Whilst this is used for the production of masks and reticles in place of previous UV light sources, direct-write-on-wafer can also be done as an alternative to E-beam direct-write. X-ray lithography is also being researched for sub-micron lithography with wavelengths in the 4–50 Å range [54], [58]. However, this is still at an early stage of development.

One particular commercial application of laser direct-write-on-wafer is the LASARRAY system [56], [59], [60]. Here, a helium–cadmium (He–Cd) laser is used to expose the photoresist, following which an etching procedure is performed to remove unwanted metal. A particular feature of the LASARRAY system is that it can be supplied complete to customers to undertake their own dedicated procedure. The system comprises three self-contained, clean-room chambers which provide wafer customization, wafer slicing, testing and packaging, the input being the wafers and the output the packaged chips. The uncommitted wafers, however, are specific to the LASARRAY process and not from any USIC vendor since they have to contain positioning information for the several stages of automatic processing. The overall floor space of the assembly is 9.03 m × 7.50 m and the direction of processing is clockwise through the three principal clean-room areas.

The overall CAD and computer control is also unique. Typical processing time for a 100 mm diameter wafer is stated to be about two hours. The cost of a complete system is however considerably more than US$1 million.

Laser cutting and laser trimming are both well-established practices, particularly in the realm of hybrid circuits where passive components have to be trimmed to close on absolute values. However, laser cutting has been employed to customize arrays which contain all possible interconnection paths on the uncommitted chips, the unwanted connections being cut by a computer-controlled laser beam. Further details of laser cutting and laser trimming may be found in the extensive bibliography listed in [56].

In all the fabrication processes previously considered, the patterning of the top interconnection layers has been a *subtractive* process, unwanted metal being removed leaving only the required connections, but with a process known as 'laser pantography' metal may be deposited in conducting strips using a scanning laser. This is thus an *additive* process.

The principle of direct-writing of conducting strips depends upon a local reaction caused by the energy of a focused laser beam in a reactive atmosphere, this reaction being produced either by the temperature rise induced by the laser spot (pyrolysis) or directly by photon energy (photolysis). Many different combinations of materials and laser sources have been studied to deposit conductors such as copper (Cu), aluminium (Al) and tungsten (W) (see Table 4.5), but one presently in commercial use uses an argon-ion laser to deposit tungsten or nickel (Ni) for the interconnection tracks.

This particular example is LASA Industries' QT–GA (quick-turn gate-array) process, a process which the company terms 'lasography' and which provides one-layer-metal or two-layer-metal gate arrays of several thousand 2-input NAND–NOR gate capacity [9], [50], [57], [59]–[62].

The complete deposition system is illustrated in Figure 4.36. Pre-processed wafers which have been fabricated up to the I/O bonding pads and the contact

Table 4.5. Some examples of conductor materials deposited by laser pantography

Possible reactive atmosphere	Possible type of laser	Conductor deposited
$Al_2(CH_3)_6$	Ar^+ (257 nm), Kr^+ (476 – 647 nm)	Al
$Au(CH_3)_2(AcAc)_2$	Ar^+ (515 nm)	Au
$CuSO_4$	Ar^+ (515 nm)	Cu
$[Pd(m - O_2(CH_3)_2]_3$	Ar^+ (515 nm)	Pd
$W(CO)_6$	Ar^+ (257 nm), Ar^+ (350–360 nm)	W

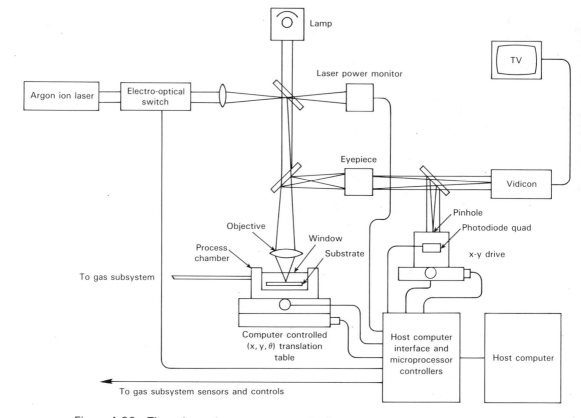

Figure 4.36 The schematic arrangement of a laser pantography system to deposit conducting paths on an uncommitted side. The die is held in a reactive atmosphere and written with a finely focused laser beam. (Source: based upon [56].)

windows (vias) are the starting point, and the initial procedures are as follows:

- wafer probe test for correct fabrication parameters;
- wafer slicing;
- packaging of individual die;
- I/O wire-bonding;
- visual inspection;
- insertion of packaged but uncommitted die into carriers for loading into the process machine.

Within the machine the uncommitted die is positioned in a process chamber and surrounded by an appropriate mixture of gases which decompose to deposit metal under the laser spot. The complete commitment procedure is as follows:

- auto-positioning of the packaged die;
- first layer of tungsten tracks deposited;
- SiO$_2$ insulating layer applied;
- laser etching of contact windows through the SiO$_2$;
- second layer of tungsten tracks deposited;
- final insulating layer applied;
- lid attached to the package.

The whole commitment operation is under dedicated computer control once the uncommitted ICs have been initially loaded.

A particular feature of this process is that because individual ICs are processed and not complete wafers and the metal deposition only takes place at a 1 μm spot under the laser beam, the size of the initial process chamber can be made very small, about 2 cm^3. The subsequent two chambers to and from which robotic arms move the IC are about $25 \times 25 \times 12$ cm for the SiO$_2$ processing and $5 \times 5 \times 5$ cm for the final packaging chamber. As a result the complete processing unit with its chemicals and computer control is entirely self-contained. Performance of the system is such that a writing speed of 1–2 cm s^{-1} is available, which allows a 1000-gate two-layer metal array to be processed in about 5 minutes, or a more complex 6000-gate array in about 120 minutes. Line widths down to 1 μm with spacings of 1 μm can be produced, with location accuracy of 0.2 μm.

As with the previous LASARRAY system and other similar company ventures [59], [63], it will be noticed that this LASA system also requires specially prepared uncommitted wafers. The cost of the system is also high, possibly US$2–3 million. Hence all these maskless fabrication procedures are unlikely to become in-house facilities for any but the largest equipment manufacturing companies or development laboratories, and even then users will be locked into using an appropriate uncommitted product only. They may, however, find a niche market for project work where very fast turn-around times are a priority but where state-of-the-art complexity or performance is not always required.

4.5 Summary and technical comparisons

Having completed this coverage of custom microelectronic techniques we may now summarize their principal attributes. We will leave the question of detailed design costs, usually known as non-recurring engineering (NRE) costs, until Chapter 7, where the choice of design style for a given project will be considered in detail. This consideration will also involve the cost of computer-aided-design (CAD) resources.

However, since the custom mask costs involved in the use of uncommitted arrays is less than that of full-custom or standard-cell ICs, initial production costs should be lower for arrays, but the smaller die size of the latter may make subsequent volume fabrication costs cheaper since more circuits may be accommodated per wafer.

Custom microelectronic techniques

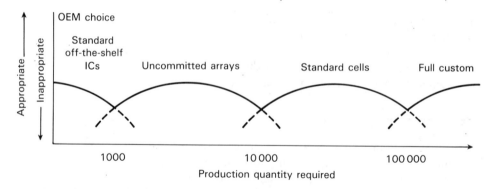

Figure 4.37 The general range of production volumes for different types of digital microelectronic realization.

The economics of this are shown in Figure 4.37 [64], [65]. For very small production quantities the use of any form of custom microelectronic IC cannot usually be justified on purely financial grounds due to the cost of the custom-design activities; for extremely large-volume requirements the smallest and most efficient design, that is, expert hand-crafted full-custom, may be the best, but in between these extremes uncommitted arrays and standard-cell designs offer their respective financial advantages.

This broad generalization remains reasonably true when detailed design costs are also taken into consideration, but the increasing size and capabilities of (a) programmable devices (see Section 3.5) and (b) channel-less arrays, is beginning to blur this picture. Programmable devices may take over the lower-volume end of this spectrum, whilst channel-less gate arrays may capture the higher-volume area at present served by standard-cell designs. These considerations will be pursued in more detail in Chapter 7.

Turning to technical details, more precise comparisons and summaries can be made. In practice it may be that technical performance, for example minimum power consumption requirements, dictates the type of microelectronic product and technology used rather than financial considerations, but in most engineering situations more than one possible design solution usually needs to be considered. (Reference may also be made back to Chapter 2, Tables 2.3 and 2.4, for technology comparison details.)

A summary of the custom technologies which are technically most appropriate for different duties is as follows:

- Analogue-only applications: bipolar
- Digital-only applications: MOS or bipolar
- Mixed analogue/digital applications: no preferred technology
- Very-high-speed operation: bipolar or GaAs
- Large number of logic gates: MOS, specifically CMOS

- High operating voltage: bipolar
- Very low operating voltage: MOS, specifically CMOS
- Low power dissipation: MOS, specifically CMOS

In general, very-high-speed digital operation requires high power dissipation, as was illustrated in Figure 2.37. This in turn can limit the number of gates which may be provided on a single custom IC. Figure 4.38(a) illustrates this trade-off.

However, both gate delay times and power dissipation in CMOS circuits are heavily dependent upon operating conditions. The delay time of a simple inverter gate vs. output load capacitance is shown in Figure 4.38(b) and (c). All these values are of course dependent upon the physical sizes of the FETs. The CMOS power dissipation is largely dependent upon the number of gate on/off cycles per second, since each gate output has to charge and discharge an output capacitance C giving a resultant power dissipation of $P = CV^2 f$ watts, where V is the output voltage swing ($\simeq V_{DD}$) and f the frequency of the on/off cycles per second. For a typical internal gate in a 2 μm gate-array or standard-cell IC this power may be 5–10 μW MHz^{-1}, while output I/Os with a 50 pF loading may dissipate 500 μW at 1 MHz or more. This a.c. power loss is considerably higher than any d.c. leakage power loss or the dissipation through the p-channel and n-channel transistor pairs when in the transition periods from on to off, and vice versa, and hence is usually considered to be the power dissipation of the whole circuit. This may typically be 150 mW for a 3000-gate USIC operating at 10 MHz.

Full-custom design has been shown to be the most efficient way to produce a circuit tailored to a specific customer's requirement, but with the disadvantage of cost and time to design. A comparison of full-custom with the two principal custom alternatives is given in Table 4.6. It will be noticed that, in general, the cheaper and faster the design route the less efficient is the final circuit for any given custom requirement, but this does not mean an unacceptable performance for the vast majority of custom needs.

Perhaps the greatest disadvantage of the full-custom and standard-cell means of producing custom circuits is that in the event of a design change after the first prototype stage a complete new mask set and fabrication run has to be undertaken. For commercial applications where the exact product specification is not absolutely fixed at the initial design stage this is a serious and potentially expensive problem. One published solution is to combine a gate-array section on a standard-cell design, the former containing the random logic or other peripheral duties which may require changes or new variants in the light of development [66]. Any such variations after the first prototype stage can be made by re-design of the final metallization mask levels only. Whether such mixed standard-cell/gate-array architectures will find a use outside major computer and other manufacturers remains to be seen.

The exact place of channel-less architectures in this overall survey is also unclear. Table 4.7 lists their capabilities in comparison with other uncommitted array architectures, from which their place at the sophisticated end of the market

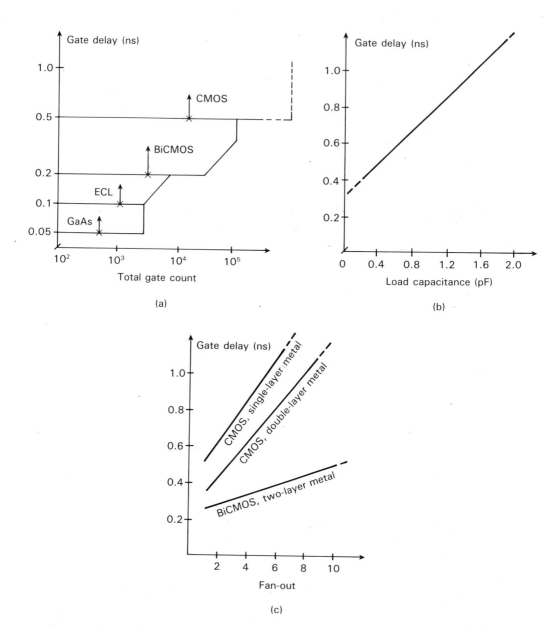

Figure 4.38 Typical technology limits for custom microelectronics: (a) minimum gate delay vs. total gate counts, controlled largely by total power dissipation; (b) the effect of output load capacitance on 2 μm CMOS gate delay; (c) the effect of fan-out loading on CMOS and BiCMOS gate delay.

Table 4.6. A general comparison between the three principal categories of custom circuits, taking full-custom as 100 in each case. Exact comparisons depend upon many factors, including commercial expertise and competition

Attribute	Full-custom USIC	Standard-cell USIC	Uncommitted array USIC
Design flexibility	100	50[a]	30[b]
Design and prototype costs	100	25	20
Design and prototype times	100	20	15
Risk factor (probability of prototype circuits not working correctly)	100[c]	50	50
Cost and time for prototype redesign	100	50	50
Die size area	100	150	200
Maximum system speed	100	75	50
Power consumption	100	120	150
Volume production costs	100	200	300

[a] Limited by available standard-cell libraries.
[b] Limited by gate-array specifications.
[c] Highest risk due to human errors and omissions.
Source: based on [64].

mentioned in Section 4.3.3 will be noted. Their particular attribute in being able to make memory and other array structures puts them in direct competition with standard-cell designs using RAM, ROM and other high-level library macros, but only requiring a reduced mask set in comparison with the full mask set of standard cell designs. They are of course confined (at present) to CMOS technology, and have no foreseeable analogue capability.

To summarize the analogue product area, uncommitted linear arrays, usually bipolar with a floorplan of standard tiles plus other components, are becoming well established. Vendors' libraries of interconnections to realize standard analogue building-blocks are increasingly available from many vendors. Routing may still be done by hand since overall chip complexity is not extreme. Analogue standard-cell products are also receiving considerable attention, possibly more for mixed analogue/digital requirements, but the number of standard-cell designs available is considerably fewer than the very extensive digital-cell libraries that most vendors

Table 4.7. Typical capabilities of five types of uncommitted array. Exact capabilities depend upon vendor expertise and target markets, and are continuously evolving

Type of uncommitted circuit	No. of 2-input gate equivalents per chip	Typical gate delay (n)	Typical number of usable I/Os
Silicon-gate CMOS (channelled)	150–20 000	1–5	20–300
Channel-less CMOS	3500–230 000[a]	0.6–0.8	60–256
Bipolar CDI	150–10 000	0.5–5.0	20–200
Bipolar ECL	100–15 000	0.1–0.5	20–300
GaAs	2000–4000	0.1	20–120

[a] Total per die; not all usable, say 40% maximum utilization.

174 *Custom microelectronic techniques*

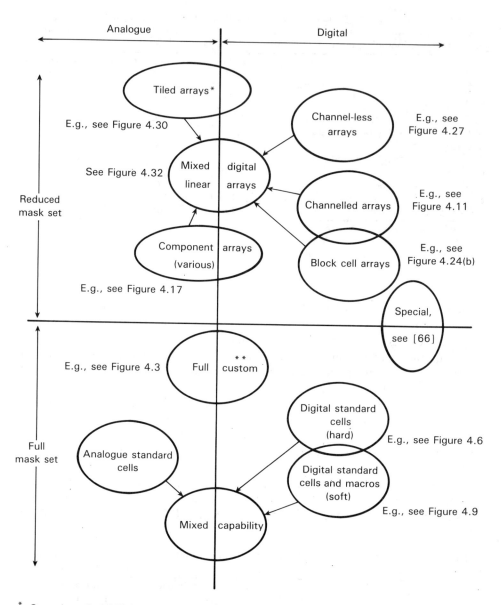

* Some (usually inefficient) digital capability
** Anything possible within the limits of the technology

Figure 4.39 A summary of the principal types of custom microelectronics, with considerable blurring of the boundaries between types. Architectures which span the reduced mask set and full mask set are at present internal to specific companies.

now make available [9], [67]. CMOS and BiCMOS technologies are generally being pursued. Figure 4.39 attempts to overview the range of custom and semicustom solutions which we have for both the analogue and the digital application areas.

In conclusion, we find the following:

- Uncommitted gate arrays, both single-layer-metal and two-layer-metal, are both widely established.
- Standard-cell USICs are also well established, although at present these are largely used internally by vendors for their own range of products.
- CMOS is currently the dominant technology for digital applications.
- ECL and GaAs cater for the very-high-speed specialist market.
- Channel-less arrays and mixed floorplan architectures are receiving considerable attention.
- Fast prototyping is becoming increasingly sought with techniques such as E-beam and direct-write and laser pantography being pursued.

All the above custom microelectronic techniques, however, depend upon good CAD resources for design, simulation and fabrication. This will therefore be the subject of Chapter 5.

4.6 References

1. Mead, C. and Conway, L., *Introduction to VLSI Systems*, Addison-Wesley, Reading, MA, 1980.
2. Pucknell, D. A. and Eshraghian, K., *Basic VLSI Design: System and Circuits*, Prentice Hall, NJ, 1988.
3. Conway, L., Bell, A. and Martin, E. N., 'A large-scale demonstration of a new way to create systems in silicon', *Lambda* (now *VLSI Design*), Vol. 1, No. 2, 1980, pp. 10-19.
4. Lyon, R.F., 'Simplified design rules for VLSI', *ibid.*, Vol. 2, No. 1, 1981, pp. 54-9.
5. Griswold, T. W., 'Portable design rules for bulk CMOS', *ibid*, Vol. 8, September 1987, pp. 62-7.
6. Lipman, J., 'A CMOS implementation of an introductory VLSI design course', *ibid.*, Vol. 2, No. 4, 1981, pp. 56-8.
7. Weste, N. H. E. and Eshraghian, K., *Principles of CMOS VLSI Design*, Addison-Wesley, Reading, MA, 1985.
8. Haskard, M. R. and May, I. C., *Analogue VLSI Design: nMOS and CMOS*, Prentice-Hall, NJ, 1988.
9. Integrated Circuit Corporation, *ASIC Outlook 1990: An application specific IC Report and Directory*, Integrated Circuit Corporation, Scottsdale, AZ, 1989.
10. B.E.P. Data Services, *Semicustom IC Year Book*, Elsevier Scientific Publications, Amsterdam, 1988.
11. Steinkerchner, S. and Rowson, J. A., 'A 75000 gate graphics chip designed in 10 weeks', *Electronic Product Design*, Vol. 10, October 1989, pp. 37-47.
12. Lin, Y. L. S., Gajeski, D. D. and Toga, H., 'A flexible-cell approach for module generation', *Proc. IEEE Custom Integrated Circuits Conf.*, 1987, pp. 9-12.

13. Rossbach, P. C., Linderman, R. W. and Gallagher, D. M., 'An optimizing XROM silicon compiler', *ibid.*, pp. 13–16.
14. Dawson, R. H., Witkosky, B. J. and Kotcherlakota, S., 'ASIC design using VITAL', *ibid.*, pp. 29–32.
15. Watkins, D., Rasmussen, R. and Chang, Y., 'Megafunctions and megacells for ASIC designs: a comparison', *ibid.*, pp. 375–8.
16. Saucier, G., Read, E. and Trilhe, J., (eds.), *Fast Prototyping of VLSI*, North Holland, Amsterdam, 1987.
17. Lau, P. K., Eden, R. E. and Lee, E. S., 'GaAs standard cell family designed with current limiting capacitor diode FET logic approach', *Proc. IEEE Custom Integrated Circuits Conf.*, 1988, pp. 101–4.
18. Gayakwad, R. A., *Op-Amps and Linear Integrated Circuits*, Prentice-Hall, NJ, 1988.
19. Rips, E. M., *Discrete and Integrated Electronics*, Prentice-Hall, NJ, 1986.
20. Barna, A. and Porat, D. I., *Operational Amplifiers*, Wiley, NY, 1989.
21. Allen, P. E., 'Computer aided design of analogue integrated circuits', *J. of Semicustom ICs*, Vol. 4. December 1986, pp. 23–32.
22. Dedic, I. J., King, M. J., Vogt, A. W. and Mallinson, N., 'High performance converters on CMOS', *ibid.*, Vol. 7, September 1989, pp. 40–4.
23. Pletersek, T., Trontelj, J., Trontelj, L., Jones, I. and Shenton, G., 'High-performance designs with CMOS analog standard cells', *IEEE J. Solid-state Circuits*, Vol. SC21, 1986, pp. 215–22.
24. Olmstead, J. A. and Vulih, S., 'Noise problems in mixed analog-digital integrated circuits', *Proc. IEEE. Custom Integrated Circuits Conf.*, 1987, pp. 659–62.
25. Dugan, T. D., 'GaAs gate arrays', *High Performance Systems*, February 1989, pp. 66–74.
26. Singh, H. P., 'A comparative study of GaAs logic families using universal shift registers and self-designed gate technology', *Proc. IEEE GaAs IC Symp.*, 1986, pp. 11–14.
27. Andrews, W., 'ECL process drops power and picks up speed and density', *Computer Design*, Vol. 28, July 1989, pp. 37–41.
28. Ruyat, S., 'CMOS power and density in ECL arrays', *Electronic Product Design*, Vol. 8, October 1987, pp. 77–81.
29. Texas Instruments, *HCMOS Gate Array Design Manual*, Texas Instruments, Inc., Dallas, TX, 1985.
30. LSI Logic, *Databook and Design Manual*, LSI Logic Corporation, Milpitas, CA, 1986.
31. NEC Electronics, *CMOS/CMOS 4A Gate Arrays*, NEC Electronics, Inc., Mountain View, CA, 1987.
32. Read, J. W., *Gate Arrays: Design and application*, Collins, London, 1985.
33. Hurst, S. L., *Custom-Specific Integrated Circuits: Design and fabrication*, Marcel Dekker, NY, 1985.
34. Edwards, C. R., 'A special class of universal logic gate, and their evaluation under the Walsh transform', *Int. J. Electronics*, Vol. 44, January 1978, pp. 49–59.
35. New, A. M., 'Statistical efficiency of universal logic elements in realisation of logic functions', *Proc. IEE*, Vol. 129, May 1982, pp. 93–8.
36. Bennet, P. S., Dixon, R. P. and Ormerod, F., 'High performance BiMOS gate arrays with embedded configurable static memory', *Proc. IEEE Custom Integrated Circuits Conf.*, 1987, pp. 195–7.
37. Nishio, Y., *et al.*, '0.45 ns 7 K Hi-BiCMOS gate array with configurable 3-port 4.6 K SRAM', *ibid.*, pp. 203–6.
38. Landman, B. S. and Russo, R. L., 'On a pin versus block relationship for the partition of logic graphs', *Trans. IEEE*, Vol. C.20, 1971, pp. 1469–79.

39. Heller, W. R., Mikhail, W. F. and Donath, W. E., 'Prediction of wiring space requirements for LSI', *Proc. IEEE 14th Design Automation Conf.*, 1977, pp. 32–42.
40. King, H., 'A statistical approach to route estimating', *Electronic Product Design*, Vol. 9, May 1988, pp. 61–5.
41. Berry, J. and Grassick, E., 'Channel-less arrays increase flexibility', *ibid.*, Vol. 8, October 1987, pp. 43–5.
42. Takahashi, H., 'A 240 K transistor CMOS array with flexible allocation of memory and channels', *ISSCC Digest*, February 1985, pp. 124–5.
43. Anderson, F. and Ford, J., 'A 0.5 micron 150 K channelless gate array', *Proc. IEEE Custom Integrated Circuits Conf.*, 1987, pp. 35–8.
44. Beunder, M., Kernhof, J. and Hoefflinger, B., 'Effective implementation of complex and dynamic CMOS logic in a gate forest environment', *ibid.*, pp. 44–7.
45. Kubosawa, H., et al., 'Layout approach to high-density channelless masterslice', *ibid.*, pp. 48–51.
46. Andrews, W., 'Small CMOS arrays flourish as channelless types emerge', *Computer Design*, Vol. 28, January 1989, pp. 47–59.
47. Bray, D. and Irissou, P., 'A new gridded bipolar linear semicustom array family with CAD support', *J. Semicustom ICs*, Vol. 3, June 1986, pp. 13–20.
48. Sparkes, R. G. and Gross, W., 'Recent developments and trends in bipolar analogue arrays', *Proc. IEEE*, Vol. 75, 1987, pp. 807–15.
49. Tektronix, Inc., *Quickchip 2 Designer's Guide*, Tektronix, Inc., Beaverton, OR.
50. Meza, P. J. and Gross, W.J., 'Standard tiles: a new design method for high-performance analog ICs', *Proc. IEEE Custom Integrated Circuits Conf.*, 1987, pp. 639–43.
51. Crolla, P., 'A family of high-density tile-based bipolar semicustom arrays for the implementation of analogue integrated circuits', *J. Semicustom ICs*, Vol. 5, December 1987, pp. 23–9.
52. AT&T, *Semi-custom Linear Array Brochure*, AT&T Technologies, Allentown, PA.
53. Ratheon Company, *RLA Series Linear Array Design Manual*, Ratheon Corporation, Mountain View, CA.
54. Sze, S. M., (ed.), *VLSI Technology*, McGraw-Hill, NJ, 1985.
55. Carter, R. C., Hardy, C. J., Jones, P. L. and Lawes, R. A., 'Customisation of high-performance uncommitted logic arrays using electron beam direct write', *J. Semicustom ICs*, Vol. 5, March 1988, pp. 23–32.
56. Leppavouri, S., 'Laser processing in ASIC fabrication', *ibid.*, Vol. 6, December 1989, pp. 5–11.
57. Elsea, A. R. and Draper, O. L., 'Advances in laser assisted semi-conductor processing', *Semiconductor International*, Vol. 10, April 1987, pp. 43–9.
58. Santo, B., 'X-ray lithography: the best is yet to come', *IEEE Spectrum*, Vol. 26, February 1989, pp. 48, 49.
59. Cole, B. C., 'Gate arrays' big problem: they take too long to build', *Electronics*, 12 November 1987, pp. 69–71.
60. LASA Industries, *LASARRAY Technical Brochure*, Lasa Industries, San Jose, CA, 1986.
61. Petach, P., 'Lasography allows ASIC gate array prototyping in hours', *J. Semicustom ICs*, Vol. 6, March 1989, pp. 11–15.
62. Cole, B. C., 'LASA's mobile ASIC fab may do the job in minutes', *Electronics*, 12 November 1989, pp. 72–4.
63. Or-Bach, Z., Pierce, K. and Nance, S., 'High density laser programmable gate array family', *Proc. IEEE Custom Integrated Circuits Conf.*, 1987, pp. 526–8.
64. The Open University, *Microelectronic Matters*, Microelectronics for Industry Publication PT504MM, The Open University, UK, 1987.

65. The Open University, *Microelectronic Decisions*, Microelectronics for Industry Publication PT 505 MED, The Open University, UK, 1988.
66. Hornung, R., Bonneau, M. and Waymel, B., 'A versatile VLSI design system for combining gate array and standard circuits on the same chip', *Proc. IEEE Custom Integrated Circuits Conf.*, 1987, pp. 245–7.
67. Design Staff Survey, 'Survey of analogue semicustom ICs', *VLSI Systems Design*, Vol. 8, May 1987, pp. 89–106.

5 Computer-aided design

From the preceding chapters it will be evident that computer-aided design (CAD) plays a key part in the field of custom microelectronics, being used in the following three principle activities:

(a) in the circuit design (synthesis) process;
(b) in simulation before any manufacture is commenced;
(c) in the fabrication process.

The CAD, or, more strictly, the CAE or CAM, used in the fabrication activities is normally the province of the IC vendor, and will not concern the OEM in any depth. We shall have no need to consider this area in any detail here; instead, we will concentrate largely on design and simulation aspects.

The necessity for CAD to be available is largely due to the volume of data involved rather than the difficulties of design, plus the need to ensure that the design is as far as possible error-free before incurring any unrecoverable manufacturing expense. Time and money are therefore the fundamental driving factors behind the need for good CAD resources, this being especially necessary for custom activities where design costs and time to market are critical.

As discussed in Chapter 1, early CAD tools for IC design were only concerned with the geometric layout and subsequent mask-making procedures, and were entirely in the hands of the IC vendors or very large manufacturing companies undertaking full-custom designs. These tools provided complex geometrical manipulation capability, with design-rule checking (DRC) to ensure that line widths and spacings and other dimensional details were all valid. Little or no help was available in optimizing the overall floorplan of the die, and no electrical-rule checking (ERC) or simulation of the final layout (post-layout simulation) was available. Also, there was no standardization of software and data formats between companies, each vendor tending to write its own in-house software programs.

Since then there has been an increasing development of CAD for circuit design and simulation, as indicated in Chapter 1, the ideal being a hierarchy of resources where the data at each level of activity is able to be handed on to the next level without any manual processing. Equally, any problems experienced at a lower level

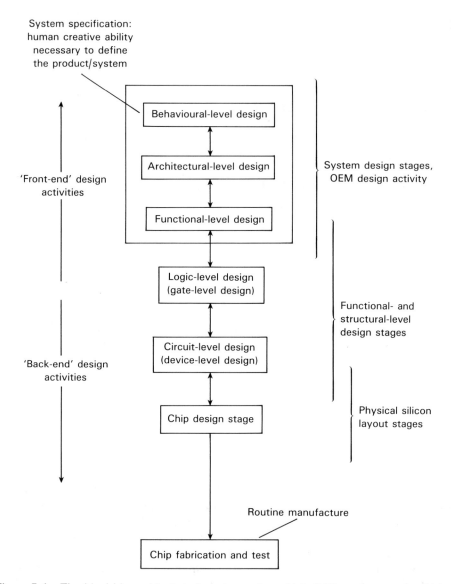

Figure 5.1 The ideal hierarchical design stages for which CAD resources should be available, with the software programs at each stage able to transmit data to, and accept data from, adjacent stages. Validation and simulation activities must also be present alongside these design activities to ensure correct-first-time designs as far as possible.

should be capable of being transmitted back to a higher level for appropriate correction by the design engineers. Figure 5.1 illustrates this ideal.

There is some blurring of distinction in the levels shown in Figure 5.1. In particular the exact demarcation lines between the following:

- the system or behavioural-level design, where the overall behaviour of the required system or product and the flow of information is being considered,
- the architectural-level design, in which the behaviour of each large block or partition and its timing requirements, etc., are being considered, and

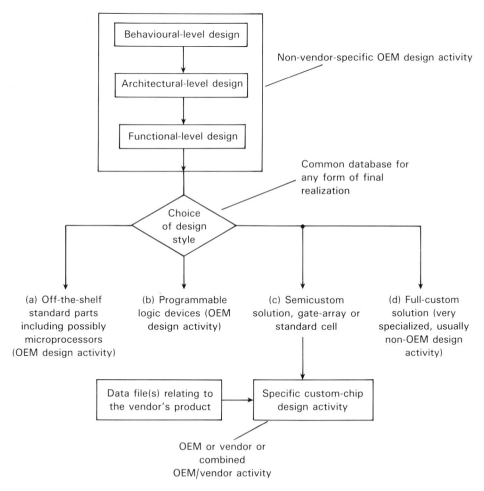

Figure 5.2 The design style choices facing the OEM, where in an ideal situation the hierarchical CAD resources of Figure 5.1 would allow the choice of design style from some common data base at functional level.

Computer-aided design

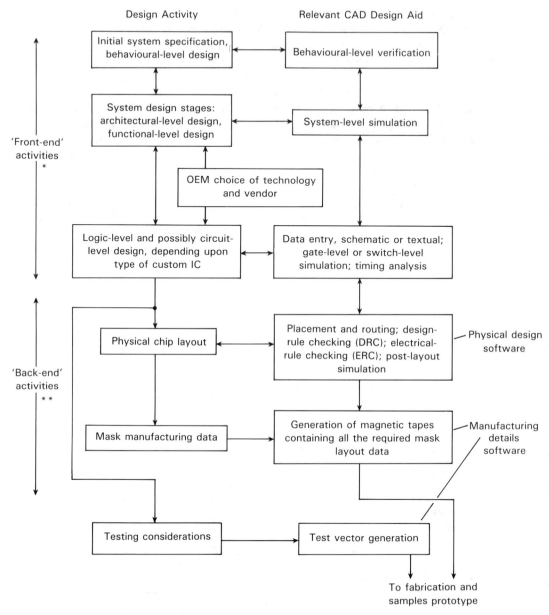

* No physical IC layout (physical design) details yet present
** Physical design details now involved

Figure 5.3 The overall IC design process for digital ICs and the corresponding CAD software resources. For all custom microelectronic design some pre-designed library details are available from vendors so that detailed geometric design at the silicon (transistor) level is not required.

- the functional-level design, in which each architectural block is further partitioned into functional boxes such as counters, gates, multiplexers, and so on,

are difficult to define. Also, because human creative ability and judgement are involved in these areas, particularly towards the top of this hierarchy, CAD support at these levels is still difficult to produce – such CAD support should employ formal languages and be knowledge-based to be of true assistance in the creative design activity.

Another important factor faces the OEM designer at the initial stages of any new product design. This is the choice of design style in which to complete the required microelectronics. The factors involved in making this choice are complex, involving both technical and financial considerations, as will be discussed in Chapter 7, but, ideally, one universal CAD suite should be available whatever the chosen design route. Figure 5.2 illustrates this further aspect facing the OEM. At present most CAD systems are not sufficiently flexible to cater for all the divergences of Figure 5.2; a fully hierarchical system may be available for, say, a programmable logic device design, but it is unlikely that this resource would be of direct use for the design of a semicustom solution.

The overall IC design process for any chosen design style with ideal CAD support is shown in Figure 5.3. This may be compared with the design activities given in Figures 4.7 and 4.12. The most well-established areas of this hierarchy – see also Figure 5.1 – are the back-end activities where there is little creative work but which mostly involve the mechanical translation of the (ideally proven) design into silicon. The least well-established CAD areas are the architectural and behavioural levels. Hence many software suites available at present start at about the functional level, or even at the gate level, which is adequate for many OEMs not concerned with designs involving complex data flows. Further general and detailed information may be found in [1]–[18].

All the preceding comments refer largely to digital logic design and not to analogue. However, in the following sections, which look more closely at the CAD tools, we will bring into the picture both analogue and mixed analogue/digital requirements as well as digital.

5.1 IC design software

From the preceding discussions, an ideal CAD system would allow a top–down design procedure, starting at the behavioural and architectural levels and moving down to the final silicon level. At each level of activity, corresponding simulation resources should be available so that design effort does not proceed too far without confirmation of correctness – these essential simulation activities will be considered in Section 5.2.

Although top–down system design is always necessary initially, there are also certain bottom–up design constraints in all forms of custom microelectronics, since the final silicon realization will invariably employ pre-designed elements from a vendor's library or other source, rather than every transistor and other component being designed in a new and unique manner. Technology constraints will also be present which limit the freedom of design at the silicon level.

For digital systems we may consider design software under the following four main headings:

(a) behavioural- and architectural-level synthesis;
(b) functional- or gate-level synthesis;
(c) physical design software; and
(d) test requirements software.

Category (a) above should be largely independent of the way the circuit is finally made; category (b) includes the design entry and schematic capture procedures which generate the data for the subsequent silicon realization; category (c) includes the placement and routing of the required circuit elements or macros in the silicon layout, and (d) includes the generation of the appropriate information by which the IC may subsequently be tested. Analogue software does not readily follow this hierarchy, and will be considered separately in Section 5.1.5. Also, software for programmable devices, being very vendor-dependent, will be considered separately (see Section 5.1.6).

5.1.1 Behavioural- and architectural-level synthesis

To cater for the increasing complexity of VLSI integrated-circuits, CAD tools are increasingly sought to handle top–down design from the behavioural and architectural levels. Such tools should link with, or encompass, the design duties at the structural decomposition level, where gates or other library parts are introduced. This abstraction above the structural level is illustrated in Figure 5.4.

The purpose of behavioural-/architectural-level CAD is to allow the required system design to be expressed in such a way as to allow the following:

- manipulation and optimization of the system design at this abstract level, with ready means to try different ways of composing the required system (the system architecture), for example serial vs. parallel processing; and
- verification of this abstract design in order to ensure correct behavioural response.

Verification is therefore an integral element of this level of CAD. However, this level cannot include detailed timing simulation, since this can only be incorporated at a lower hierarchical level when gate or silicon layout details are known. Ideally, this high-level CAD should be independent of any subsequent choice of

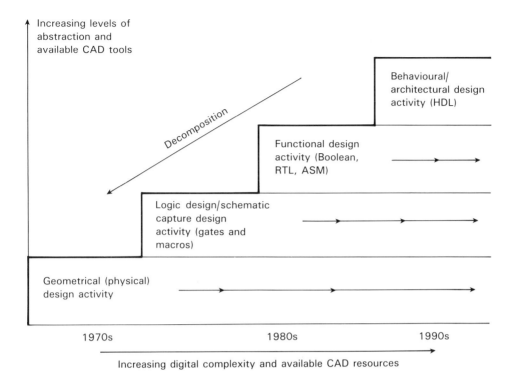

Figure 5.4 The ascending levels of abstraction in the digital system design process, where the behavioural- and architectural-level CAD is ideally independent of the form of final realization. RTL = register transfer language, ASM = algorithmic state machine.

implementation and be technology independent, thus allowing a system designer to investigate a number of alternatives before becoming involved in gate-level design and other details.

These high-level CAD tools do not, however, relieve the designer from having to undertake the initial creative activity of specifying what the required system has to do in terms of recognizable events and functions. CAD is only an aid to verification of the designer's first ideas, clarifying any key interface requirements, and generally optimizing a system structure before more detailed design work begins.

To model and simulate a high-level description of a digital system or circuit requires a behavioural-level computer language. The particular languages used in IC CAD are termed *hardware description languages* (HDLs). These are often derived from high-level programming languages, and hence HDL models often resemble a program written in a high-level language such as Pascal or C. Hardware description languages should contain many of the procedural constraints found in high-level languages [19], [20], such as arithmetic operators, logical operators, if-then-else, while and case statements. These constraints allow the behaviour of a digital system to be modelled in an unambiguous English-language-like manner, and allow for computer simulation of the circuit model's response. Time relationships based upon clock periods can be built in. Hence, HDLs can support the optimization and verification of the architectural details of a proposed design, but they do not overcome the necessity of a good architectural proposal in the first place or the designer's skill in suggesting alternatives; nor do they guarantee that the design can be finally made within the limits of a chosen technology.

A number of hardware description languages have been developed. The one most widely quoted is VHDL, which was developed from the US Very High-Speed Integrated Circuit (VHSIC) government-funded research programme, and is now an IEEE and US Department of Defense standard. All custom IC designs for the Department of Defense have now to be described in VHDL. Other HDLs are ELLA™ (Electronic Logic Language), originally developed in the United Kingdom by the Royal Radar Research Establishment over the period 1978–83, Verilog HDL from Gateway Design Automation, HHDL (HELIX™ HDL) from Silvar-Lisco, M-HDL from Silicon Compiler Systems, and others. (ELLA is the registered trademark of the Secretary of State for Defence, UK, and is marketed by Praxis Systems plc, Bath, UK; HELIX is the registered trademark of Silvar-Lisco, Belgium.) Unfortunately, unified HDL standards have yet to be established, although, due to Department of Defense requirements, VHDL is currently an *ipso facto* US standard (IEEE Standard 1076-1987) in spite of any imperfections.

Whilst HDL can be used to define elements as small as individual gates or functions using Boolean-type expressions, for example

SIGNAL, x, y, cin, cout, sum : BIT
sum < = x XOR y XOR cin
cout < = (x AND y) OR (x AND cin) OR (y AND cin)

which describe a full-adder with no timing clause (default to 0 ns), the essential power of a hardware description language is to allow both:

(a) the textual description of a system at the abstract behavioural level using control flow information; and
(b) the textual description of a system at the architectural level, using data flow descriptions of the architectural partitions.

Once the general functionality of a system has been verified at the behavioural level, then the functionality of the partitioned system can be simulated and compared with the former for agreement. This hierarchical partitioning may be extended to smaller and more detailed partitioning if appropriate. The ability to use (instantiate) a block (a design 'entity') many times in an overall architectural description, and to give generic data to allow families of devices to be constructed, plus assertions which must be true at all times, are further features of HDLs, together with necessary commands such as WAIT, GUARD and AFTER. Table 5.1. shows a short VHDL description.

IC design software 187

The general procedure using a hardware description language is given in Figure 5.5. A textual description of the behaviour of the required system is first entered, from which the HDL system compiles and checks a behavioural description file. Details of the system input signals are also entered to build up the HDL stimulus file. Simulation of the behavioural or architectural system can then be made, with the output being printed out or plotted for checking and verification. If both

Table 5.1. VHDL description of a sorting circuit which accepts a stream of up to 256 16-bit positive integer numbers and shifts them out in ascending order of magnitude: (a) the entity declaration, which defines the input and output signals of the macro; (b) the behavioural model which defines the required action of the macro; (c) the architectural model, which describes one possible design of the macro

```
entity sorter is
   port (din:      in vlbit_1d(15 downto 0);   -- data in
         dout:     out vlbit_1d(15 downto 0);  -- data out
         phi1, phi2,                           -- clocks 1 and 2
         reset,                                -- reset control signal
         read,                                 -- sort dir'n ct1 signal
         double,                               -- double-word ct1 signal
         newword: in vlbit);                   -- word boundary ct1 signal
end;
```

(a)

```
procedure insert  (variable data:  inout  vlbit_2d(0 to
                                          255, 0 to 15);
                   constant size:  in     integer;
                   constant din:   in     vlbit_vector) is
   variable wix:   integer;
begin
   -- find location for new data word, shifting greater words up:
   wix := size - 1;
   while wix /= 0 and data(wix - 1) › din loop
      data(wix) := data(wix - 1);
      wix := wix - 1;
   end loop;
   -- insert new data word in correct location:
   data (wix) := din;
end insert;
```

(b)

```
architecture behavior of sorter is
   constant width:   integer := 16;   -- width of words to sort
   constant length:  integer := 265;  -- number of words to sort
begin
--
--concurrent assertion statement:
--flag attempt to use unimplemented feature after initialization:
   check: assert not (double = '1' and now> 0ns)
                           report "Double-word sorting is not implemented.";
--
--main sorting process:
--insert each new word in order in data bank and emit sorted words:
main: process
   variable data:       vlbit_2d(0 to length - 1, 0 to width - 1);
                        --data words in process of sort
   variable inwix,
       outwix:integer := 0;-- word indices for input, output
begin
wait until newword = '1';-- wait for new word boundary
   for clockperiod in 1 to width - 1 loop
      wait until phi1 = '1'; wait until phi2 = '1';
   end loop;                           -- wait for last clock period
   wait until phi1 = '1';              -- wait for start of clock period
   dout <= data (outwix);              -- emit next data word
   wait until phi2 = '1';              -- wait for second phase
   if reset = '1' then
      init: for wix in 0 to length - 1 loop
         data (wix) := X"FFFF";        -- initialize data bank
      end loop;
      inwix := 0;                      -- initialize data indices
      outwix := 0;
   else
      if read = '1' then               -- increment input index and
         inwix := (inwix + 1) rem length;
         insert (data, inwix, din);    -- insert next data word in order
      else                             -- increment output index
         outwix := (outwix + 1) rem length;
      end if;
   end if;
end process;                           -- return to top of process
end behavior;
```

(c)

Note: An additional procedure may follow to define a sequence of input test signals (logic 0 and logic 1) and determine the output response in order to verify (c), see Table 5.3. (Source: Viewlogic Systems, MA.)

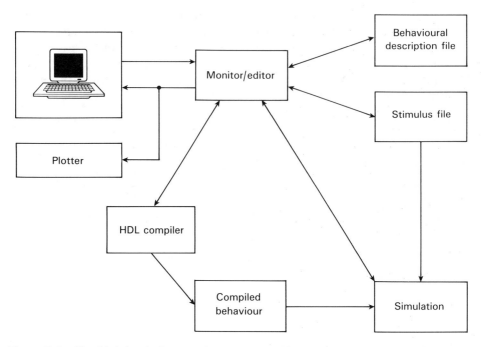

Figure 5.5 The high-level abstract design and verification structure of HDL software.

behavioural- and architectural-level models are entered, the HDL system may automatically output any functional errors between the two.

Details of VHDL, which, like all other hardware description languages, is a complex programming language, may be found in the following publications: [1], [7], [9], [21]–[25]. A useful worked example of a circuit which implements a parallel hardware sorting algorithm is given in [26], which emphasizes that estimated timings only can be given at the HDL stage, and that the final implementation may deviate significantly from these high-level estimates. In-depth details of other HDLs, particularly the proprietary ones, are not so widely available [1], [6], [8], [21], [22], [27], [28].

At some stage in the design process the HDL design data has to be passed on to the structural level of design, and here there may be a gap between the HDL data and the data format acceptable to the more detailed CAD tools used at these later design stages. Figure 5.6 indicates the hierarchical links which are required. Vendors' tools at this lower level of design are thus needed to accept the HDL input data.

One problem is that an HDL behavioural-level description implies no particular hardware realization, and hence there has to be an implied or accepted architectural partitioning for the detailed design synthesis to be able to begin. Nevertheless, many CAD/CAE vendors are beginning to provide interface links to HDL, for example

IC design software 189

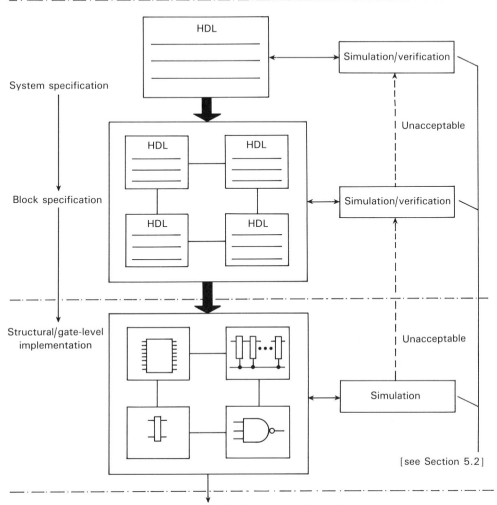

Figure 5.6 The hierarchical levels between HDL synthesis and verification and the lower design levels.

(i) the Design Compiler from Synopsis (USA) which will accept a Verilog HDL input, (ii) LOCAM from Praxis (UK) which will accept ELLA; and (iii) Viewdesign from Viewlogic Systems (USA) which will accept VHDL. These CAD links may, however, impose some restrictions on the HDL design level.

Although VHDL is now an approved US standard, the present concepts of hardware description languages are still under criticism from a wide body of opinion, which is perhaps a measure of the continuing evolution of this top-end

CAD. Some published criticisms are as follows:

- Whilst it may be suitable for system documentation, is a computer language a suitable medium for system design?
- It is difficult if not impossible to fully describe a microprocessor or other complex macro in HDL.
- Diagrams have a superior readability over text for system design purposes, and therefore some computer means of handling high-level schematic input data without complex textual encoding would be preferred.
- High-level synthesis and high-level optimization should be considered as two design activities that are best served by two separate CAD tools rather than one.
- Since there is no established repertoire of synthesis tools and methodologies to help in the translation of initial behavioural descriptions into architectural descriptions, it is premature to try to standardize on a single HDL language for design purposes.

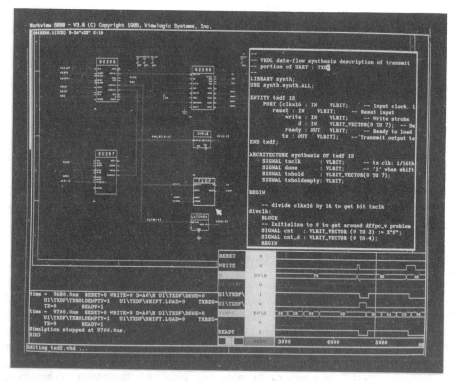

Figure 5.7 The multi-window display available with Viewdesign CAD, where VHDL synthesis data is on the right, with corresponding schematic and waveform simulation details in the other two windows. (Courtesy of Viewlogic Systems, Inc., MA.)

In total, HDLs at present provide an unfamiliar working environment for system designers who are engineers rather than computer scientists, and further evolution or education may be required to achieve wider acceptance as a design tool. In some respects it mirrors early CAD resources at the layout design level where the required gate-level details and their interconnections had to be entered manually in text form into the computer system, whereas schematic capture with automatic compilation of netlists from this graphical data is now universal.

CAE vendors are, however, aware of these problems, and are beginning to provide a more familiar working environment by combining HDL and schematics. For example, Viewlogic Systems' Viewdesign CAD links VHDL for high-level synthesis with simultaneous schematic representations, as illustrated in Figure 5.7. Other CAE vendors, such as Silicon Compiler Systems, provide similar integrated facilities [22]. However, the US Department of Defense requirements in respect of VHDL do not require custom ICs for defence applications to be *designed* using VHDL, but only that the design be *documented* in a VHDL behavioural description. Hence there are interface tools such as the VHDL Interface from VLSI Technology, Inc., which translates conventional IC design data into the IEEE 1076 standard VHDL language. Third-party system simulation can (in theory) be undertaken on this VHDL data, which may be useful to verify that the IC design is correct within a complete system environment. It may, therefore be that the present VHDL standard becomes a documentation standard rather than a preferred synthesis tool.

5.1.2 *Functional- and gate-level synthesis*

In contrast with the above behavioural and architectural levels, the specific structure of the final circuit realization has to be introduced at the functional and gate levels. This involves the types of gates and other macros which are physically available for the final silicon design, rather than the possibly abstract building-blocks at behavioural level.

The current most common synthesis procedure at this level for VLSI circuits is to take each block of the architectural level, and manually partition it until it is finally described at the available gate or macro level. There is, therefore, no specific CAD synthesis tool involved. Simulation is essential, however, and hence the major CAD resources at this design level are for (i) schematic assembly of the circuit ('schematic capture'), and (ii) simulation, rather than synthesis. Most CAD/CAE vendors' software covers this area, since it is the first level at which their library of available cells is formally introduced into a top–down design procedure.

However, it is still possible to describe the required system at this level using a somewhat lower-level CAD model than HDL. This is the register-transfer language (RTL) model, which is a data-flow type of model that can describe hardware blocks in detail rather than the abstract HDL functional modelling. It may or may not be vendor-independent. It is the highest level of modelling which can currently be simulated in a hardware accelerator (see Section 5.2.6). There is clearly an area of overlap between the use of HDLs and RTLs for digital system modelling, with protagonists of each saying that only one should be developed to cover all hierarchical levels from gate level upwards, but at present RTL is largely hardware- and vendor-specific whereas HDL is more abstract.

The RTL model representations can be used for synthesis purposes to try out design variations or simplifications, but it remains a modelling rather than a true synthesis tool. Details of RTLs may be found in [1], [3], [22], [29].

There are, however, some true CAD synthesis tools available for functional- and gate-level design processes. These tools are possibly most prominent in the proprietary CAD resources made available for programmable logic devices (see Section 5.1.6) since there is often considerable emphasis upon optimization and minimization of LSI-size designs in order that they may fit into the smallest and, hence, cheapest PLD. These tools may be considered in two parts:

(a) those which deal with the states of the system and attempt to optimize the state assignment design; and
(b) those which deal with the associated combinatorial logic, particularly the random or 'glue' logic, and attempt to minimize this logic component.

Logic design is a subject outside the general coverage of this text, but it currently involves algorithmic state machine (ASM) synthesis, as well as the more familiar areas of truthtables, Boolean equations, state assignments, minimization and other standard digital logic concepts. Details of all these areas may be found in many standard texts and elsewhere [2], [30]–[33]. However, logic minimization, which is classically taught using Karnaugh maps and the Quine–McClusky minimization algorithm, is currently best performed for large functions by a recent series of efficient CAD programs, particularly ESPRESSO [34], [35]. Others, such as TAU [36], provide similar fast minimization procedures. These resources usually form part of a vendor's CAD suite, rather than being purchased individually.

Hence, outside the PLD area, the synthesis of complex ICs is not yet well served by available CAD tools, and is still largely dependent on human ability to do the creative work, with CAD help in optimization and verification. Further information on the present and future status may be found in [5], [8], [20], [37].

CAD tools do, however, come into their own at this hierarchical level of design in the following two non-innovative areas:

(a) to cater for 'soft' cells and macros in a vendor's library, where the exact cell specification has to be matched to the circuit designer's requirements; and
(b) for schematic capture purposes, where the detailed circuit schematic is completed using the CAD graphics.

The concept of soft cells and macros was introduced in Chapter 4, Section 4.2.2; it caters for the need to be able to make available different capacity circuits such as ROMs, RAMs and PLAs, or to optimize cell performance to meet a given timing specification. The CAD tools which do this are, without exception, IC vendors' tools since they are intimately concerned with the library components, and as a result the software details are usually proprietary. The use of soft macros does, however, slightly blur the interface between this functional and gate-level activity and the following physical design activity, since in the subsequent placement and routing it may also be advantageous to revise the ordering of inputs and outputs or otherwise manipulate the actual physical instance of such cells, rather than taking them as fixed entities from the synthesis level. Functionally, however, the soft cells and macros will not change at the placement and routing stages.

On the schematic capture side, where the CAD resources provide a non-creative role to capture the design as the detailed synthesis builds up, CAD tools from IC and CAD/CAE vendors are readily available which enable the designer to:

- build up the circuit schematic on the CAD screen from a library of symbols,
- interconnect the resulting schematic diagram,
- replicate completed sub-assemblies such as shift registers, counters, etc.,
- change or correct ('undo') the schematic as necessary,

see, for example, Figure 1.5. In addition, the schematic capture CAD will:

- ensure that no invalid connections, such as a gate output connected to a power supply, are made;
- ensure that no input nodes are left unconnected;
- ensure that all connections are terminated at both ends;
- check that the fan-out from each output node is within the permitted maximum;
- compile an inventory of all the cells and macros used in the final schematic; and
- compile a list of all the connections, and the cells to which they are connected.

Schematic capture software thus provides (i) electrical rule checking (ERC), and (ii) data in the form of the netlist of interconnections for final pre-layout simulation (see Section 5.2.3) and for the following physical design activities.

5.1.3 *Physical design software*

Once schematic capture has been done and simulation completed to verify the design, all original creative design work is complete. (If subsequent post-layout simulation reveals timing or other problems, then of course some iteration of the design activity may become necessary. There is, therefore, always a loop back from a lower to a higher hierarchical level necessary to cater for unforeseen difficulties.) The remaining design procedures at the silicon level are largely complex mechanical procedures and can therefore be handled easily by computer, although human intervention and ingenuity is still necessary to provide guidance or solve difficulties which the CAD software cannot resolve.

As discussed in Chapter 4, the two principal design activities at this level are as follows:

(a) placement, where the required functions are allocated positions on the chip floorplan; and
(b) routing, where the detailed geometric pattern of all the required interconnections between functions and I/Os is completed.

This is indicated in Figure 5.8. With uncommitted arrays the floorplan is fixed and the routing constrained within the available wiring channel areas, the principal flexibility here being in the size of the chosen array, which, as a general rule, should have a total gate count at least 30% higher than the total gate count of the schematic design. (Such gate counts are normally in 2-input NAND gate equivalents.) For cell-library and full-custom realizations this constraint is not

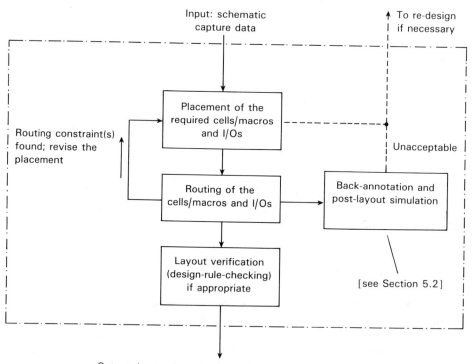

Figure 5.8 The physical (geometrical) design level, where the circuit layout and interconnection routing are defined, the placement and routing activities being interactive if necessary.

present, and the usual objective now is to make the smallest final floorplan area (die size) for the circuit.

Placement

The principal aim of placement is to facilitate and optimize the subsequent geometric routing. Unfortunately, the results of placement cannot be fully tested until the detailed routing is undertaken, and hence most procedures involve a provisional placement design followed by some re-placement if unacceptable or impossible-to-route situations are subsequently encountered.

The computation involved in investigating every possible floorplan placement is completely impractical, and hence most placement procedures involve some initial global placement followed by local iterative improvements (For m cells there are $m!$ possible adjacent placements, ignoring rotation and mirror-imaging of each cell, but as half of these will be global mirror-image placements of the other half, there are effectively $\frac{1}{2}m!$ possible different floorplan layouts without cell rotation and mirror-imaging.) The suggested placements are based on a global consideration of the wiring requirements rather than on individual paths, with a target of minimizing the total interconnect length, or averaging the wiring densities, or some other chosen wiring criteria. In the case of gate arrays a placement is acceptable once the routing can be accommodated within the wiring channels and provided that no unacceptable interconnection propagation delays are present.

Placement procedures may be either heuristic, based upon the designer's knowledge of the circuit schematic, or algorithmic. The following three classes of placement algorithms may be found:

1. Constructive initial placement: in which an analysis of the degree of connectivity between cells is first made, from which a first placement solution is generated.
2. Branch-and-bound methods: whereby cells are initially grouped into large blocks, each of which is then repeatedly sub-divided so as to form a decision tree of associated cell groupings.
3. An interactive re-placement procedure: in which cells are interchanged, possibly at random or possibly under some connectivity indicator, the result of each action being observed.

In practice there may be combinations of these classes within a particular place-and-route CAD program.

In many placement procedures a measure of the 'goodness' of the initial placement is the minimization of the total wiring length. However, with channelled architectures there is what is termed a Manhatten interconnection topology, with the Manhatten distance M between two points x_1, y_1 and x_2, y_2 being defined by

$$M = \{|x_1 - x_2| + |y_1 - y_2|\}$$

Alternative definitions of length are as follows:

- the summation of distances along the wiring channels only, the 'trunk length', ignoring individual cell connections across the wiring channels and all perimeter conditions;
- one-half of the perimeter of the smallest rectangle which encompasses all terminal points of a complete net.

Note that the latter definition is not necessarily the same as the summations of the lengths of the individual connections which go to make up a complete net, but it is equivalent to the lowest bound tree length in orthogonal routing between all points of a net, provided the net routing is not blocked by intervening obstacles. In all cases, weighting of connection distances may be made in order to emphasize critical connections at the expense of less important connections.

Hence if a chip consists of T total connections, then using, say, the Manhatten length the total on-chip length L_M is given by

$$L_M = \sum_{i=1}^{T} M_i$$

and if each connection is given a weighting factor w_i then the total weighted Manhatten length is

$$L_{MW} = \sum_{i=1}^{T} w_i M_i$$

Minimization of this parameter may be considered as a goal for optimal placement.

Other suggested 'goodness' parameters include minimization of the total number of interconnect crossovers or minimization of the total number of 90° turns in the completed routing, assuming north–south/east–west routing directions only in the latter.

The placement algorithm often employed for constant-height cells is linear placement. In linear placement it is assumed that all cells are first arranged in a single horizontal row. Permutation of cell positions is then undertaken in order to minimize the total Manhatten or other distance summation. However, since it is not feasible to consider all possible permutations of cell positions in this minimum summation search, some 'clustering' algorithm is desirable in order to provide a first global placement, following which local perturbation of cells or cell clusters may be invoked to fine-tune the final placement.

Clustering aims to bring together strongly associated cells at the expense of dispersing those which have little commonality. The connectivity V_{PQ} between two cells P and Q may be defined as

$$V_{PQ} = \left\{ f(P) \frac{C_{PQ}}{D_P} + f(Q) \frac{C_{PQ}}{D_Q} \right\}$$

IC design software

where

$f(P), f(Q)$ are empirical factors related to the size of cells, P, Q respectively
C_{PQ} = number of connections between P and Q
D_P = number of connections from P not routed to Q
D_Q = number of connections from Q not routed to P

For example, in Figure 5.9 taking $f(P), f(Q), f(R)$ and $f(S)$ all as unity, we have

$$V_{PQ} = \left\{\frac{4}{1} + \frac{4}{3}\right\} = 5.33$$

$$V_{QR} = \left\{\frac{1}{6} + \frac{1}{5}\right\} = 0.37$$

$$V_{RS} = \left\{\frac{3}{6} + \frac{3}{0}\right\} = \infty$$

$$V_{PR} = V_{PS} = V_{QS} = 0$$

From this it will be noted that loosely connected cells have a low connectivity value, being zero if no interconnections exist, whilst any cell which is 100% connected to another has infinite connectivity value.

In this example it is obvious that cell S should be adjacent to cell R. Equally, there is no requirement for cells P and R, cells P and S, and cells Q and S to be specifically paired with respect to each other. The clustering algorithm therefore proceeds as follows:

1. Compute the connectivity value of all pairs of cells.
2. Combine the two cells with the highest connectivity factor; call this a new cell.
3. Re-calculate all cell connectivity factors.
4. Repeat (2) and (3) until all cells have been clustered.

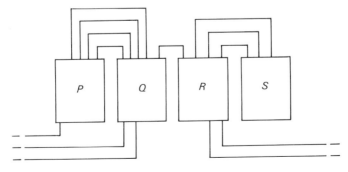

Figure 5.9   The clustering of cells and connectivity parameters.

This is the outline procedure. Problems arise, however, as 'cells' become large in this clustering procedure, since they tend to dominate the connectivity values. This is where the introduction of the $f(P), f(Q), \ldots$ factors becomes necessary in order to counteract the increasingly dominated effects of the large macro clusters. Other difficulties may arise due to connections running between three or more cells, which may require further empirical adjustments. However, a reasonably sensible clustering of the original cells results, but whether it differs markedly from an initial placement which represents a logical layout of the given circuit schematic is debatable.

Having produced some initial linear cell clustering, then perturbation of the cell clusters can be undertaken if desired in order to improve the total placement. Finally, the single row of placed cells has to be folded into rows to produce a floorplan of the required shape.

More general placement algorithms which are not necessarily restricted to constant-height cells and channelled architectures include min-cut placement, stochastic placement, and simulated annealing placement. These alternative placement algorithms are appropriate for channel-less sea-of-gates uncommitted arrays and standard-cell realizations.

In the min-cut placement algorithm all the cells and macros of the circuit are initially considered as one block (as shown in Figure 5.10(a)). A cut $C1$ is applied to divide the assembly into two blocks $B1$ and $B2$, the interconnect lines crossing this cut being termed 'signal-cuts'. The min-cut placement algorithm then seeks to minimize the number of signal-cuts by swapping cells between $B1$ and $B2$, which is equivalent to trying some other cut $C1$ to divide the two parts. This procedure is then repeated by the introduction of a second cut (see Figure 5.10(b)), with the minimization of signal-cuts across $C2$ then being sought. Further cuts and repeats of the minimum signal-cut search continue until a satisfactory solution has been reached, which may be at the ultimate individual cell per block level or terminated at some preceding stage.

This min-cut procedure is an application of a previous partitioning algorithm

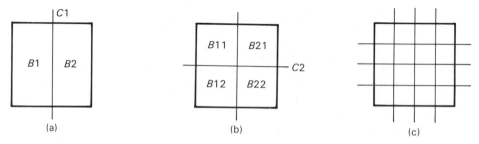

Figure 5.10 The Min-cut placement algorithm: (a) first cut $C1$; (b) second cut $C2$; (c) subsequent cuts. Note that the partitions after each cut may not be the same physical size on the layout.

known as the Kernighan–Lin partitioning algorithm [38]. However, applied to the IC placement problem there is the practical situation that the physical size of cells being swapped across the min-cut boundaries may not be the same, and hence a final chip layout to implement placement based only upon the minimization of signal-cuts may not be satisfactory.

Stochastic placement requires the construction of a larger number of solutions for comparison, using random variables to select different block placements, the usual measure of 'goodness' being the total interconnect length. After a range of solutions has been generated, the best positions for major blocks are accepted and local iterations are applied to fine-tune the results. This method uses more computer time than, say, the min-cut algorithm, but does not converge upon a solution which may not be globally optimal.

In the simulated annealing program, the procedure is modelled upon the physical process which takes place during the cooling of a molten substance such as glass. Here the stresses between molecules are minimized by the random movement of molecules during the cooling down process, such that each molecule tends to find a position at which there are minimum resultant forces acting on it, the amount of movement which each molecule is allowed to make being dependent upon the molten material temperature. Thus in the placement algorithm, a variable termed the 'temperature' is set to some initial value which defines the amount of movement that each logic block can have in searching for an improved position where there is minimum 'force' acting on it from other blocks, this force being based upon the number and length of the block interconnections. A Monte Carlo technique is used to generate random block displacements from an initial layout, and the 'temperature' of the process is gradually reduced to zero to give a final placement solution.

This procedure requires large computer processing power, but can produce good results, particularly as it does not suffer from being locked into locally optimum solutions at the expense of a global solution.

Other algorithmic procedures which may be found, particularly in the iterative improvement phase of placement, include the following:

1. Iterative pairwise interchange: in which pairs of modules are interchanged; if a reduced total interconnect length results, then the interchange is accepted; if not, the original placement is retained.
2. Relaxation methods: in which each connection is thought of as a stretched elastic string containing a force f proportional to the length l of the connection; relaxation techniques to minimize the total force Σf between all cells by re-placement are then applied.
3. Force-directed pairwise interchange: which is a combination of (1) and (2) above.
4. The use of 'seed cells': that is, an arbitrary cell is chosen and all cells closely connected to this seed cell are then drawn into proximity; further seed cells are chosen until all cells are finally grouped.

However, in spite of the availability of such CAD programs, placement still involves human activity, possibly relying upon an initial heuristic placement based upon knowledge of the circuit architecture and functional partitioning, followed by interactive intervention to accept solutions or suggest alternative placements. The subsequent routing and post-layout simulation may also reflect back upon this placement, and dictate revisions to the floorplan if problems are encountered later.

For further theoretical information on placement see Rubin [10], Goto and Matsuda in [14], Hanan and Kurtzberg in [39] and Schwartz [40]; for further information on particular algorithms or practice see [41]–[51].

Routing

Routing is the final phase of the physical design process before any manufacturing costs are involved. It features prominently in all vendors' suites of CAD software, and is still a subject of intense development. (Both placement and routing theory go back to pre-LSI days, having origins in the design of printed-circuit boards containing discrete components or SSI/MSI integrated circuits. Indeed, placement and routing of PCBs may now be extremely complex due to the use of multi-layer boards.)

However, between an initial placement procedure and the detailed geometric routing there may be a global routing (or loose routing) procedure, the objective of which is to produce a routing plan in which each net is defined in one or more segments, allocated to an appropriate interconnect level, and given a category of priority for the subsequent detailed routing [51], [52]. In the case of simple single-layer-metal gate arrays, this loose routing may not be necessary; on the other hand with complex circuits it is sometimes regarded as an integral part of the placement procedure since it strongly influences the placement efficiency, particularly as the correlation between the efficiency of placement based upon total interconnect length only and the subsequent ease of routing is debatable [53]. Global routing may therefore be inseparable from the placement procedure and be strongly linked to the type of product which has to be routed.

A common problem in global routing is to find the 'best' way of connecting a net which links more than two nodes, each segment of which is subsequently separately routed by the detailed routing procedure, see later. A net with n nodes may be linked by $n-1$ segments, but there are n^{n-2} ways in which this may be done, neglecting the possibility of intermediate connecting nodes. The problem of selecting the optimum segments so that the total interconnect length is minimum is the 'travelling salesman' problem, but if intermediate nodes are also allowed, then the problem escalates. The latter is called the Steiner tree problem and, like the travelling salesman problem, is NP-complete [51], [54], [55]. (NP-complete (non-deterministic polynomial) are problems which cannot be solved by any known polynomial algorithm. If a correct solution does exist it can, however, be verified [55].)

Consider the four nodes shown in Figure 5.11. The travelling salesman solution is given in Figure 5.11(a), with a minimum spanning tree solution shown in Figure

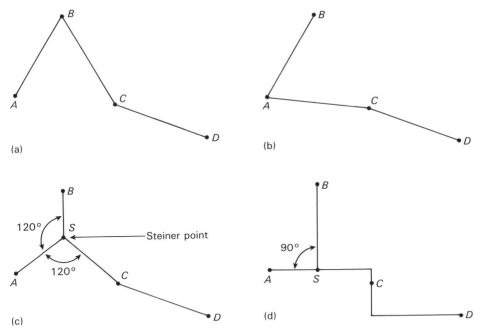

Figure 5.11 Minimum spanning trees for a 4-node net: (a) the travelling salesman (minimum distance with no backtracking) solution; (b) minimum spanning tree; (c) minimum Steiner tree in non-rectilinear geometry; (d) minimum Steiner tree in rectilinear geometry.

5.11(b) [56], [57]. However, allowing intermediate nodes, the alternative Steiner spanning trees shown in Figure 5.11(c) and (d) become relevant [39], [51].

The multiple-node routing problem can, however, be tackled by the detailed routing procedure by giving the routing algorithm a source node and then defining all other nodes as potential target nodes. The first target node reached is accepted to give a completed segment. A second source node is then chosen with the remaining nodes as targets. Routing is complete when all the segments have been determined.

The further tasks of global routing, namely allocating of interconnect level and priorities for detailed routing, are highly dependent upon the particular product being routed. However, the routing priorities may have a marked effect upon the difficulties of the subsequent routing, since high-priority routes such as clock lines or busses may hamper the routing of remaining nets [39], [52], [58].

The detailed geometric routing is the final phase of the physical design process before manufacturing costs are incurred. Like placement it features prominently in all vendors' suites of CAD software.

Geometric routing algorithms may be divided into two main classifications, namely (i) general routers which deal with the routing between any two points A

and B not confined to specific wiring channels, and (ii) channel routers which, as their name suggests, are relevant for channelled arrays and similar floorplans. The former category is represented by the following two main types of router:

(a) the wavefront or maze-running type of router, sometimes also known as a 'grid-expansion router';
(b) the line-search or line-expansion type of router.

Channel routers may be divided into the following three principal types:

(a) the left-edge router,
(b) the dog-leg router, and
(c) the 'greedy' router,

although many practical routers may contain combined features, plus additional heuristics introduced to cater for specific layout requirements or problem areas.

The most widely known type of general router is Lee's wavefront router, first introduced in the early 1960s [59]. Here a routed path follows a rigid grid pattern from a source node to a target node, moving from one cell in the grid to an immediately adjacent one. For example, consider the situation illustrated in Figure 5.12(a), where the white cells indicate space available for interconnection routing, and the black cells indicate obstructions due to prior routing or other obstructions. If we require to route from source cell position S to target cell position T, we first number the available white cells as follows:

1. All cells immediately surrounding S are numbered '1'.
2. All cells immediately surrounding the cells numbered '1' are numbered '2'.
3. Continue numbering cells surrounding every already-numbered cell with the next higher integer number until the target T is encountered.

This is shown in Figure 5.12(b).

Lee's routing algorithm then selects a routing path between S and T by backtracking from T towards S, always moving from one grid square to a next lower-numbered cell, such as those shown in Figure 5.12(c). The advantages of Lee's algorithm are as follows:

- Provided the initial cell numbering can be extended from S to T, then a routing path between S and T is guaranteed.
- The path chosen will always be the shortest Manhatten distance between S and T.
- It can readily cope with nets involving more than two points S and T.

Disadvantages, however, include the following:

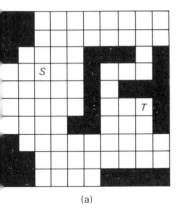 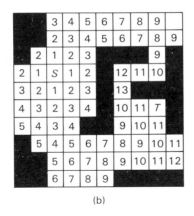 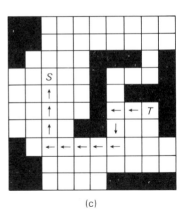

Figure 5.12 Lee's wavefront routing algorithm: (a) grid division of the floorplan, with S the source node and T the target node; (b) grid numbering spreading out from S towards T; (c) possible routing back from T to S.

- Complications occur where there is a choice of cells in the back-tracking, since in theory any cell numbered $(x-1)$ may be chosen to follow cell x.
- The choice of a particular route between S and T may preclude subsequent netlist connections from being routed, with rip-up and re-routing (by hand) of previously completed nets being necessary for 100% completion.
- A considerable and unnecessary area of the available wiring space may be covered and enumerated by the grid radiating outwards from S, with high computer memory requirements and long CPU running time.

Various techniques to refine the basic procedure have been proposed [60], [61], in particular to avoid the labelling of matrix squares which are generally running in an inappropriate direction from the target cell.

The alternative line-search algorithm of Hightower [62] does not require the comprehensive numbering of grid squares radiating from S towards T. Instead, north–south and east–west lines starting from S are established, from which a search for the nearest point from which a perpendicular escape route towards T avoiding constraints can be made. This is a new source point S_1. A similar procedure is started at T, aiming generally towards S, giving a new source point T_1. This is illustrated in Figure 5.13(a).

This procedure is repeated from the new source and target points S_1 and T_1 until, eventually, a line which traces back to S cuts a line which traces back to T. Hence a Manhatten connection path between S and T is established. This is shown in Figure 5.13(b). The advantages of this and similar algorithms include the following:

- The algorithm tends to find the path with the minimum number of changes in direction.

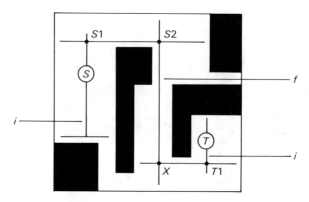

Figure 5.13 Hightower's line-search algorithm, with *i* the initial routes from *S* and *T*, and *f* the final routing for connection path $S_1S_2XT_1$, *X* being the meeting point in the routing paths.

- Very greatly reduced storage and computing times are needed compared with Lee's algorithm.
- The final routing path is less likely to block subsequent routing paths compared with Lee's algorithm.

Against these points, the minimum Manhatten distance may not necessarily be found and, unlike Lee's method, no guarantee of 100% routing is available. However, modifications to improve the algorithm have been reported which guarantee 100% routability if a path exists [63].

In practice a large number of the required interconnections are relatively simple straight connections, with a minimum of bends in each route. Hence, many commercial routers include the means to first route simple (possibly local) connections and critical paths which should be kept as short as possible, followed by, say, a Lee-type router to complete the more complex routes.

In channel routers, rather than completely routing each long net in turn over possibly long distances, the wiring requirements along each individual channel are first undertaken. The perimeter wiring around the central area may subsequently be considered as additional wiring-channel routing, or as a final general routing requirement.

With channel routers, we normally assume that the channel may be represented as a rectangular grid with uniform spacing. Assuming that the rows of cells and wiring channels are horizontal, then the cell terminals are spaced at uniform grid positions along the wiring channel, with the connections to be routed running as horizontal paths along the channel, connecting with orthogonal polysilicon fingers from the cell terminals.

The best-known channel router is the left-edge channel router of Hashimoto and Stevens [64]. Its development is shown in Figure 5.14. It starts by selecting the

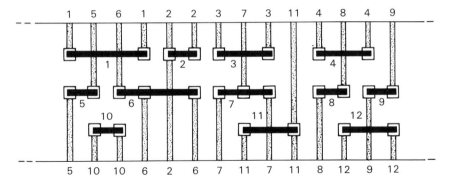

Figure 5.14 The left-edge channel router, consecutively routing horizontal nets 1, 2, 3, ...

extreme left-hand cell connection and routes this at the top of the channel. In the case of two directly facing cell connections, it first selects the top one of the two. The next connection routed is the first cell connection to the right of the routing just completed, which can be routed at the same track level in the wiring channel as the first connection. This procedure repeats until no more connections can be routed at this level, following which it returns to the left-hand edge and selects the next available pin for routing at the next-level track position in the channel, thus building up parallel routing along the channel.

It will be observed that this left-edge channel routing algorithm is subject to the following two fundamental constraints:

(a) a horizontal constraint, in that a horizontal overlap of nets on the same track level cannot be allowed; and
(b) a vertical constraint, in that a vertical overlap of directly facing connections from cells cannot be allowed.

When these two constraints are combined, then we may encounter a cyclic constraint which precludes routing a required net. This problem is shown in Figure 5.15, and may be overcome by re-placement or by using a second level of interconnect (polysilicon or second-level metal) to circumvent the difficulty [64], [65].

The inclusion of dog-legs in the channel router may also overcome cyclic constraints, but at the expense of a more complex routing procedure. Deutsch's dog-leg router [66] allows channel routing to be completed by means such as those shown in Figure 5.16. Branch-and-bound techniques are also available [65], but may involve excessive computer run-time if many iterations of the branching are investigated.

Other variants of channel routing algorithms may be found, including Zonal [67], Greedy [68] and Hierarchical [69]. A bench-mark for these various

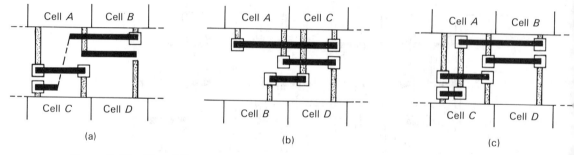

Figure 5.15 The difficulty encountered by the left-edge channel router: (a) the route between cell *B* and cell *C* cannot be completed with a horizontal segment; (b) re-placement of cells *B* and *C* to achieve a horizontal run; (c) use of a polysilicon crossunder to achieve the route with the original placement.

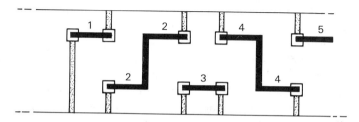

Figure 5.16 The dog-leg router of Deutsch, which may circumvent cyclic restraints.

algorithms has also evolved over recent years, this being a series of netlists which the channel routing algorithm is asked to route. The most complex of these netlists is Deutsch's 'Difficult' example, which has a theoretical lower bound track density of nineteen tracks across the wiring channel to route completely; early channel routers did not succeed in achieving the lower bound, but later algorithms, such as the hierarchical router, do achieve this minimum [69], [71].

Channel routers such as Deutsch and others can effectively route to all four sizes of a rectangular wiring area, as distinct from the longitudinal routing only of simple channel routers. They are therefore sometimes referred to as 'block' or 'switchbox' routers, and the areas they route as 'four-sided channels' [10], [70], [71]. They may be applied to other than channelled architectures provided the area to be routed is a four-sided rectangle; if the area is not rectangular, for example between large macro blocks of different size, then partitioning into rectangular blocks and switchbox routing of each block may be done. However, in spite of the sophistication of many switchbox and channel routing algorithms, it is often also necessary to provide, say, a Lee router to solve particularly difficult interconnection paths which the other routers cannot solve. In the case of single-layer-metal channelled floorplans with fixed via positions to the cells, 100% automatic routing can rarely be achieved; human intervention and re-placement usually becomes necessary to achieve the final routing. For additional discussions on routing see [10], [14], [52], [71]–[75] and the further references these list, particularly in [75].

As a final point concerning the expediency of routing, it will be evident that as die size increases, CPU routing times will also increase. Times given in the mid-1980s for typical maze routing and channel routing algorithms were as shown in Table 5.2, which gave approximate relationships of $t \propto \{\text{No. of gates}\}^2$ for maze routing and $t \propto \{\text{No. of gates}\}^{1.2}$ for channel routing. The values given in Table 5.2 are now likely to be appreciably less, but the relationships are still relevant.

The completion of routing constitutes the end of the physical design activity. However, before the final physical design is translated into manufacturing data it is usual to undertake final checks and post-layout simulation. This is perhaps more important for circuits involving a full mask set than for very simple single-layer-metal gate arrays, but some final verification must always be undertaken.

As implied in Figure 5.8, this verification involves some or all of the following:

- layout verification (design rule checking or DRC) to ensure that all geometric widths and clearances, etc., meet the process design rules;
- circuit verification to ensure that the layout is a correct implementation of the required circuit;
- timing verification to ensure that the circuit response time is acceptable with the detailed routing of the final layout.

The last two checks require the extraction of circuit details from the layout information. This may involve the re-creation of the full circuit diagram from the layout details or the interconnection netlist, which can then be compared with the input design data. Timing verification involves re-simulation of the circuit using the actual resistance and capacitance values extracted from the layout, instead of the nominal values which had to be used before the physical design details were completed. This will be referred to again in Section 5.2.4.

These activities are normally undertaken by the vendor or independent design house, and the results sent to the customer (OEM) for approval. Should the post-layout timing simulation prove unacceptable, then some re-design or re-placement of the circuit is necessary in order to achieve the required performance.

Table 5.2. Comparative CPU times for different algorithms

Chip complexity	CPU time t	
	Maze routing	Channel routing
1000 gates	1 minute	1 minute
10 000 gates	1.6 hours	16 minutes
100 000 gates	144 hours	2 hours

Following design approval, the final physical design activity is the conversion of this layout data into the format required for subsequent mask-making or direct-write-on-wafer. This vendor activity will use an agreed data format, such as the following:

- Gerber format, a standard first established for plotters and similar instruments;
- Calma GDSII format, developed specifically for LSI and VLSI geometries;
- Caltech Intermediate Form (CIF), also developed for LSI and VLSI geometries;
- Electronic Design Interchange Format (EDIF), which is becoming an international standard for the exchange of electronic design data.

The latter has a hierarchical format which is capable of descending to the geometric level to describe floorplan geometries. However, it is at present more usual to employ GDSII or CIF or other proprietary formats rather than EDIF for this level of data transfer [10]. (Further details on standard formats will be covered in Section 5.5.6.)

The vendor or independent design house may do a further check on this final data by requesting a print of each mask layout, perhaps 100 times full size and known as 'colour-keys' or 'blowbacks', which can be used as a visual check that nothing has gone grossly wrong during transfer of this data, for example the use of an incomplete or wrong magnetic tape.

Further details of IC layout and mask-making may be found elsewhere [1], [10], [76].

5.1.4 *Test programme formulation*

An essential part of the front-end design activities must be a consideration of how the final integrated circuit will be tested. This becomes critically important as the size and complexity of the IC increases, and above, say, 5000 gates, special means to ease the task of testing become increasingly necessary.

The problem of testing is complex because of the following two factors:

1. Only the input and output pins of an IC are available for test purposes, probing of the interior nodes of the circuit not being possible under normal test conditions.
2. The time to undertake a test procedure is limited.

If time were not a constraint, it would be possible to test any circuit exhaustively through all its possible modes of operation, and check for correct functionality under all conditions. However, consider even a simple 16-bit accumulator IC which can add or subtract a 16-bit input number from any 16-bit number already stored in the accumulator. In total there may be 19 input signals (16 bits plus 3 control signals), plus the 16 internal latches. Thus the number of different input combinations which can be applied is 2^{19}, and the number of different input numbers which may be stored in the accumulator is 2^{16}. This gives a total of $2^{19} \times 2^{16} = 3^{35} = 34\,359\,738\,368$ possible different logic conditions for this simple circuit. Hence to test this circuit exhaustively would theoretically require over 34×10^9 input test vectors, which at an applied rate of one million per second would take almost ten hours to complete. In general, any n-input circuit with s internal storage elements (latches or flip-flops) theoretically requires 2^{n+s} test vectors for exhaustive testing.

Clearly, this problem must be addressed at the design stage in order that:

- some appropriate non-exhaustive test sequence shall be defined which checks the correct functionality within certain confidence limits; and/or
- the circuit design shall incorporate special circuit details which facilitate testing by providing a test mode whereby the circuit is partitioned or otherwise modified from its normal mode of operation under test conditions.

The latter is termed 'design-for-test' and the techniques involved are known as DFT techniques.

Because of the great importance of test, it will be the sole subject of Chapter 6. CAD resources will be seen to be available to aid the designer in developing an appropriate test strategy and to determine appropriate test vectors. However, this is additional to the hierarchical design activities which we have considered in the preceding pages, and does not alter the concepts already covered in any way.

5.1.5 *Analogue and mixed analogue/digital synthesis*

The subject of analogue system synthesis spans the design of the basic individual building-blocks such as amplifiers, oscillators and converters which go to make up a complete system, plus the system design which is the interconnection of these building-blocks to meet the required system specifications.

Basic analogue design is a skill which must be learnt, and many excellent texts are available [77]–[85]. However, few CAD resources are available to help in the initial creative work, but CAD assistance is extensively used in simulation, see Section 5.2.5. Analogue design is further complicated by the difference between designing using discrete components, where accurate value, close tolerance passive components are widely available, and designing monolithic circuits where these attributes are not present, plus the diversity of requirements which may be needed (see Figure 4.29).

Most OEM designers will therefore use some form of pre-designed analogue circuit blocks since he or she cannot afford to become an expert analogue designer at the silicon level. Assuming that off-the-shelf ICs and other components are not used, the custom choices are as covered in Chapter 4, namely a vendor's library of proven circuits which can be made from uncommitted component-level arrays, or a vendor's library of analogue cells for a standard-cell solution.

The following CAD help which is available in the design phase is largely graphical:

- symbol generation;
- capability to draw the circuit on the VDU screen as it is being built up;
- schematic capture and automatic formatting of the design data for simulation.

The essential part, however, is the simulation capability, the results of which may be presented as appropriate graphical plots [17], [86]–[89]. This will be illustrated further in Section 5.2.5.

CAD help is, however, available in one particular design area, that of *filter design*, although even here simulation forms the essential background. In this area we have the nearest analogue equivalent to the silicon compiler used for digital synthesis purposes (see Section 5.3). With filter synthesis programs it becomes possible to enter the required filter characteristics into the CAD system, and the program will attempt to determine the appropriate component values [80], [81], [90]–[92]. Nevertheless, it still involves designer expertise to suggest the appropriate type of filter to meet the specification, and to accept or reject the results of the synthesis.

Mixed analogue/digital design is not well served by CAD resources at present. The usual practice is to design the two parts separately, utilizing human expertise to handle the interface. No single simulation resource is currently available which can handle the detailed device-level simulation necessary for the analogue sections and the gate or higher hierarchical level of simulation necessary for the more populous digital parts. Much research work is being undertaken in this area, but in the absence of a single comprehensive design theory covering the wide range of functions used in analogue systems it remains difficult to formulate CAD tools covering this widely diverse area. Human expertise and creative talents backed up by good CAD simulation and verification are likely to remain the general means of analogue and mixed analogue/digital design [80], [86], [88], [93], [94].

5.1.6 *Programmable logic devices*

In contrast to the sparsity of CAD tools for analogue synthesis, CAD tools are widely available for all the design activities associated with PLDs, being part of the resources marketed by PLD vendors and distributors. Digital-only devices are involved at present.

The CAD support usually includes the following:

IC design software 211

- design entry, possibly incorporating vendor's library data;
- design consistency checks;
- simulation, either before of after the following step or both;
- automatic placement and routing of the design on a particular PLD – the selection of the optimum size of PLD from the vendor's range of devices may be part of this program;
- down-loading of this design data into the appropriate programming unit for the commitment (dedication) of the PLD; and
- design documentation.

At the design entry phase one or more of the following hierarchical levels of design may be used:

- high-level logic statements;
- state machine statements, specifying states and conditional branching and also inputs and outputs of the state machine;
- Boolean equations;
- truthtables;
- schematic capture, possibly using a vendor's menu of primitives or macros;
- netlist defining the required system.

This is indicated in Figure 5.17.

The high-level logic statements which may be used in the design of the larger and logically more powerful types of PLD are usually written in a language such as Pascal or C, but are not necessarily as comprehensive as in the general-purpose, high-level description languages (HDLs) introduced in Section 5.1.1. This is because functional simulation is not always done at the architectural level, but, instead, the emphasis is more upon a top–down design route targeted upon specific PLD architectures with simulation at the PLD level. An example is Altera's high-level language AHDL, which is a VHDL derivative with enhancements to aid the initial design entry using library macros which can be the equivalent of the 74[**] series ICs, and where the target PLD architecture determines the hierarchical decomposition [22], [95]. The supporting graphics editor allows multiwindow representation of the synthesis procedure, as shown in Figure 5.18.

There are, however, many CAD suites available which support a range of vendors' products. Among them are ABEL, PALASM, CUPL and PLPL; details of these and others may be found in vendor's literature and elsewhere [96]–[103]. The syntax of these software tools is unfortunately not standardized; for example

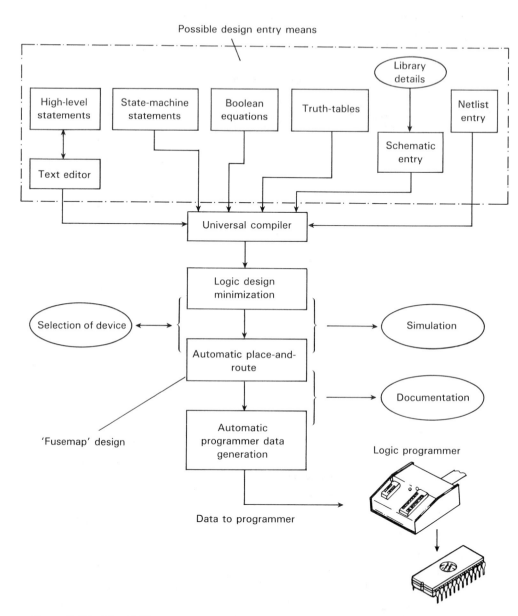

Figure 5.17 The CAD resources available for programmable logic devices. Not all the possible design entry means may be used for the less sophisticated range of PLDs. Also place-and-route in this context means the mapping of the required functions on the chosen PLD and determination of the programmable interconnect pattern.

IC design software 213

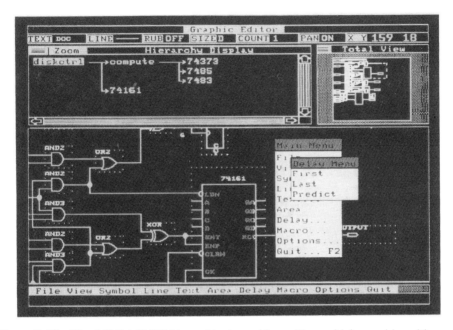

Figure 5.18 The MAX + PLUS hierarchical graphics editor, which provides either top–down or bottom–up design facilities. (Courtesy of Altera Corporation, CA.)

in CUPL and PLPL the three common logic operators are

	CUPL	PLPL
logical AND	&	*
logical OR	#	+
logical negation	!	/

although translators to convert from one format to another have been written.

Currently, the largest and most complex user-programmable product available is the logic cell array (LCA), see Chapter 3, Section 3.6. The vendor's CAD tools to support LCA design also support design entry interfaces from several other schematic entry formats, including ABEL, PALASM and others, for example as shown in Figure 5.19, which allows designs already undertaken for any type of PLD to be translated into an LCA realization [101].

For the more simple types of PLD which do not have such dense logic as the more complex products, simple schematic capture or Boolean equations may be entirely adequate. An essential part of all PLD design software is Boolean minimization and factorization in order that an efficient and complete design tool is available for OEM use to fit the architecture of the device [96]. Such CAD resources targeted on the final IC realization thus provide the most complete means currently available for an OEM to progress from design concept to an efficient IC realization without recourse to any outside design help or participation.

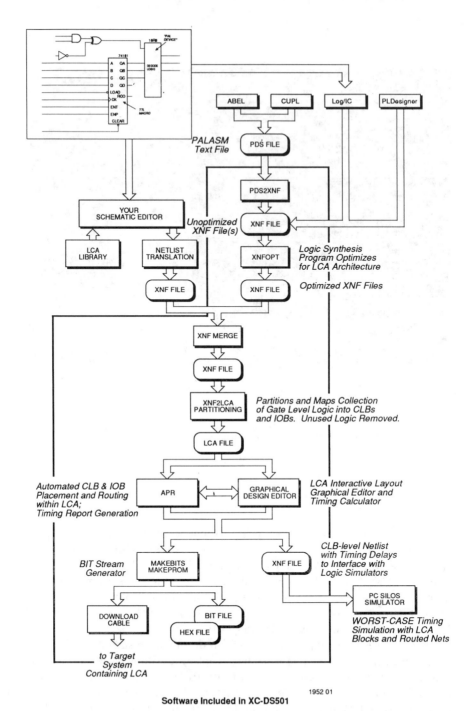

Figure 5.19 The Xilinx CAD system which supports inputs from ABEL, CUPL, LOG/IC and PLDesigner plus schematic netlists and text files. (Courtesy of Xilinx, Inc., CA.)

5.2 IC simulation software

It will be apparent from the preceding sections that simulation is the essential component of CAD. Whilst there has been much talk of 'correct-by-construction' in higher levels of design, verification is still needed to ensure that all the design details are correct as far as possible before manufacture is commenced. This importance is reflected by the percentages given in Figure 5.20.

The level of simulation which may be considered in digital IC design activities are as follows:

- behavioural level;
- architectural level;

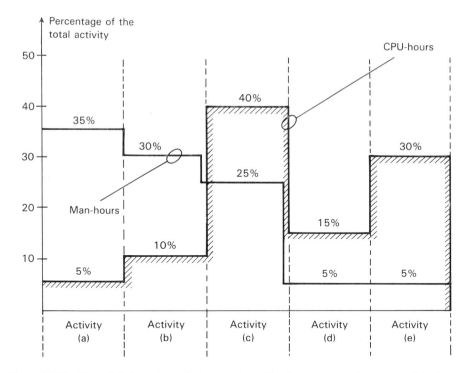

Figure 5.20 The digital system design and verification process in terms of typical total man-hours and CPU hours, with simulation and verification being the largest CPU component: (a) product specification and system design activities; (b) logic and circuit design activities; (c) simulation and verification activities; (d) physical layout activities; (e) layout verification and test-pattern generation. (Source: based upon [37].)

- functional level;
- gate level;
- switch level;
- circuit and device level;
- silicon fabrication level.

Behavioural and architectural levels are concerned with the system design before it is partitioned at the functional level into specific macros and logic functions; gate level is concerned with the individual gates which go to make up the functional level, whilst switch level is an alternative to the gate level where gates are modelled as switches. The latter is particularly relevant for MOS circuits. Circuit and device level is concerned with the process level of the design, whilst the silicon fabrication level models the physics of the actual silicon fabrication processes. We shall have no need to consider the transistor level [11] in this text. Analogue circuit simulation does not involve this hierarchical structure, and the usual simulation level is at the circuit and device level only.

The lower the level of simulation the more accurate are the simulation results, but the amount of computer simulation time increases rapidly. On the other hand, at the top hierarchical level either broad confirmation that a complex system design is correct or easy means that evaluation of alternative architectures is all that is required, not detailed timing and other verification which can only be done when the design has progressed further to the macro or gate level. For more simple circuits, say with fewer than 5000 gates, functional-level simulation may be the highest level of simulation necessary, since this may be done with an acceptable amount of computing.

All levels of simulation involve computer application of input signals to the model of the circuit, and checking of the output response. This output response may be converted into graphical waveform representations, which are easier to interpret than print-outs of logical values. As well as functional verification, the detection of spikes and hazards at the lower levels are an essential part of simulation.

5.2.1 *Behavioural-, architectural- and functional-level modelling*

Since it is difficult to draw a clear line of distinction between these levels of simulation, we consider them together. All involve a programming language which describes the system and its appropriate degree of partitioning, with varying degrees of timing information from zero delays to nominal or average values. The terminologies

- mixed-level simulation,
- multilevel simulation, and
- mixed-signal simulation

Table 5.3. A continuation for verification purposes of the VHDL example given in Table 5.1; here a sequence of input vectors is applied to the architecture and the output response generated. No detailed cell architecture or timing is involved at this stage

```vhdl
architecture action of driver is
   constant PHI_PW:         time := 100ns;
   constant P:              time := 400ns;
   constant PHASE:          time := P/2;
   constant NEWWORD_DLY:    time := (P - SKEW -
                                    PHI_PW)/3;
   constant NEWWORD_PW:     time := 30ns;
   constant NBITS:          integer := 16;
   constant NWORDS;         integer := 256;
begin
--
--
   timing: block
   begin
      phi1     <=  '0'   after PHI_PW when phi1 = '1' else
                   '1'   after P - PHI_PW when phi1 = '0' else
                   '1'   after P - SKEW - PHI_PW;
      phi2     <=  '0'   after SKEW when phi1 = '0' else
                   '1'   after SKEW when phi1 = '1' else
                   '0';
      newword  <=  '0'   after NEWWORD_PW when newword = '1' else
                   '1'   after NBITS * P - NEWWORD_PW
                         when newword = '0' else
                   '1'   after NEWWORD_DLY;
   end block timing;
--
-- setup test
--
   test: process
      begin
--
-- initialization
--
      reset <= '1';
      read <= '1';
      double <= '0';
      din <= b"0000000000000000";
      wait for NBITS * P;
--
-- load initial data
--
      reset <= '0';
      for nword in 1 to NWORDS loop
         nextval(nword, din);
         wait for NBITS * P;
      end loop;
--
-- shift out data
--
      read <= '0';
      for nword in 1 to NWORDS loop
         wait for NBITS * P;
         checkval(nword, dout);
      end loop;
      wait;
   end process init;
end action;
```

may be found to cover procedures which span several categories [6], [12], [29], [104]–[106], although this must be distinguished from simulation systems which attempt to cover both analogue and digital circuits and which may be referred to as mixed-mode or multimode simulators [80], [106]. Unfortunately, the use of these adjectives is sometimes unclear, with mixed-mode sometimes being used for mixed-level digital-only simulation.

Unlike analogue simulation, which requires the accurate modelling of continuously variable voltages and currents, digital simulation can use models with the binary values 0 and 1, with time as a further parameter. At the more detailed lower levels of simulation, other concepts such as impedance must be incorporated, giving rise to unknown or undefined logic states or further distinctions such as low or high impedance source/drive conditions, etc., but this degree of detail is not required at the behavioural, architectural and functional levels.

Simulation at these high levels requires the choice of an appropriate hardware description language such as is considered in the preceding sections. VHDL is possible, but at present the most widely used are HLDs whose structure is directly relevant for additional simulation activities, such as Verilog HDL, Silvar-Lisco HHDL and others, or even those designed specifically for simulation rather than synthesis, such as HILO–GHDL and SIMULA [21]. The OEM is largely dependent on what is commercially available from chosen vendors, with cell library data and a good graphics interface being key features.

The VHDL listing to verify the functionality of the sorter circuit given in Table 5.1 is shown in Table 5.3, where PHI_PW defines a 100 ns positive clock pulse for clock lines phi1 and phi2, the clock pulses being equally spaced within a cycle period P of 400 ns. No propagation delay timing data is included in this VHDL description, as may be present in other proprietary HLD descriptions.

System designers normally prefer graphical rather than textual inputs and outputs, and hence most commercial HDL systems provide waveform outputs such as those shown in Figure 5.7. Windowing is a powerful advantage in this area, as it allows the designer to give priority to the different aspects of the design and simulation activities.

5.2.2 Gate- and switch-level simulation

Some of the high-level HDLs which may be applied to the architectural and functional levels also descend to the gate level for simulation purposes, for example GenRad's HILO, Silvar-Lisco's HELIX and others. These general-purpose tools have the capability of incorporating many different digital signal conditions, for example logic 0, logic 1 and logic X (unknown), each at either strong, weak or high impedance. This gives a total of nine possible signal conditions. Others have five (or more) different strengths available, giving fifteen (or more) different possible conditions. Many of these conditions may be necessary for full-custom design activities where many transistor sizes may be present, but may not be so relevant for semicustom gate-array and cell-library design activities.

However, gate-level simulation has been extensively developed in its own right, since it forms the cornerstone of the design verification process. Many of the algorithms and practices developed at gate level are incorporated in higher-level simulators, and it is thus appropriate to consider this and the similar switch level in some detail.

Gate-level simulation involves the following four inter-related activities:

(a) check of the data input;
(b) re-formatting of this input data into an appropriate form for simulation;
(c) the simulation procedure; and
(d) output display formatting, either textual or graphical or both.

Input data
Although the input data may have been graphical as far as the designer was concerned, this has to be converted by the software into textual form with statements defining each of the primitives and their interconnections, plus statements defining propagation delays and other signal information. In this 'flattened' form, all complex macros are partitioned into their constituent primitives.

This data is then checked for possible errors, for example duplication, omission of interconnections, illegal primitives, and so on, but this check cannot of course detect wrong design data such as a NOR gate being specified rather than a NAND, or an unwanted Inverter gate being included. Indeed, finding such errors is one of the purposes of simulation.

Re-formatting
The 'model compiler' then generates an appropriate model of the circuit for simulation purposes from the above syntactically correct description, either (i) by generating a compiled code model in which the circuit description is organized into a logical sequence which progressively mirrors the data flow of the circuit, or (ii) by grouping the circuit description into a set of interlinked data tables which are subsequently operated upon by the simulation program.

In the compiled code model the order in which the gates is compiled is significant since the output from a given gate must be updated before this signal can be used in a following gate input. This process allows very fast simulation to be accomplished if a single-pass simulation can be achieved without revision of the ordering.

In the case of the table-driven format, each table contains information relating to the connections from one table to other tables, as well as full information relating to the associated primitives, their input signals and their timing characteristics. By this means, a selective trace simulation can subsequently be implemented, allowing the simulation to follow pointers from one table to another as the simulation proceeds. This table-driven format allows more simulation information to be included in each table than in the compiled code model, and changes to be made in a table without having to re-compile the whole circuit [107].

Simulation procedure

The basic simulation algorithm performs the following actions:

1. Set all gate outputs at time $T = 0$ to X (unknown).
2. Apply the first input stimuli to the model.
3. Evaluate the effects of this stimulus and schedule the changes in gate outputs in an appropriate event queue or timing schedule.
4. Continue to propagate this simulation until no further changes take place.
5. Repeat with further input stimuli.

The effect of gate input changes on gate output is given by look-up tables which document the effect of all input stimuli, including timing information. Thus no detailed (analogue) computations are involved in gate-level modelling, although the vendor's original generation of this look-up data may have involved such simulation (see Section 5.2.3).

The most important part of this simulation process is determination of the method of scheduling the changes as the simulation run proceeds. This may be in the form of an events queue, in which the results of the simulation at time t are placed in the event queue at time ($t + delay$) for the subsequent simulation activity. However, with dissimilar propagation delays an event in later time may cause a subsequent event earlier than one previously scheduled. Hence a continuous update and revision of the simulation next-events list is necessary in order to avoid mis-scheduling, which may involve appreciable CPU time. On the other hand, no waiting time occurs in moving on from one event in the list to simulation of the next scheduled event.

The alternative to a next-events list is a time-mapping events schedule. Here, every propagation delay is expressed as an integer multiple of a small unit of time Δt, the simulation program scanning each Δt period for the next simulation activity required. The advantage of this form of scheduling is that there is no re-scheduling of priorities as with a next-events list, but, against this, CPU time is wasted in scanning empty Δt time slots where no event is scheduled.

These principles of scheduling are detailed more fully in Lightner [29] and elsewhere [40], [107].

As circuit size increases, simulation time may be reduced by further techniques. For example, in large networks only a very small percentage of the gates change state as a result of a preceding change, and hence it is only necessary to simulate the changes in this fan-out area. This 'selective trace' may be incorporated to exploit the latency of a large network. Parallel simulation of parts of the network which do not interact may also be done. All these forms of event scheduling for gate-level simulation are in contrast to most circuit level simulators (see later) where

IC simulation software

all circuit nodes are examined at each time increment, whether they have or have not changed state.

Output display formatting

The output display format can be textual in the form of logic 0 and 1 lists for each input and output node. This occurs at each simulator time interval. It is very tedious for the human designer to check, except perhaps to highlight very short duration spikes or glitches which show up as a logic 0 (or 1) in a stream of logic 1s (or 0s), and it is thus now common practice for the output of gate-level simulation to be converted into logic waveforms for general checking purposes. The overall picture of gate-level simulation is therefore as indicated in Figure 5.21. Further details may be found elsewhere, although the fine details of most vendors' programs are proprietary [1], [3], [12], [13], [29], [40], [107].

Switch-level simulation is specifically relevant for MOS logic circuits, since it overcomes certain problems in trying to model MOS circuits at the gate level. In switch-level simulation each transistor may be treated as a conducting or non-conducting perfect switch having the truthtable shown in Table 5.4. As in gate-level

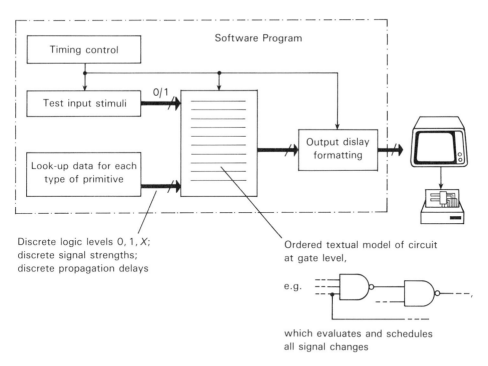

Figure 5.21 The schematic of gate-level software simulation, where the electrical performance of the gate is defined by discrete values with the input/output relationships given in look-up tables and not computed.

Table 5.4. The conducting/non-conducting and unknown truthtable for MOS switch-level simulation

Source (input) logic signal	Gate (control) signal	Drain (output) logic signal	
		n-type device	p-type device
0	0	Z	0
1	0	Z	1
X	0	Z	X
0	1	0	Z
1	1	1	Z
X	1	X	Z
0	X	X	X
1	X	X	X
X	X	X	X

simulation X is an unknown (undefined) logic signal, but the additional signal Z introduced here is 'open circuit', that is, the switch is non-conducting and hence the output state is not defined by the input node. Also, device threshold voltages are not considered in this simple truthtable.

Further reasons why switch-level simulation is often more appropriate than gate-level simulation are as follows:

- Since CMOS logic gates contain only p-channel and n-channel devices the conventional logic gate configurations may always be regarded as networks of switches, see Figure 5.22(a).
- Many 'transmission-gate' circuit configurations which do not have a direct Boolean NAND or NOR model are found in vendors' libraries, for example see Figure 5.22(b).
- Similarly, complex logic functions (for example, see Figure 5.22(c)) do not have a direct NAND/NOR model. And, finally,
- MOS devices may be used as bidirectional switches so that logic signals flow both ways through them, again something that cannot easily be modelled by Boolean gates.

Thus, as all MOS/CMOS logic circuits contains only MOS devices and no resistors, the switch-level model is an obvious possibility for simulation purposes. (This remodelling of CMOS gates as switches will be referred to again in Chapter 6 when considering fault modelling of logic circuits.)

Switch-level simulation of MOS logic circuits therefore consists of a network of nodes interconnected by bidirectional switches. This topology can be extracted directly from the circuit layout *without reference to the functionality of the circuit*, and hence forms a powerful further check on the circuit and physical design.

The actual simulation activity can be a behavioural check on the circuit without any detailed timing considerations. In this case the effect of an input simulation is evaluated by tracing the signal changes through the perfect switches until all

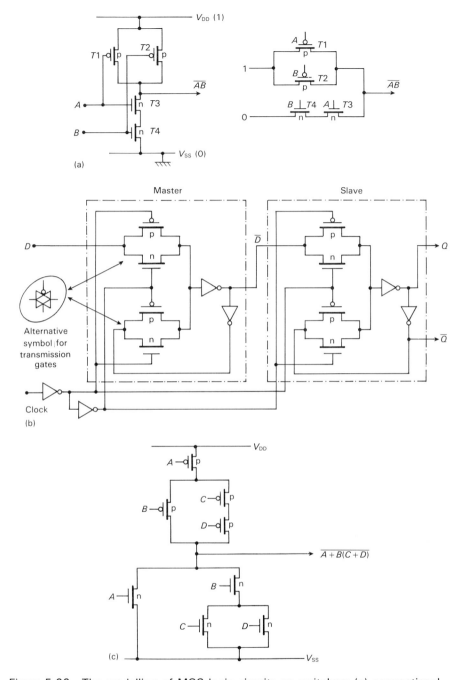

Figure 5.22 The modelling of MOS logic circuits as switches: (a) conventional CMOS NAND gate re-modelling to emphasize the transistors acting as on/off switches; (b) CMOS D-type master–slave circuit using transmission gate configurations; (c) complex CMOS logic gate realizing the function $\overline{A + B(C + D)}$ which may be remodelled as in (a).

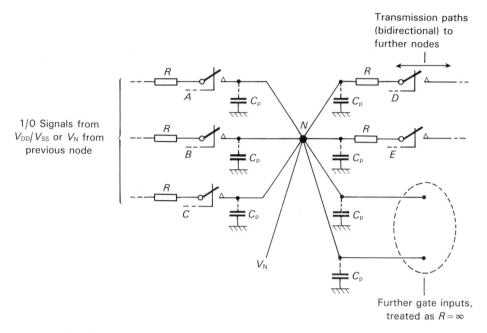

Figure 5.23 The basic concept of switch-level simulation where perfect switches switch voltages to a summing node N and thence to further nodes. Kirchoff's law must hold at every node. For timing simulation parasitic capacitance values C_P may be introduced. Note that transmission paths are bidirectional, but closed paths from, say, V_{DD} to V_{SS} must be design errors.

changes have been evaluated (see Figure 5.23), following which the next test input is evaluated. One problem which may be encountered is where an undefined logic value at a node is encountered, as would be the case if all transistor switches connected to this node were open together. The problem with an undefined node is that this undefined state is likely to propagate through the rest of the circuit, and hence the simulator may stop when this condition is encountered and flag a possible design fault in the circuit being simulated.

Alternatively, the switch-level model may simulate timing by including nominal circuit resistance and capacitance values which provide a measure of circuit propagation times. Because these parameters are at transistor level, the CR times to charge or discharge nodes can give good results, but without the wide range of signal levels, circuit impedances and timing details that may be necessary for gate-level modelling. It is also generally impossible to detect race conditions and hazards, but against this the simplicity of switch-level modelling allows very large MOS circuits to be simulated very rapidly.

In total, switch-level synthesis provides a fast means for MOS circuit simulation, between the very detailed circuit-level simulation (see Section 5.2.3) and gate-level simulation using vendor's data on complete primitives. The scheduling techniques such as those used in gate-level synthesis are also used for switch-level procedures. Unfortunately, both gate-level and switch-level CAD tools are largely vendor-specific, without the ability to be integrated easily into a hierarchical simulation environment [104]. This highlights the need for standardized data interchange formats such as EDIF (see Section 5.5.6), which design engineers are increasingly demanding. For further details see [107] and elsewhere [12], [13], [29], [40], [108].

5.2.3 Silicon-level simulation

Silicon-level, or circuit-level, simulation is the most detailed and accurate level of simulation, but, because of this, detail cannot realistically be applied to circuits consisting of more than a few hundred transistors. It is therefore appropriate for basic building-blocks of SSI, MSI and perhaps LSI complexity, but not for VLSI.

The almost universally used program for silicon-level simulation is SPICE ('Simulation Program, Integrated Circuit Emphasis'), and particularly its very many commercial derivatives such as PSPICE, HSPICE, etc. (see Section 5.5.2). SPICE was first developed at the University of California at Berkeley in the early 1970s, and is based upon the small-signal a.c. equivalent circuit representations of active and passive devices. These in turn are based upon the physical properties and terminal characteristics of each device, with particular emphasis on the non-linear characteristics of the active device (transistors). Basic details of small-signal equivalent circuits have been summarized in Appendices C and D. It should be appreciated that as the operating point (current and voltage) in a non-linear device changes, so the small-signal equivalent circuit values also change, and hence it is necessary for a simulation program such as SPICE to perform a large number of simulation calculations with different operating points as the circuit currents and voltages vary.

SPICE requires the following information to be given:

- the circuit interconnection details;
- R, C, L values of all linear devices and perhaps interconnections;
- details of all non-linear devices (transistors, diodes and perhaps monolithic capacitors) and their parameter values at some chosen operating point;
- the d.c. supply voltage(s);
- the input(s) to the circuit;
- the output(s) or other details which the program is to calculate.

Some of the above will not arise in digital circuits, e.g. discrete inductors or capacitors, monolithic capacitors (see Figure 2.8), etc., but may be present in analogue circuit simulation, see Section 5.2.5.

A particular feature of the SPICE program, and an indication of its power, is that it is not necessary to specify the operating point of each non-linear device, nor is it necessary to give the small-signal parameter values over a range of operating points. Instead, the programme will (i) first calculate the d.c. operating point of each device from the d.c. equations that it knows for the device and from the circuit information that it has been given, and then (ii) work out the small-signal parameter values at this point from the values given at some other point. In order to perform such calculations, the equations relating the small-signal equivalent circuit values at one operating point to those at other operating points are built into the program, which will perform this re-valuation of parameter values many times in a simulation run as circuit currents and voltages change.

The method employed in SPICE for determining an operating point is an iterative procedure, whereby the program first suggests certain current and voltage values and checks to see whether these values are consistent with the circuit topology and the device d.c. equations. (For example, for bipolar transistors we have $I_C = h_{FE} I_B$, $I_E = I_B + I_C$ and $I_E = I_S(e^{40 V_{BE}} - 1)$, see Appendix C. For MOS transistors we have the d.c. equations given in Appendix D.) If not, the values are repeatedly disturbed until they converge on consistent values. Unfortunately, in some earlier versions of SPICE convergence did not always take place, particularly if zero values had been entered for some of the less important parameter values, and as a result the simulation could not proceed. Later versions, particularly the commercial derivatives, should not have such convergence difficulties.

All these procedures are hidden in the SPICE program, and the user does not have to be conversant with the details. Further resources within the program are default values should the user not wish to specify certain parameters. For example, the forward-biased diode equation of

$$I_D = I_S e^{40 V}$$

see Appendix C, is given a further enhancement to cater for particular physical constructions, being

$$I_D = I_S e^{40 V/n}$$

where n is some constant whose value is between 1 and 2 [109], [110]. If not specified, the program uses the default value of 1. Similarly, if not specified, the value for I_S is taken as 10^{-14} A for silicon. Table 5.5 shows some of the bipolar and unipolar transistor parameters used in SPICE, together with their default values.

Provision for simulating circuit performance at temperatures other than $25°C$ is also available within the SPICE program, which involves appropriate modifications to the normally used equations.

To run a SPICE digital simulation program it is necessary to provide full details of each primitive and all interconnections. The internal structure of individual types of gate, for example a 3-input NAND gate, may be listed once and then picked up as a module in the rest of the circuit netlist, or it may be available in the simulator

Table 5.5 Some of the transistor parameters and their values which are used in SPICE simulations: (a) epitaxial npn bipolar; (b) n-channel enhancement mode unipolar

Description of parameter	Symbol	SPICE designation	Typical value	SPICE default value
Saturation current	I_{CS}, I_{ES}	is	10^{-15} A	10^{-16} A
Transconductance	g_m	–	$40 I_C$ A V^{-1}	$40 I_C$ A V^{-1}
Forward-current gain	β	bf	200	100
Reverse-current gain	β_R	br	0.5	1
Base resistance	r_b	rb	100 Ω	0
Series emitter resistance	r_e	re	1 Ω	0
Series collector resistance	r_c	rc	50 Ω	0
Forward transit time	τ_t	tf	4×10^{-10} s	0
Early voltage	V_A	vaf	200 V	infinity
Emitter capacitance, $V_{BE} = 0$	C_{te0}	cje	2 pF	0
Emitter–base contact potential	ψ_{BE}	vje	0.75 V	0.75 V
Emitter–base grading factor	m_{je}	mje	0.33	0.33
Collector capacitance, $V_{CB} = 0$	C_{tc0}	cjc	1 pF	0
Collector–base contact potential	ψ_{BC}	vjc	0.75 V	0.75 V
Collector–base grading factor	m_{jc}	mjc	0.33	0.33

(a)

Description of parameter	Symbol	SPICE designation	Typical value	SPICE default value
Body-junction saturation current	I_S	is	10^{-15} A	10^{-14} A
Transconductance parameter	x	kp	3×10^{-5} A V^{-2}	2×10^{-5} A V^{-2}
Body threshold parameter	γ	gamma	0.37	0
Channel length modulation	λ	lambda	0.02	0
Zero bias threshold voltage	V_T	vto	1 V	0
Series drain resistance	r_d	rd	1 Ω	0
Series source resistance	r_s	rs	1 Ω	0
Oxide thickness	t_{ox}	tox	10^{-7} m	10^{-7} m

(b)

library data. The individual transistor models use library or default values, although for MOS transistors the W/L ratio may be specified individually – this is particularly necessary for full-custom designs where transistor geometries may be fine-tuned. Input logic signals may be given zero rise and fall times, or any required linear ramp from logic 0 to logic 1, and vice versa.

An example of the netlist for a single 2-input TTL NAND gate simulation is given in Figure 5.24(a). The five transistors $q1-q5$ are all of the same type, 'mod 1', and, similarly, the diodes are all of type 'mod 2'. Details of these two models have to be given. Input B is being held at 5 V, whilst input A is a pulse input falling from 5 V 4 ns after $t = 0$, with a fall time of 5 ns, dwelling at 0 V for 45 ns and rising again to 5 V with a rise time of 7 ns, this pulse input being repeated every 90 ns. The output response is to be calculated every 1 ns for 90 ns at output node 3.

228   Computer-aided design

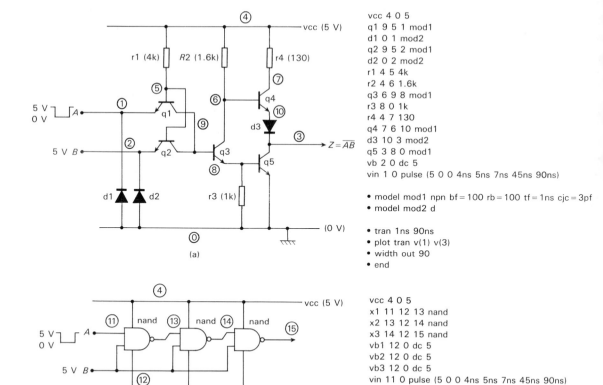

Figure 5.24   SPICE simulations of a 2-input TTL NAND gate: (a) the gate circuit with all internal nodes identified, and the resulting netlist. The diode model 'mod2 d' is the default model; (b) simulation of three such NAND gates in cascade, with the signals at input node 11 and output nodes 13, 14 and 15 to be printed out. The full netlist for the simulation must begin with a definition of 'nand', = the details given in (a) above. (Note, a SPICE netlist may use all lower-case letters for component identification as here or all CAPITAL letters but not both. Also the precise syntax in some commercial SPICE programs may be slightly dissimilar.)

The response of three such NAND gates in series may be requested, as shown in Figure 5.24(b). Similar netlists can be compiled to cover more complex digital networks but, clearly, the amount of detailed simulation involved at this circuit level escalates rapidly as the number of gates and transistors increase.

Details for a simulation of a single 2-input CMOS NAND gate are given in Figure 5.25, with the same input waveform as shown in Figure 5.24. The two p-

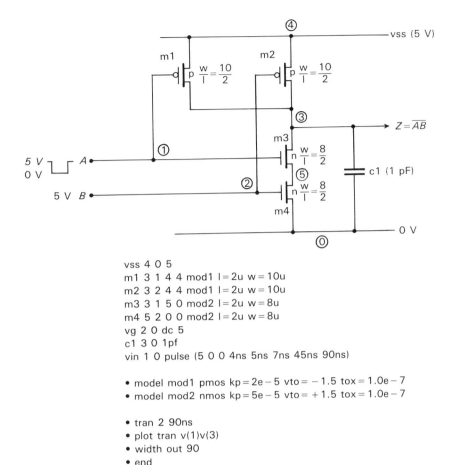

Figure 5.25 SPICE simulation every 2 ns of a 2-input CMOS NAND gate, where the W/l details of each transistor must be specified. Note that the substrate connection of each MOS transistor also has to be specified, = most +ve rail for p-channel devices and most −ve rail for n-channel devices.

channel transistors $m1$ and $m2$ are of type 'mod 1' with length and width geometries $L = 2$ μm and $W = 10$ μm, respectively, whilst the two n-channel transistors are of type 'mod 2', with $L = 2$ μm and $W = 8$ μm.

Because the full electrical performance of the active devices, together with all other circuit parameters, can be built into the SPICE simulation, very detailed simulation results can be obtained, the accuracy depending solely upon the accuracy of the models and parameter values. Every node of the circuit is evaluated for changing conditions at each simulation interval, which means a great deal of unnecessary calculation in digital circuits with a high degree of latency, for example where a long counter is involved, and hence there is the need for the higher but less detailed switch- or gate-level of simulation for most digital circuits of LSI/VLSI complexity.

Further details of SPICE, exactly how it computes d.c. operating points and the fine details of the transistor models, may be found in [109]–[112]. Since SPICE is a general-purpose simulation covering linear, non-linear and transient analysis, we will refer to it again in Section 5.2.5.

5.2.4 *Back-annotation and post-layout simulation*

As has been noted previously, it is not possible to include actual resistance and capacitance values associated with gate, macro and I/O interconnections in any simulation activity until the physical design has been completed. Prior to this, estimated or nominal R and C values have to be used, for example to consider each gate output having to drive a capacitance of $n \cdot C_G$ where n is the fan-out of the gate and C_G is some nominal value of capacitance associated with one gate input. However, in the actual chip floorplan there may be certain long interconnect paths and/or the use of a number of polysilicon underpasses which could materially affect the maximum speed of the circuit.

The layout extraction tools providing this detailed information on interconnection paths are invariably vendors' tools. The OEM may use them in some circumstances, but generally it is part of the final activity undertaken by vendor or independent design house before manufacture is commenced. The extraction tools provide the following:

- details of each interconnection routing, and hence the interconnect R and C values;
- identification of long interconnect paths or other critical paths; and
- the annotation of the netlists previously used for simulation with this detailed parametric data – hence the term 'back-annotation' which indicates this addition to the previously compiled netlists.

This is indicated in Figure 5.26 [113]–[116].

The post-layout simulation using this layout data is the closest simulation to the final product that can be performed before manufacture. The problems which post-layout simulation may find and which were not evident at early simulation stages invariably concern long interconnections in critical paths of the circuit, so that the circuit speed does not meet the required specification. This could be a fault of the original specification working too close to the limit of performance of the chosen technology and vendor's product, but whatever the reason a design revision is necessary either to:

- revise the placement in order to improve these critical interconnect paths; or
- revise the circuit design to reduce gate fan-out, and hence increase speed, or some other design revision; or

IC simulation software

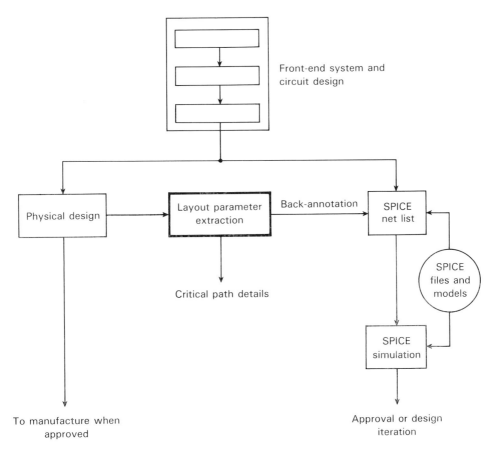

Figure 5.26 The concept of back-annotation whereby the layout extraction tool feeds back parametric data on the routing to a pervious simulation netlist. SPICE simulation is shown here, but, equally, gate- or switch-level simulation may be involved, particularly with VLSI layouts.

- relax the original specification to meet what can be achieved; or
- change the technology or product to improve the performance.

The most powerful post-layout tools can also create the circuit netlist from the final placement and routing, and thence the circuit diagram. If this extraction is then simulated instead of back-annotating data into the original design netlists, the most comprehensive check of all the design stages, including the physical design activity, is available. But whatever the checking procedure undertaken after placement and routing, the OEM will finally have to give the vendor or design house a formal approval and sign off the design as acceptable before any further work is undertaken.

232 Computer-aided design

5.2.5 Analogue and mixed analogue/digital simulation

For analogue-only circuits, the almost universal simulation tool is SPICE and its many commercial variants. The simulation accuracy of SPICE over a wide range of voltage, current, frequency and temperature conditions makes it particularly relevant. Its disadvantage, that is, the inability to handle circuits of VLSI complexity due to the detailed calculations involved, does not arise in analogue-only circuits, since it is rarely the case that more than a few hundred transistors, at the most, are involved.

The use of SPICE for analogue simulation is described in many texts [13], [77]–[80], [84], [109], [111]. However, most commercial SPICE-based tools provide a graphical as well as a textual output, with the capacity to display frequency characteristics, etc., in the conventional manner with which a designer is familiar (for example, see Figure 5.27). The ease of changing component values or

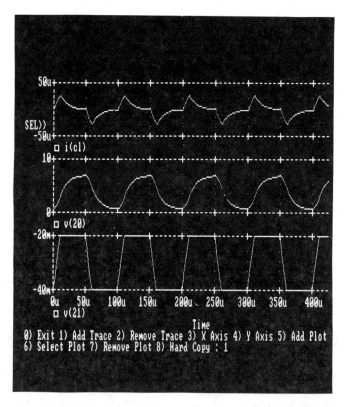

Figure 5.27 Screen display of a SPICE simulation of an analogue circuit with nodal simulations being performed for each 1 μs interval. For obvious reasons such graphical displays are sometimes referred to as 'oscilloscope' outputs. (Courtesy of MicroSim Corporation, CA.)

tolerances in a simulation and observing the results makes analogue design a very much faster procedure than having to make and measure a physical breadboard layout, but this does not of course relieve the designer from having to provide all the creative input.

For mixed analogue/digital simulation, it is exceedingly difficult to produce a single comprehensive CAD tool, largely because of the fundamentally different requirements of fine detail but low-component count of the analogue parts and the coarser detail but high-component count of the digital parts. This is sometimes referred to as the 'little-A, big-D' situation. The present solution is either (i) to have separate analogue and digital simulators and manually transfer data between them, or (ii) to have one system containing two separate simulators within one simulation environment, with the user programming which one to apply to the different parts of the circuit. For example, SPICE derivatives may be used for the analogue parts and switch- or gate-level simulation for the digital parts. (For 'little-A, little-D' situations, for example uncommitted analogue arrays which also have some limited digital capability, then SPICE may be relevant for simulating the whole circuit provided sufficiently comprehensive CAD hardware is available.) In some commercial packages the ability to move from one simulation tool to the other, for example to initialize circuit conditions before subsequent switch- or gate-level simulation is started, or to analyse digital macros in greater detail than is provided by gate-level simulation, may be particularly useful, capitalizing upon the feature that a circuit simulator analyses every individual node in a simulation time step whereas a digital-level simulator processes event changes only [105].

However, it is the method of organizing the interface between the two simulators that distinguishes different vendors' approaches. This is made more complex if there is feedback between the digital and analogue parts, for example in phase-locked loops, than in the case where there is only a one-way transfer of information, for example, in A-to-D or D-to-A conversion. The five principal methodologies which have been adopted to handle mixed simulation are as follows:

(a) unidirectional coupling of separate analogue and digital simulators, running one at a time;
(b) bidirectional coupling of separate analogue and digital simulators either running in parallel or running alternatively;
(c) unified coupling with the appropriate analogue or digital simulator being automatically called for by sub-routines;
(d) single integrated simulator, sometimes termed a 'core' or 'extended' simulator, consisting of a core analogue simulator with provision for analogue behavioural models of digital cells and macros; and
(e) similar to (d), but with the core simulator being a digital simulator with analogue behavioural models for the analogue cells and macros.

In the first three of the above five categories an appropriate boundary interface must be provided to link the two quite separate types of simulator. This data link tends to be vendor-specific, as are the behavioural models used in the last two of the above categories. Among the difficulties in linking analogue and digital simulations is the dissimilarity of the information handled in the two domains. For example, an analogue circuit driving a digital gate requires knowledge of the gate input impedance characteristics before the analogue output signal can be accurately determined, but this is not always available in digital modelling. Conversely, in the digital simulation, discrete information on signal driving strengths (see Section 5.2.2) and unknown (X) conditions is required, which is foreign to analogue simulators.

A single, mixed-mode simulator with one unified modelling language which can handle both analogue descriptions and digital descriptions, and with hierarchical capability to cover primitive to architectural levels, would clearly be advantageous. Considerable research effort is being expended in this area, but the fundamental difference between analogue and digital simulation will always remain [105], [117]–[120].

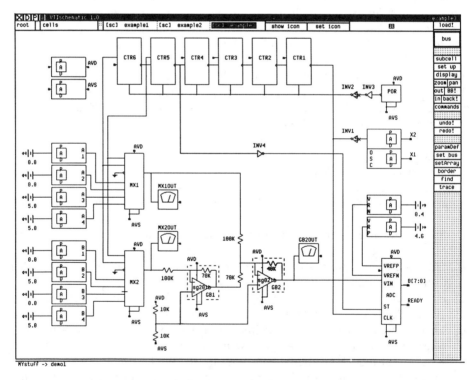

Figure 5.28 The mixed analogue/digital simulator MIXsim of Sierra Semiconductor, which provides schematic entry of mixed analogue and digital macros from a fully characterized standard-cell library, with meters added to the schematic to indicate analogue signal nodes of particular importance. Other highlighting can also be added. (Courtesy of Sierra Semiconductor Corporation, CA.)

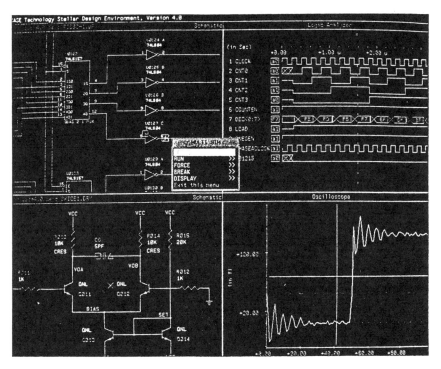

Figure 5.29 The multiple-window display availability with MultiSim, which supports both analogue and digital simulators working on the same mixed-schematic circuit diagram. (Courtesy of Teradyne, Inc., MA.)

The graphics available for mixed analogue/digital simulation, however, are already significant. Figure 5.28 shows a schematic entry of a mixed circuit which can subsequently be simulated to give the analogue and digital behaviour. Top–down schematic capture is available to this macro level, the analogue cells being characterized in a standard-cell analogue library so that detailed SPICE-level netlists do not have to be entered [117].

Textual output or graphical output of mixed simulation activities is available. In particular, the inclusion of windows to display both analogue and digital results side-by-side is increasing. Figure 5.29 indicates the measure of capability which may be available in a graphics-in/graphics-out resource.

5.2.6 Simulation accelerators and system simulation

Gate- and switch-level simulation is appropriate for most digital ICs, but as the complexity increases to greater than, say, 10 000–20 000 gates, simulation run times become unacceptably long and computer memory requirements very great. Quoted simulation times include over 20 hours for a 15 000 gate design running on a 68030-based workstation, with over 4 Mbyte of memory being required for data. Such CPU times also depend upon the accuracy of the gate-level model, that is, how many logic states and impedance levels are involved, as well as on the computer power being used.

The use of mixed-level simulation reduces CPU time at the expense of accuracy, and most vendors provide this facility in their digital simulation tools. However, there still remains the difficulty of accurately modelling very complex macros, particularly microprocessors, which, due to their complexity and software dedication, are difficult to model and may require many thousands of lines of code. This in its turn makes simulation using such complex models very slow. Architectural- and behavioural-level simulation, on the other hand, may not be sufficiently accurate.

A further problem, which we have not mentioned in the preceding sections, is that a custom IC is often only a part of a more complex system, and may have to work in association with other off-the-shelf ICs of LSI or VLSI complexity. The microprocessor may, for example, be a separate IC rather than a macro on the custom chip, and therefore the simulation of the complete system may be particularly important to confirm the details and performance of the custom design.

The following two approaches may be found to speed up the simulation of large digital networks:

(a) the use of computers designed specifically for simulation activities; and
(b) the use of actual physical circuits (e.g. a working microprocessor) as a part of the simulation tool.

The former is usually referred to as a 'simulation accelerator', and the latter as a 'hardware modelling' system.

Simulation accelerators support event-driven simulation at gate level and possibly at functional and architectural levels, but not at circuit level, which is conventionally handled by SPICE-like simulation tools. The improvement in simulation speed is achieved (i) by dispensing with a normal general-purpose computer architecture and designing the accelerator hardware to match the simulation algorithms, and (ii) by the use of parallel or pipelined processing to increase the throughput. A conventional host computer used for design activities, and perhaps circuit-level simulation, is still required to pre-process and post-process the data, but does not take part in the accelerated simulation process. This is indicated in Figure 5.30(a). Simulation throughputs of over 1 million events per second with one million gates simulation capacity can be provided by current commercial products. Further details may be found in [7], [104], [107], [121]–[123] and in vendors' literature.

Hardware modelling, however, sometimes referred to as 'physical modelling', is unlike any of the previously considered simulation methods in that it combines

IC simulation software 237

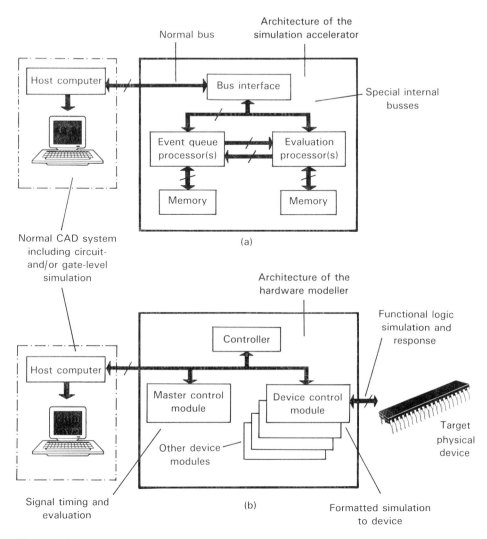

Figure 5.30 Multi-simulator, multi-platform simulation environments, with design data downloaded from the host computer for fast simulation: (a) the simulation accelerator with high-speed custom-designed processors; (b) the hardware modelling system with its target physical VLSI device.

both software and target hardware in one CAD tool. By using an actual physical device, e.g. a microprocessor or other VLSI component, its behaviour to the simulator inputs provides a fully functional response back to the simulator in real time.

Like simulation accelerators, a hardware modelling system is a resource additional to the normal CAD tools. The system comprises circuitry to configure the simulation signals to the physical device and process the returned signals, and to provide all the other necessary interface requirements for the rest of the CAD system. This is illustrated in Figure 5.30(b). The vendors of such products provide library details to interface to a range of standard LSI and VLSI devices, so that the OEM can use the system with a working device which acts as an accurate model for the rest of the simulation activity. Guaranteed working devices may also be supplied by the vendor, but, in any case, diagnostic test procedures may be built into the hardware modeller in order to confirm continuing fault-free operation of the physical devices [7], [104], [120], [124].

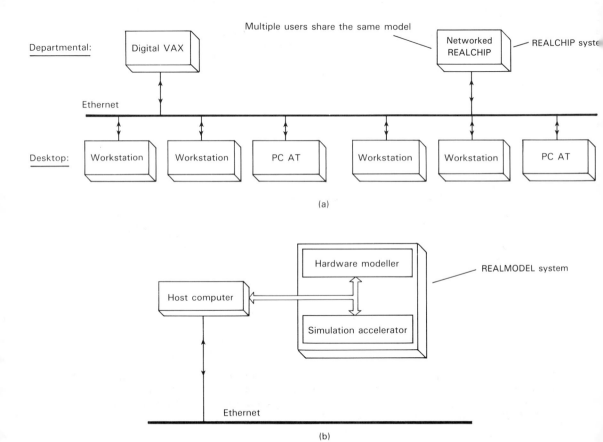

Figure 5.31 Example networking of enhanced simulation resources; see also Figures 5.34 and 5.36: (a) the REALCHIP[TM] hardware modelling system of Valid Logic, and its possible networking — normal simulation tasks are undertaken on individual workstations and REALCHIP is addressed when the simulation of a VLSI part is required; (b) the REALMODEL[TM] product of Valid Logic which combines both a hardware modelling system and a simulation acceleration in a closely coupled resource.

The cost of both simulation accelerators and hardware modelling systems is high, running into many tens of thousands of dollars, and hence such resources would normally be provided for each design office as one-off shareable resources, being networked to multiple users on an Ethernet or other local area network (LAN) facility, as indicated in Figure 5.31(a). Additionally, both these types of accelerator may be combined into one system simulator which enables them to communicate directly with each other instead of always having to go through the outside host computer (see Figure 5.31(b)).

These techniques for complex digital network simulation are beyond the needs of many OEMs, particularly first-time designers of custom ICs. They do, however, have an important role to play in association with larger USIC products, for example the very large channel-less products now increasingly available, possibly speeding up simulation by a thousand times in comparison with normal gate-level simulation, and for system simulation where it is particularly important to confirm a USIC design which is surrounded by further LSI and VLSI circuits.

5.3 Silicon compilers

In its fullest sense a silicon compiler is a CAD tool which would take a behavioural-level description of a required circuit and from it generate the silicon floorplan physical design details for manufacture. If this is done without human intervention, then the final design should be correct by construction, and, in theory, should not require any separate simulation and verification procedures. (The term 'compiler' is of course taken from computer programming, where the software writer details the required computer program in a high-level language which is then automatically compiled down to the final machine code. The final code is correct by construction and any malfunction must be a result of error at the human level. The final code may not be as compact and efficient as could be produced by a human programmer, but the amount of effort required to program at low levels is prohibitive for all except very small programs of less than, say, a few hundred lines of code. A further discussion on the semantics of software compilers vs. silicon compilers may be found in [1].)

However, unlike computer programming where, for a specific machine, specific operating system and specific compiler, there is one final result of the compiler operating on an input program, a high-level system description can, theoretically, have an infinite number of ways in which it can be made at the silicon level due to the flexibility of digital logic design. Guidance must therefore be given to a silicon compiler to tell it what choices to make in the hierarchical expansion down to the physical level, or it must be designed and used for a particular restricted type of digital system or form of realization. In many cases a more accurate description may therefore be a 'silicon assembler' if pre-determined blocks (macros) are used in this hierarchical design activity.

The 'silicon compilers' which have so far been developed fit into the following categories:

- compilers for memory arrays (ROM and RAM), A-to-D and D-to-A converters, arithmetic logic units and other macro functions with known architectures;
- compilers for programmable logic devices where the silicon architecture is fixed and the design process consists of mapping a behavioural specification on to the constrained device topology;
- compilers for digital signal processing (DSP) applications, for example digital filters, where the circuit configuration is known or defined;
- compilers for telecommunications, control and datapath applications, again where the general circuit configuration of the final design is known or defined;
- certain analogue applications, using a menu of standard analogue functions but requiring the computation of component values to meet the behavioural specification; and
- compilers for digital circuits, where the designer first has to decide from an available menu which types of building block (macros) to use in the design, the compiler assembling and optimizing them for the particular applications.

The first category above is more often referred to as a 'module generator' rather than a silicon compiler, since it usually deals with one block only in a much larger total system. Module generators usually only optimize the size and silicon layout of a soft macro design to meet a required specification [10], [125], but may be associated with a larger compiler dealing with the whole system design.

Compilers for PLDs, however, are true silicon compilers in that no human intervention is necessary between the processes of entering the system requirements in truthtable or state table or other form, and the design output stating how the target device should be personalized. This is possible because there is little freedom to innovate as a result of constraint by the PLD topology and performance, and because the size of most PLD circuits is relatively small so that only modest computer power is necessary. Correct-by-construction should thus always result provided the correct design requirements are entered by the designer [96]–[102], [126].

Compilers for DSP, telecoms and other applications, including analogue, have the advantage that the general circuit configuration is known or is entered by the designer, and that the rules for calculating all component values and other variables for the given silicon technology are known. Hence the CAD tool can be entirely rule-based, not requiring any innovative procedures once the design requirements have been fully specified [80], [92], [127]–[129]. One of the advantages claimed for compilers used for analogue purposes is that the designer does not need any in-depth knowledge of analogue design; this is perhaps too simplistic, since considerable innovative work may be necessary to finalize the specification, and the CAD tool may be unable to complete the design specification, whose resolution again requires human expertise.

However, it is in the field of non-programmable digital VLSI circuits that developments in silicon compilation are most usually associated, the objective being to encompass all the tedium of a large digital design between behavioural and layout levels. The first development in this area was the result of work carried out at the California Institute of Technology (CALTECH) using their 'Bristle-Block' approach. Here, the required digital system was designed as a data path architecture, involving registers, arithmetic logic circuits and data-shifting circuits linked by appropriate data highways (busses). These 'core' cells were defined as parameterized soft macros stored in outline but not detailed layout form, and the CAD tool called up the required cells and stacked them along several parallel data paths so as to form a roughly rectangular layout. Cell 'stretching' or 'compression' was then undertaken to equalize cell heights along each linear row.

This compilation technique is appropriate for data-path circuits where signals are passed linearly from one cell to another, but not for random logic, which cannot be accommodated in such a topology. Thus some PLA-type structure was also usually associated with a Bristle-Block compilation. Further details may be found in [71], [130], but, overall, this technique had restricted application and produced a final silicon layout which could be 50–100% greater in area than could otherwise have been produced.

Commercial silicon compilers have been marketed by several vendors, but most are more accurately described as 'silicon assemblers' since the design engineer has to define the necessary macros and also possibly the general target floorplan in order to restrict the architectural choices that the CAD tool has to make. Among commercial products are Seattle Silicon Corporation's CONCORDE VLSI compiler [131] and Silicon Compiler Systems Corporation's GENESIL compiler [132]. The use of module generators within the compiler and vendor-specific high-level languages and algorithms is a feature of most compilers.

Further details of the theory and practice of silicon compilation may be found in [130], [133]–[135]. The efficiency of certain compilers in comparison with a hand-crafted, full-custom design for the 8-bit 65C02 CMOS microprocessor may be found in [136], which shows that there was a considerable silicon area penalty to be paid for the silicon compiler design solutions.

In summary, silicon compilers currently require a design to be described at the functional level, with the tool automating much of the design activity from there on, and hence they are only somewhat more powerful than fully automatic place-and-route tools. Simulation still remains a separate and necessary activity in order to verify that the design entry data is correct, and that the final physical layout design will meet the required performance specification. The development of new and more powerful languages for logic synthesis and the incorporation of some form of artificial intelligence [134] may help to ensure that future silicon compilers realize more closely the ideal aim of automatic correct-by-construction design from a behavioural level input, although a target floorplan architecture will probably always need to be specified.

5.4 CAD hardware availability

The general evolution of CAD hardware and software resources for IC design has been introduced in Chapter 1, Section 1.3. Here, we will amplify this information and indicate the present maturity of these resources.

The use of large mainframe computers is now no longer relevant to the actual design activities, since the evolution of the workstation and personal computer (PC) has enabled sufficient computer power to be brought to each designer's desk for most of the design activities. However, the powerful mainframe resource, with its mass storage capacity, may still be required by large design offices as a host computer for the following activities:

- to undertake long simulation runs from data downloaded from individual designer's workstations;
- to coordinate and handle the data transfer between different areas of activity and generally take charge of scheduling and housekeeping activities; and

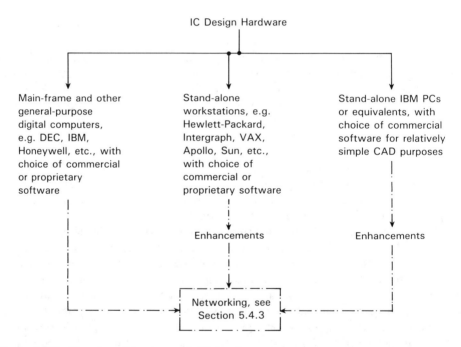

Figure 5.32 The diversification of CAD hardware resources, with the PCs generally being for small- to medium-size OEMs, the workstations for companies with greater experience and needs, and mainframe computer power for vendors and large design offices.

- to act as the principal data bank for all the company design and manufacturing details, with restricted access to ensure that no part of a product design can be changed without authority and without appropriate wider consultation.

This overall CAE product control may possibly be considered as part of managerial or product engineers' duties rather than of the IC designer, but the latter must be conscious of how his or her role fits into the complete picture [137], [138].

The general diversification of CAD hardware is as shown in Figure 5.32, with the workstation and the PC being the usual interactive resource used by IC designers. We will therefore consider these two categories in greater detail.

5.4.1 Workstations

The term 'workstation' may be broadly defined as a hardware assembly consisting of the following:

- a central processor, currently a 32-bit processor;
- a VDU, usually 15" or 19", monochrome or colour;
- input devices, usually keyboard plus mouse;
- working (main) memory;
- bulk memory, invariably several hundred megabytes;
- communications (networking) facility to printers, other workstations, host computer, hardware accelerator, etc.

Figure 1.5(a) illustrates a typical product.

Although there have been workstations designed by CAD vendors who then marketed a unique package of matching hardware and software for IC design activities, for example the breakthrough developments of Daisy Corporation and Valid Logic Systems, Inc., more recent developments have seen the adoption of general-purpose workstations by most IC CAD vendors, enabling them to concentrate on the development of increasingly sophisticated software, see Section 5.5. (It should be appreciated that workstations are not exclusively associated with IC design activities, but, rather, are general-purpose CAD/CAM tools suitable for civil, mechanical and electrical design activities, being dedicated to a particular activity by appropriate software. Good graphics capability is a common attribute for all these areas.) This separate specialization of hardware and software suppliers is an indication of the intensity of continuing development and evolution wherein one company cannot easily remain expert in both areas.

The suppliers of workstations currently include Digital Equipment Corporation, with their range of VAXstation™ and DECstation™ products, Hewlett-Packard, Sun Microsystems, Inc., Apollo Computers, Inc., and Intergraph Corporation. (Apollo has recently become a subsidiary of Hewlett-Packard, but both names are currently retained. Also, Daisy Systems Corporation and Cadnetix merged to become a single company, Daisy/Cadnetix, and in turn have become a subsidiary company of Intergraph Corporation. This is an indication of the fluidity of the CAD/CAM/CAE market.) A great deal of commonality in the choice of central processing unit and operating system may be found between products of the same generation, since most workstation suppliers themselves have to rely upon specialist IC vendors such as Intel, Motorola and (increasingly) Japanese suppliers for state-of-the-art VLSI circuits. This close dependence is indicated in Figure 1.7. Table 5.6 shows that the 25 MHz Motorola 68020 CPU assisted by the 68881 floating-point processor and the 68220 memory management circuit was widely adopted. However, every 2–3 years there is usually a new wave of improvement.

The following two fundamental developments in the more recent generation of workstations may be observed:

(a) the almost universal adoption of the UNIX operating system, originally developed by AT&T; and
(b) the increasing use of Reduced Instruction Set Computing (RISC) processors.

The RISC/UNIX combination was pioneered by Sun in their series 4 workstations, which first employed Fujitsu CMOS gate arrays, with faster ECL versions being promised. Intergraph Corporation were also pioneers in this area, using the high-speed CLIPPER chip set originally designed by Fairchild. It is probable that this RISC/UNIX combination will be a feature of most workstations through the 1990s.

It is thus of no great significance to quote performance specifications of existing workstations, since maximum performance is continuously increasing and costs are generally falling for any given performance level. Figure 5.33 shows comparative performance figures, with workstations at the lower end of the picture compared with supercomputers at the top end. The y-axis is in million floating point s^{-1} (MFLOPS) rather than million instructions s^{-1}, since graphics processing relies heavily upon floating point operations, typically requiring as a rule of thumb ten times as many MFLOPS as MIPS to reach a given standard of user response [139]. It is of interest to note that all the hardware platforms shown in Figure 5.33 incurred costs of around $100 000 per MFLOPS, although it is not obvious why this should have been so. The mid-1990s stand-alone workstations may possibly provide a performance of about 20 MFLOPS, although at lower costs than are indicated in Figure 5.33.

5.4.2 *Personal computers (PCs)*

The personal computer, first introduced by IBM in the late 1970s, has increased in power and application so that it is now becoming difficult to draw a clear dividing line between workstations and PCs [139], [140]. This blurring of capabilities has

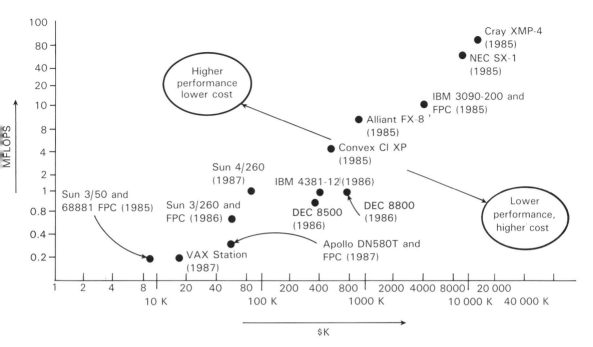

Figure 5.33 The general capability of MFLOPS vs. cost of workstations and mainframe computers in the second half of the 1980s. Future tendencies will be for performance to increase but performance per dollar to decrease; FPC = floating point co-processor. (Source: based on [139].)

increased with the second generation of machines, using, for example, Intel 80386 CPUs. A typical IBM PC/AT is illustrated in Figure 1.5(b).

In terms of operating speed and memory capacity the PC generally compares with workstations as follows:

- PC: 1–2 MIPS, 1 Mbyte RAM, 10–20 Mbytes bulk memory (floppy disk);
- workstation: 4 MIPS, 2–4 Mbyte RAM, 100 or more Mbytes bulk memory (hard disk);
- newer RISC/UNIX workstations: up to 10 MIPS, 4–8 Mbyte RAM, several hundred Mbytes bulk memory.

These are typical figures only, and should be read as comparative data between the groups. Table 5.6 gives additional data.

The problem with the first-generation PCs was largely with the MS-DOS operating system, which is excellent for simple housekeeping and clerical duties but not so relevant for CAD/CAE activities since multitasking and windowing cannot be supported. Additionally, the original MS-DOS could only address 640 K of RAM. The PC's strength lay in their very wide global adoption, with numerous

Computer-aided design

Table 5.6. Typical specifications for a range of past and present PCs and workstations, with continuing evolution being a very prominent factor

Product	CPU	Floating point co-processor	Clock rate (MHz)	MIPS	Operating system	Mbytes RAM (standard or minimum)	Mbytes hard disk (standard)
IBM PC/AT	80286	None	6	N.A.[d]	MS-DOS	0.64	20
IBM PC/AT	80286	80287	8–10	N.A.	MS-DOS	0.64	20
IBM PS/2 Model 70/80	80386	80387	10–25	N.A.	OS/2	4.0	60–640
IBM System 6000[a]	Own	on-chip	20	29.5	AIX	8.0	120–640
Compaq 386	80386	80287	16	N.A.	MS-DOS	1.0	140
Apollo DN3000	68020	68881	12	N.A.	UNIX	2.0	70
HP Apollo 9000/400 dl	68030	68882	50	12	HP/UX	8.0	200–400
HP Apollo 9000/425t	68040	on-chip	25	20	HP/UX	8.0	200–400
Sun 3/260	68020	68881	25	3	Sun UNIX	8.0	280
Sun SPARKstation SLC[a]	SPARC	SPARC	20	12.5	Sun UNIX	8.0	104–2007
Sun SPARKstation IPC[a]	SPARC	SPARC	25	15.8	Sun UNIX	8.0	104–2007
Sun SPARKstation 470[a]	SPARC	TI 8847	33	22.5	Sun UNIX	32	669–8000
Intergraph 2000 series	Clipper +80386	Clipper	25	12.5	UNIX	16	200
Sony NWS 1500	68030	68882	25	4	UNIX	8.0	240
DEC VAXstation 3100	Own	Own	25	N.A.	UNIX[b]	32	208
DEC VAXstation 3100[a]	Own	Own	25	N.A.	UNIX	32	208
DEC VAXstation 8000[c]	Own	Own	25	N.A.	UNIX[b]	32	208

[a] RISC/UNIX based workstations.
[b] VMS also available.
[c] Enhanced graphics capabilities.
[d] NA = not published or not available.

clone versions, which encouraged the widespread development of CAD software from very many vendors which worked as efficiently as possible within these constraints. The later introduction by IBM of their OS/2 operating system and new bus structure ('Micro Channel Architecture' or MCA) was intended to recover the uniqueness of the IBM product and prevent clone versions being marketed by competitors, but the subsequent acceptance of UNIX has forced IBM and most workstation vendors to adopt this as the operating system for CAD/CAE purposes – this has also forced DEC to provide its version of UNIX as well as VMS for its range of workstations.

The other strength of the PC was the ability to plug in additional expansion and processor boards on the PC bus to increase the performance for specific duties. This has been the principal reason why PCs, even with the MS/DOS operating system, have been so successful for many CAD/CAE activities, and, conversely, why the later IBM, OS/2 operating system, which with its new MCA bus structure was incompatible with these existing enhancement boards, has not been widely welcomed.

A further player in the field below the workstation has been the Apple Macintosh. The strength of Apple products has been in their graphics and intensive adoption of icons and windows, making them the most favoured hardware platform of the 1980s for scientific word-processing, desk-top publishing and 3-D pictorial representations. However, the advantage of graphics, icon and windowing has now been learnt by the engineering CAD/CAE community, and it remains to be seen whether Macintosh hardware based upon the 80386 or subsequent CPUs will expand to find a place in the electronics CAD/CAE area.

However, with company rationalizations and take-overs, it seems probable that there will be four major suppliers of CAD/CAE hardware, namely IBM, DEC, Hewlett-Packard and Sun, with the demarcation between PCs and workstations becoming more difficult to define. Software vendors (see Section 5.5) will increasingly have to provide software which will run on a range of available hardware platforms.

5.4.3 *Networking*

Isolated stand-alone PCs and workstations are appropriate for very small design offices, where one designer can do all the required in-house design activity per product, and pass his or her design file(s) over to, say, a vendor for further design, simulation and fabrication activity. However, for organizations with more than a small number of machines it usually becomes either necessary or desirable to link resources so that they can exchange data or share expensive common hardware. This applies equally to PCs and workstations.

The means adopted to link distributed in-house resources is a serial bit stream transmitted over a local area network (LAN). A number of LAN designs are commercially available, but the two most common ones are:

(a) the Token Ring, originally produced by IBM; and
(b) Ethernet, which is a bus-type interconnect.

The Token Ring is a loop linking all the hardware stations (see Figure 5.34(a)), the interconnect medium being a fibre-optic cable or a metallic-screen twisted-pair. Access to the ring is controlled by an electronic token which is transmitted continuously in one direction around the ring, being set to indicate whether it is alone or accompanied by a destination address plus information for this address. The token is interrogated at each station on the ring, and then passed on to the next one until it finally reaches its destination address if accompanied by information. Here it is accepted, together with the information that accompanies it. In the absence of information the token continues to be circulated.

When a station wishes to send a package of information it checks to see if the received token is alone, and if so adds its information and required destination address. The Token Ring originally offered a transmission bit rate of 5 Mbits s^{-1},

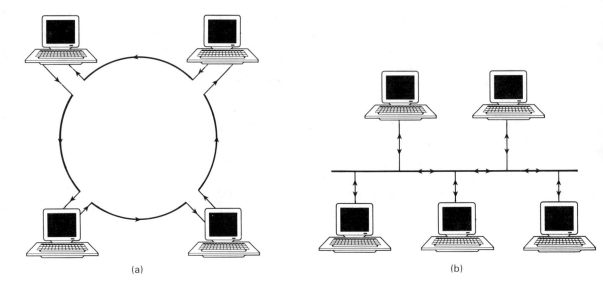

Figure 5.34 The Token Ring and the Ethernet local area networks: (a) Token Ring LAN; (b) Ethernet LAN.

but new interface boards with faster ICs introduced in 1988 increases the rate to 16 Mbits s^{-1}.

Ethernet differs fundamentally from the Token Ring, and does not have a ring interconnection pattern between stations. Instead it is bus-type interconnect as shown in Figure 5.34(b), with all stations connected to the single interconnect bus. The stations on an Ethernet network contend for access, and if two attempt to transmit simultaneously, then protocols determine the priority. Every station receives data on the network, but ignores the data if this does not contain its station address.

Ethernet can transmit at up to 10 Mbits s^{-1}, although speed of actual data transfer can fall with a high level of traffic due to the priority protocols then involved. Both Ethernet and Token Ring are specified in the IEEE Open Systems Interconnect (OSI) standards which specify electrical characteristics, protocols, and cabling and connection features of local networks. Ethernet, IEEE specification 802.3, is more usually used in engineering environments, for example see Figure 5.31, with Token Ring, IEEE specification 802.5, possibly being more common in non-technical business environments.

LAN networks are typically limited to a total length of about 1 km because of cable losses and interference problems. However, in order to use the higher bandwidth and reduced losses of modern fibre-optic cables and connectors, and their inherent immunity to interference, a new high-speed network technology has recently been introduced. This has become increasingly desirable because of the increase in file sharing, whereby a Network File System allows workstations on the

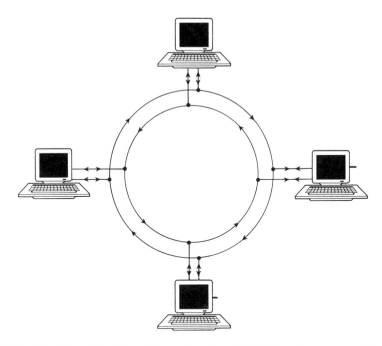

Figure 5.35 The Fibre Distributed-Data Interface (FDDI) local area network, with fibre optics interconnections.

network to share one set of disks located on a host computer. The new system is the Fibre Distributed-Data Interface (FDDI), and consists of a dual ring interconnection topology, as shown in Figure 5.35. Both rings may be used simultaneously, giving a total bandwidth of 200 Mbits s^{-1}, using a token method of station selection similar to the Token Ring. Further attributes of the FDDI are that it can span a much wider territory, up to 100 km, making it a metropolitan area network (MAN) rather than a LAN if required, and can tolerate a break in one of the two rings without impairing the network catastrophically [141].

Where data has to be transferred in real time between different CAD resources at the full operating speed of the equipment, then LANs are not appropriate. Here a local bus interconnect is required which can link the actual operating busses of the two (or more) equipments so as to provide the desired distributed processing capability. Among the systems available are VMEbus, Multibus and Futurebus+, with variants being marketed by individual vendors. These systems can transfer data at rates of from 6 Mbytes s^{-1} upwards, and may be combined with LANs (as shown in Figure 5.36) for the transfer of off-line data.

Bus interconnect systems are normally employed for linking equipment up to a few tens of metres apart, although in theory there is now no theoretical reason why longer interconnect distances should not be available using multiple-core fibre-optic cables. However, the biggest problem is how to link different types of systems,

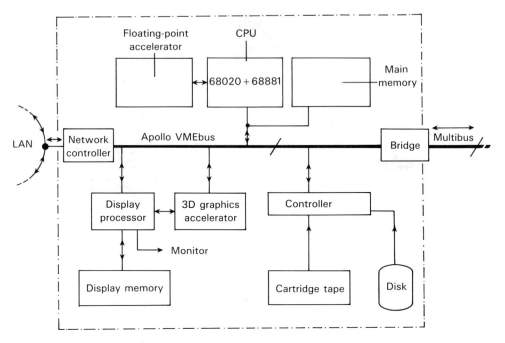

Figure 5.36 The integrated architecture of the HP Apollo Domain workstation environment, with local modules communicating with each other at bus speeds and serial LAN communication to further resources.

in particular VMEbus and Multibus, which have been widely used in the 1980s, to Futurebus+, which may become the standard for the 1990s [142]. The problem is that these busses utilize completely different event formats, making transfer of data an indirect procedure requiring 'bridge' and 'slave' circuits to provide the transfer.

5.5 CAD software availability

The choice of CAD hardware and its continuing evolution is compounded by the even more diverse and rapidly changing range of design software. Managerial decisions on what should be purchased for OEM in-house activities are therefore complex and will be considered further in Chapter 7.

Another problem with the present status of CAD software is that data formats have not yet become fully standardized. The effects of this are two-fold, as follows:

1. Although individual software programs may be available for all the hierarchical levels of activity shown in Figure 5.3, they may not easily be 'bolted together' to form an integrated hierarchical suite without encountering interface problems.

2. Data formats may differ between vendors, so that even with the gate-array choice of design style one vendor's software may not be compatible with another's, and possibly neither will be compatible with, say, a PLD vendor's software.

These problems largely arise because of the diverse origins of CAD software, compounded by the fact that vendors may have had vested interests in not encouraging the free interchange of design data to other vendors' CAD systems. The OEM, however, would clearly prefer vendor independence for all the CAD resources which he or she may use.

The CAD software which the OEM or independent design house may use comes from three principal sources, (i) the IC vendors themselves, (ii) workstation vendors, and (iii) independent software vendors. This is illustrated in Figure 5.37. Software from academic institutions may also be involved, although this usually requires some commercial company to make it more robust and provide documentation and updating.

The advantages and disadvantages of these various sources of original CAD software are as follows:

- The software from an IC vendor will usually be specific for his range of product devices, and in general will not cover the design of any other vendor's product. However, it will cover all the likely OEM design activities, will be proven for

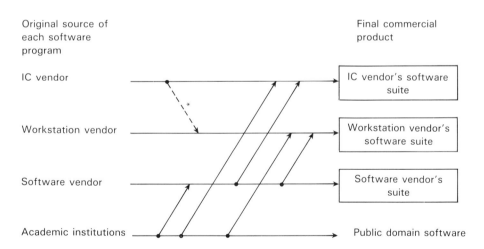

* For vendor-specific data on chip layouts, etc.

Figure 5.37 The interleaving sources of IC CAD software, omitting very large OEM companies who may write or compile their own software for internal use only. One software vendor's program may also be sold to other software vendors in order to complete a suite. (Source: based upon [17].)

its particular duties, and is likely to be inexpensive for the OEM to purchase or hire.
- The software from a workstation vendor may be tailored to the vendor's own design of workstation, and may not run on any other hardware platform. It will not be specific in matching the products of any particular IC vendor, and will therefore require further information in order that design data can be translated into a particular vendor's gate-array or standard-cell product. It will usually be an expensive but comprehensive suite of programs, suitable for independent design-house use as well as use by large OEMs.
- The software from independent software companies is likely to run on a range of hardware platforms, but, like workstation vendor's software, will not be specific for any particular IC vendor's products. It may be a comprehensive suite of software or a single software package to perform one task, such as simulation or routing, very efficiently.
- Finally, the software from academic institutions is likely to be of an innovative nature when first released, not necessarily completely proven and documented, and possibly requiring considerable further work to turn it into a fully fledged product.

The simulation package SPICE is the classic example of microelectronic software originating from an academic institution and now globally used in IC CAD work.

Because software development costs are becoming so high and companies specializing in the hardware platforms are becoming so competitive, increasing rationalization must now be included in the picture painted above. The growing tendency is for specialized electronic design automation (EDA) companies to market comprehensive suites of software, including their own, but increasingly the best software currently available from elsewhere, for specific duties, a complete package being capable of running on several industry-standard hardware platforms such as PC/ATs and DEC, Sun and Apollo workstations. The complex engineering must now achieve the following:

- link different software packages together to form a comprehensive hierarchical suite with one common database, thus enabling an OEM to purchase progressively as much as needed without encountering incompatibilities;
- provide, as far as possible, different versions of the suites to run on different hardware platforms with different operating systems such as MS/DOS, UNIX or VMS (UNIX is being used increasingly, but currently not all versions of UNIX are compatible; the use of MS/DOS is likely to decline for CAE purposes);
- liaise with a large number of IC vendors to provide design data on these vendors' products, often known as 'ASIC design kits', thus providing routes to silicon;
- provide networking facilities so the different hardware platforms and other hardware resources may be linked, for example see Figure 5.31; and
- to provide maintenance and update of all these interlinked resources.

This is not a task which any OEM would undertake lightly in order to assemble his suite of CAD resources.

5.5.1 Design and layout software

As discussed in Sections 5.1 and 5.3, the upper levels of IC design are not as fully represented in available software tools as the lower design levels such as schematic capture, placement and routing. The status is generally as follows:

- *Silicon compilation software*: available from specialist vendors, and requiring a workstation (or equivalent) hardware platform.
- *Schematic capture*: very widely available, running on PC or workstation platforms.
- *Placement and routing*: very widely available, running on PCs or workstations for gate arrays but usually requiring the increased memory capacity of workstations for standard-cell designs – this depends upon the complexity of the cell library and the amount of additional memory which can be added to PCs.
- *Physical silicon design software*: again widely available but mainly for vendor or design-house use rather than the OEM who may not wish to descend to the silicon level. Usually workstation-based.

5.5.2 Simulation software

Since it forms a key part in all forms of IC design, simulation software has seen greater emphasis and development than any other associated CAD tool.

High-level digital simulation software running on workstation (or equivalent) platforms and using HDLs (see Section 5.2.1) are being marketed increasingly, but are not yet generally accepted by the OEM designer. However, the status of lower simulation levels of digital simulation is generally as follows:

- *Gate and switch-level simulation*: very widely available, previously running on workstations but increasingly available for circuits not exceeding, say, 5000 gates on PCs.
- *Circuit level simulation*: here the almost universal simulation tool is based upon SPICE, with every CAE vendor providing such a resource. Among the very many commercial derivatives of SPICE from different vendors are CSPICE, DSPICE, HSPICE, MSPICE, PSPICE, SSPICE, USPICE, VSPICE and ZSPICE, plus other versions tailored for more specialized applications, such as microwave devices [143].

Whilst many will have been produced to derive the d.c., a.c. and transient behaviour of any given integrated circuit over a range of temperatures, whether bipolar, MOS or GaAs, many others such as PSPICE from MicroSim Corporation are now providing mixed analogue/digital simulation capability, the analogue simulation being a fully detailed SPICE determination with a digital simulation program running separately from the SPICE program to handle the (possibly larger) digital part.

Most SPICE programs require the capability of a workstation in order to simulate circuits containing hundreds of transistors in a reasonable time, but an increasing number of vendors are providing PC (or equivalent)-based versions which require enhancement of the PC capabilities. Again, taking PSPICE as an example, versions are available for the following hardware configurations:

- Sun, Apollo and DEC workstations, running on a variety of operating systems;
- PC/AT or equivalent platforms with 0.64 Mbytes of RAM, type 8087, 80287 or 80387 floating point co-processor, up to 16 Mbytes hard disk, MS/DOS operating system;
- 80286 and 80386-based PCs with OS/2 operating systems; and
- Macintosh II platforms with 6881 floating point co-processor.

The PSPICE digital simulation option requires additional fixed disk memory capacity on the PC/AT, and appropriate additional memory on the latter two hardware options. The latest developments, involving windowing and graphics-in/graphics-out capabilities, and output processing such as Fourier analysis and distortion analysis of the simulation outputs, are increasingly available from specialist software programs. Graphics display adaptor boards are required with PC hardware platforms to give the necessary graphics capability.

5.5.3 *PLD design software*

Software for all the OEM design and programming activities associated with programmable logic devices is commercially available from both vendors and distributors of such devices.

The most widely used design tools include:

- ABEL,
- CUPL,
- PALASM,
- PLDesigner,
- PLPL,

and others, which are used by many vendors to build up a complete CAD package for their programmable products. Enhancements such as in Advanced Micro Devices' AmCUPL, may also be incorporated. For the larger capability logic cell arrays (LCAs), Xilinx provides its own development system XACT, which (as shown in Figure 5.19) has provision for accepting input data from PLD design tools.

The hardware platforms for PLD and LCA software invariably include the PC/AT, together with the more powerful workstations should the OEM possess such resources. However, PC-based platforms require considerable hard disk back-up memory to hold all the design data, a typical specification being as follows:

- 80286 or preferably 80386 PC/AT or equivalent;
- 0.64 Mbytes RAM (minimum);
- 20 or, preferably, 40 Mbytes hard disk;
- MS/DOS version 3.1 or higher;
- graphics display adaptor board.

The size of memory required for program and device data depends upon the types of PLD and the comprehensiveness of the design and test software. The Xilinx program for LCA design, for example, requires 3.5 Mbytes for the XACT program and 6.5 Mbytes for 9000-gate LCA details, which must be available in addition to the memory used for all the other resident software and for the running requirements.

All the above figures should be interpreted only as being illustrative of the widely available PLD software resources [95]–[102].

5.5.4 *Printed-circuit board (PCB) design software*

Like PLD design resources, CAD tools for PCB design are readily available from a wide range of specialist vendors, usually as stand-alone design tools. However, most CAE vendors who package a complete hierarchical suite of programs with a common data base for custom-design activities also include a PCB design tool in their suite of programs, or an interface to one or more of these tools from specialist companies. As well as PCB layout design, such tools can often also design the layout for wire-wrapped assemblies.

PCB design software pre-dates custom microelectronics, being originally developed for PCBs containing discrete components and SSI/MSI integrated circuits. Available tools now range from the relatively simple, catering for up to 100 or so ICs on single-sided or double-sided boards, to very complex programs for multilayer boards with perhaps eighteen to twenty layers and containing many hundreds of IC packages. The latter may require the computing power of a VAX or similar mainframe computer to solve the placement and routing in an acceptable time. Among the accepted industry-standard PCB layout tools are P-CAD, OmniCad, CADDS, Scicards, and others.

For the majority of OEM and custom microelectronic needs, the lower end of the above spectrum of capability is usually appropriate. The hardware platform on which the large majority of these design tools will run is again the PC, although, increasingly, 80386 processor-based machines are being specified. A co-processor is not usually required, but at least 20 Mbytes hard disk is usually needed [144]–[147].

5.5.5 Housekeeping software

Documentation of complex system designs is a critical factor in the design, manufacture and update aspects of the product, and, increasingly, a common company database of all aspects of the product is necessary. In the absence of good housekeeping, there is the danger of a company having individual stand-alone islands of CAD/CAE/CAM which do not communicate electronically with each other, and which then require human effort to exchange data and maintain overall consistency.

Figure 5.38 indicates the commonality which should exist between design activities and documentation. The precise complexity will depend upon company activities, and could range from a very simple to an extremely complex network of resources. Typically, some or all of the following resources will be involved:

- PC or workstation-based engineering design stations for IC design activities;
- similar design stations for PCB and other electrical and mechanical design activities;
- computer controlled test facilities;
- laser printers and desk-top publishing resources for documentation, illustrations, maintenance publications, etc.;
- overall stock control;
- LAN network and central mainframe computer with mass storage for central data files.

This overall company organization is beyond our consideration here. Software to provide all these housekeeping facilities is available with advice from specialist consultants on building up a company structure, but it remains a managerial decision within each manufacturing company as to the scope of, and necessity for, in-house computer-based housekeeping resources [137], [138], [148], [149]. The software rather than the hardware usually incurs the greatest problems, with compatibility of data formats still the principal cause of difficulties.

5.5.6 Data formats and standards

From all the preceding considerations it is clear that there is an increasing need for software tools to be able to communicate with each other, without the time-consuming and highly error-prone need of human intervention and re-entering of

CAD software availability 257

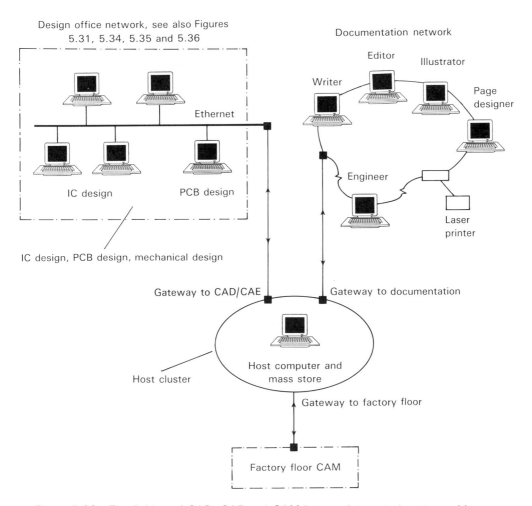

Figure 5.38 The linking of CAD, CAE and CAM into an integrated system with one common database, sometimes referred to as CIM (computer integrated manufacturing), although this designation is more usually reserved for the factory-floor post-design activities. The particular arrangement of Ethernet or equivalent bus to connect closely coupled resources and LAN or equivalent networking will vary for each installation.

data from one tool to another. The following are thus required:

(a) standardization of data formats in CAD software for specific duties from different vendors, e.g. formats for mask-making or automatic test machines, etc.; and
(b) interchange formats to allow data interchange between tools doing different parts of the design and manufacturing activity.

Unfortunately, it is not possible to achieve global standardization of data formats between all parties for particular design or manufacturing activities. Instead, several standards have emerged from specialized vendors, which are currently accepted by other companies or machines which use this particular level of data. For interchange between different duties translation to some standard data format such as EDIF (see below) is then required. This is illustrated in Figure 5.39.

Data formats which have received acceptance through widespread use for particular duties include the following:

- *Gerber format*: developed by the Gerber Scientific Instrument Company for plotters, particularly appropriate for the graphics artwork of PCB and wire-wrap layouts.
- *EBES (Electronic Beam Exposure System) format*: developed by Bell Labs., which is used for the control of E-beam machines for direct-write-on-wafer.

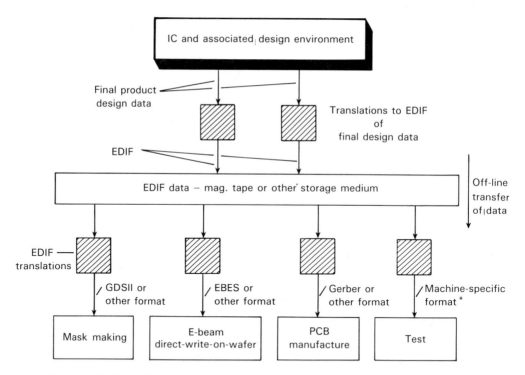

* No standard machine language available

Figure 5.39 The principle of the translation of non-EDIF design data into standard EDIF format for transfer to other in-house or vendor activities. Note that translation to EDIF may not always be necessary to provide off-line transfer of data, but it provides a common format when multiple-sourcing or vendor- or machine-independent interchange of data is desirable.

- *CADDIF format*: developed by Schlumberger-Factron, which is a possible standard for testing machines but not yet extensively adopted. (Many commercial IC testers, however, can accept the simulation data generated during the IC design process as input data.)
- *GDSII format*: developed by the Calma Corporation, which is the most widely adopted standard to describe the geometric details of an IC design for mask-making.
- *CIF (Caltech Intermediate Format)*: developed at the California Institute of Technology, which is an alternative to GDSII for mask-making.

The first two of the above formats are machine-specific, and can therefore be fed directly into the appropriate machine. The others are non-machine specific, and have to be converted into the actual binary data required by a particular machine. Further details of these *ipso facto* standards may be found in the literature [10], [150]–[152].

Standard interchange formats, which can act as a link between individual tools with their own individual formats, currently include the following:

- *IGES (International Graphics Exchange Specification)*: a standard originally developed in the late 1970s for the interchange of data between mechanical CAE systems, particularly relevant for numerically controlled machines and production design data.
- *EDIF (Electronic Design Interchange Format)*: a still-evolving standard covering all activities in the IC design area.
- *SDIF (Stimulus Data Interchange Format)*: aimed at providing a standard format for all IC testing activities.

EDIF is becoming widely accepted in the IC design field, with many CAD vendors now providing EDIF interfaces to their design tools. EDIF adopts a textual format, and is therefore readable by design engineers. It resembles a LSIP programming language and has a hierarchical structure enabling any form of information on circuit topology, geometric details and behaviour to be incorporated. An EDIF file contains a set of libraries, each of which contains a set of cells. Each cell can be described by 'views', with the following seven types of view currently being available:

(a) netlist, for circuit topology details;
(b) schematic, for logic symbol structures;
(c) symbolic, for more abstract representations;
(d) mask layout, for layout geometries;
(e) behaviour, for functional descriptions;
(f) document, for general textual information; and
(g) stranger, for any information which does not fit into the above six views.

Examples of these and the general hierarchy may be found in [10].

To some degree, EDIF overlaps both IGES and SDIF. It can be used for PCB layout data as well as IC layout, and can therefore take over a duty which is also covered by IGES; at the other extreme the SDIF interchange format has been developed to address the difficult data problems involved in automatic testing of ICs [153], [154], having a similar syntax to EDIF but being more specific to the definition of testing requirements. Further developments of EDIF may therefore incorporate more of the attributes currently provided in SDIF, making the future of SDIF less certain. Additionally, a new generation of commercial testers, ideally with the capability of testing both analogue and digital ICs, may promote the development of one standard tester format, which can then be linked more easily to a standard data interchange format.

EDIF is therefore the increasingly recognized international-standard data interchange format for all IC design activities, with continuing developments to enhance its scope and capabilities. As well as the link to testing, work to link with VHDL, and possibly other hardware description languages, is also being considered. Further details may be found in [10], [104], [155], [156].

5.6 CAD costs

It is not feasible to put hard and fast costs against the wide range of CAD hardware and software which is available for electronic design purposes for the following reasons:

- The market is highly competitive, with new or improved products being announced continuously.
- Take-overs and mergers between CAE vendors frequently occur, which modifies the sales basis for many products.
- Non-realistic prices may be charged in order to penetrate new markets or maintain the existing market share for certain products.
- The introduction of new microprocessor chip sets modifies the cost/performance ratio of hardware.
- Software may be provided by IC vendors at non-commercial rates if this is a means of obtaining USIC production orders from an OEM.

What is generally true, however, is that CAD hardware cost have fallen dramatically in the past two decades, first with workstations in competition with mainframe and mini computers, and, later, PCs with additional plug-in enhancements challenging an appreciable part of the market captured by workstations.

It must also be appreciated that hardware and software CAD costs are only one part of the expenditure which an OEM may incur in the design of a microelectronic-based product. Other non-recurring engineering (NRE) costs, such as mask-making costs and staff time, are involved – a consideration of these important NRE costs will form part of the managerial discussions of Chapter 7.

The following paragraphs provide a typical guide to CAD costs. The financial figures quoted should only be taken as being broadly typical, but they do give a general picture of CAD costs [17], [18], [157]–[159].

The first introduction of workstations, by Daisy, Mentor and Valid, involved costs of around $100 000 per seat. (The term 'per seat' is frequently used when quoting CAD/CAE costs; it is the cost of the resources available to one designer, and may be a single stand-alone workstation or the cost of a network of stations divided by the number of stations.) This figure was itself lower than prices previously charged for large Calma or Applicon graphics stations or mainframe or mini computers used for design purposes. By 1987 these costs had reduced to the figures shown in Table 5.7; further reductions in costs have now brought prices down to less than the $10 K mark for some of the base-line versions.

The cost of PCs has also fallen since their introduction. However, the addition of co-processor and graphics enhancement boards plus additional mass storage memory has meant that basic PC costs have to be at least doubled to include these necessary enhancements. Total costs of $5 K–$10 K per seat may thus be involved, blurring the cost distinction between enhanced PCs and workstations [139].

Table 5.7. 1987 costs per seat of workstations, on the basis of a network of four stations per design office and central data bank. These figures are for illustrative purposes only to show how per-seat costs may be calculated; current figures will show a reduction on these 1987 prices

Apollo		DEC		Sun	
Workstation resource					
DN3000-C		VAXstation 2000		3/110LC	
68020 CPU		MicroVAX II CPU		68020 CPU	
68881 FPC		78132 FPC		68881 FPC	
4-Mbyte memory		4-Mbyte memory		4-Mbyte memory	
15-in colour VDU		15-in colour VDU		15-in colour VDU	
Apollo Token Ring or Ethernet		Ethernet		Ethernet	
Domain IX and AEGIS		LAV or NFS		NFS	
		VMS or Ultrix		SunOS	
	$8900		$7900		$15 900
	×4 = $35 600		×4 = $31 600		×4 = $63 600
Central resource					
DSP4000		VAX 8600		3/260S	
68020		with 4-Mbyte memory		68020	
68881				68881	
8-Mbyte memory		4-Mbyte memory add-on		8-Mbyte memory	
348-Mbyte disk		456-Mbyte disk		280-Mbyte disk	
60-Mbyte cart. tape		95-Mbyte cart. tape		60 Mbyte cart. tape	
	$29 400		$470 300		$49 500
Total:	$65 000		$501 900		$113 100
	Price per seat: $16 250		Price per seat: $125 475		Price per seat: $28 275

Source: based upon [139].

The cost of software for custom IC design purposes varies from as low as a few hundred dollars for individual software programs running on PCs, for example SPICE-based simulation programs for simple analogue design purposes, to more than $100 K for the outright purchase of a fully integrated hierarchical suite of programs from a vendor covering all the design activities from behavioural level downwards (see Figure 5.3), including PCB and mask-making data. To a large extent the OEM purchases as much or as little as required, and thus costs have to be individually determined. The average company experience is that purchased software costs are usually at least equal to hardware costs.

CAD for specific duties such as PLD and PCB design is usually PC-based and is available from specialist vendors or distributors as a complete hardware/software package. PLD design package costs may be of the order of $10 K–15 K complete; PCB design package costs can be as low as $3 K–5 K for a simple system, rising to $10 K–15 K for more sophisticated resources. For extremely complex multilayer PCB boards, which are beyond the needs of most OEMs, the total costs of a non-PC-based system can rise to $50 K–100 K or more [101], [144].

Finally, it should be noted that both hardware and software incur annual maintenance charges. Software maintenance and updating is particularly significant since continuous updating is done by the supplier to correct errors or ambiguities, or to improve details of performance. The per-annum costs of CAD/CAE maintenance and updating is usually found to be 10–20% of the purchase cost, and thus the OEM is likely to spend possibly half as much again on maintenance and updating over the life of the resources on top of the original purchase costs.

These financial matters will be considered in Chapter 7, but again it should be stressed that the actual costs quoted here are illustrative only and are subject to continuous change, usually downwards in terms of real value.

5.7 Summary

The viability of custom microelectronics depends upon the availability of proven silicon fabrication methods and upon good CAD resources to enable custom ICs to be accurately and expeditiously designed. The various levels of the latter resources are broadly summarized in Figure 5.40, with the front-end design activities involving creative activities not yet being so well served as the back-end standard manufacturing activities. Simulation forms an essential part throughout the design process, ideally preventing the design activity proceeding from one level to the next without first proving the correctness of the design details.

CAD hardware has developed into the two principal resources, namely the workstation and the PC (see Figure 1.5) with the dividing line between the two blurring rapidly. Future developments are heavily reliant on new processor chip sets produced by the major IC manufacturing companies as off-the-shelf catalogue parts.

Summary

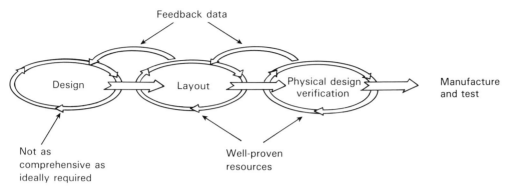

Figure 5.40 A broad summary of the custom IC design activities which are supported by appropriate CAD/CAE resources, with simulation being a key element at each stage.

IC design software remains the greatest problem area for the OEM in deciding what and how much is necessary in terms of in-house activities Certain software packages such as:

- GARDS layout system from Silvar-Lisco for the placement and routing of gate arrays,
- CADAT simulation from HHB Systems for switch, gate and behavioural simulation,
- HILO from GenRad for high-level and macro simulation and for fault simulation,
- the many commercial derivatives of SPICE for detailed transistor-level simulation,

may be considered as representative of the areas considered in the preceding sections, but an OEM would not normally attempt to compile his own software suite from multiple sources due to data interchange difficulties. Completely compatible systems using proprietary software, and also third-party tools, where appropriate, are available from CAE vendors such as Valid Logic, Silvar-Lisco, Viewlogic Systems and others, providing comprehensive resources such as those shown in Figure 5.41.

Data interchange standards remains one of the greatest needs, with EDIF being the principal means so far available. Continuing work to extend the range of EDIF and increasing acceptance by CAE vendors is of particular significance. *Networking* of resources with bus interconnect and local area networks (LANs) is also an increasing feature of design environments, with continuing developments also taking place in these areas.

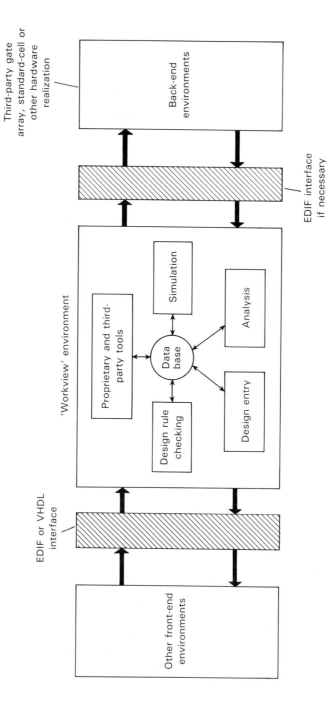

Figure 5.41 The integrated design environment provided by Viewlogic, with facilities to link to other design environments via EDIF and VHDL read/write interfaces, and kits to link to the input requirements of gate-array, standard-cell, PLD, LCA and PCB vendors. (Courtesy of Viewlogic Systems Inc., MA.)

Finally, every IC design has to be tested after fabrication. This also forms a part of the CAD activities, particularly if design-for-test features are built in at the system design stage. However, because of the increasingly important aspect of test in the complete design and fabrication activity we have deliberately left this vital area for the following chapter to give separate coverage commensurate with its importance.

5.8 References

1. Russell, G. (ed.), *Computer Aided Tools for VLSI System Design*, IEE Peter Peregrinus, London, 1987.
2. Hicks, P. J. (ed.), *Semi-Custom IC Design and VLSI*, IEE Peter Peregrinus, London, 1983.
3. Russell, G., Kinniment, D. J., Chester, E. G. and McLauchlan, M. R. (eds.), *CAD for VLSI*, Van Nostrand Reinhold, Wokingham, UK, 1985.
4. Rohrer, D. A., 'Evolution of the electronic design automation industry', *IEEE Design and Test*, Vol. 5, December 1988, pp. 8–13.
5. Newton, A. R. and Sangiovanni-Vincentelli, A. L., 'CAD tools for ASIC design', *Proc. IEEE*, Vol. 75, June 1987, pp. 765–76.
6. Johnson, P., 'Mixed-level design tools enhance top–down design', *Computer Design*, Vol. 28, January 1989, pp. 85–8.
7. Harding, W., 'System simulation ensures chips play together', *ibid.*, Vol. 28, August 1989, pp. 70–83.
8. Dettmer, R., 'Logic synthesis: a faster route to silicon', *IEE Review*, Vol. 35, December 1989, pp. 427–30.
9. VLSI System Design, *User's Guide to Design Automation*, CMP Publications, Manhasset, NY, 1988.
10. Rubin, S. M., *Computer Aids for VLSI Design*, Addison-Wesley, Reading, MA, 1987.
11. Engl, W. L. (ed.), *Process and Device Modeling*, Advances in CAD for VLSI Book Series, Vol. 1, North-Holland, Amsterdam, 1986.
12. Hörbst, E. (ed.) *Logic Design and Simulation*, *ibid.*, Vol. 2, 1986.
13. Ruehli, A. E. (ed.), *Circuit Analysis, Simulation and Design*, *ibid.*, Vol. 3 (2 parts), 1986.
14. Ohtsuki, T. (ed.), *Layout Design and Verification*, *ibid.*, Vol. 4, 1986.
15. Goto, S. (ed.), *Design Methodologies*, *ibid.*, Vol. 6, 1986.
16. Hartenstein, R. W., *Hardware Description Languages*, *ibid.*, Vol. 7, 1986.
17. The Open University, *CAD for Custom Chips*, Microelectronics for Industry, Publication PT505CAD, The Open University, UK, 1988.
18. The Open University, *Computers in the Design Process*, Microelectronics for Industry, Publication PT506COM, The Open University, UK, 1989.
19. Marcus, C., *Prolog Programming*, Addison-Wesley, Reading, MA, 1987.
20. Kelley, A. and Pohl. I., *C by Dissection: The essentials of C programming*, Addison-Wesley, Reading, MA, 1987.
21. Meyer, E., 'VHDL strives to cover both synthesis and modeling', *Computer Design*, Vol. 28, October 1989, pp. 42–5.
22. Harding, W., 'Logic synthesis forces rethinking of design tools', *ibid.*, December 1989, pp. 51–7.
23. IEEE Publication, *VHDL Language Reference Manual, Draft Standard 1076/*B, IEEE Computer Society, Publication Department, Los Angeles, CA, 1987.

24. Shahdad, M., Lipsett, R., Marshuer, E., Sheehan, K., Cohen, H., Waxman, R. and Ackley, D., 'VHSIC hardware description language', *IEEE Computer*, Vol. 18, February 1985, pp. 94–103.
25. Hookway, R. J., 'System simulation using VHDL', *Journal of Semicustom ICs*, Vol. 5, September 1987, pp. 23–9.
26. Sullavan, R. and Asher, L. R., 'VHDL for ASIC design and verification', in *Semicustom Design Guide*, CMP Publications, Inc., Manhasset, NY, 1988.
27. Dettmer, R., 'ELLA, a language for VLSI', *IEE Electronics and Power*, Vol. 31, July 1986, pp. 517–22.
28. Robertson, S., 'Hardware representation and design control using the ELLA™ system', *Journal of Semicustom ICs*, Vol. 5, March 1988, pp. 18–22.
29. Lightner, M. R., 'Modelling and simulation of VLSI digital systems', *Proc. IEEE*, Vol. 75, June 1987, pp. 786–96.
30. Clare, C., *Designing Logic Systems using State Machines*, McGraw-Hill, NY, 1972.
31. Green, D., *Modern Logic Design*, Addison-Wesley, Wokingham, UK, 1986.
32. Muroga, S., *Logic Design and Switching Theory*, Wiley, NY, 1979.
33. McClusky, E., *Logic Design Principles*, Prentice-Hall, NJ, 1986.
34. Brayton, R., Hachtel, G., McMullen, C. and Sangiovanni-Vincentelli, A., *Logic Minimization Algorithms for VLSI Synthesis*, Kluwer Academic Publishers, MA, 1984.
35. Rudell, R., *EXPRESSO IIC User's Manual*, University of California at Berkeley, Dept of EECS, 1986.
36. Poretta, A., Santomauro, M. and Somenzi, F., 'TAU, a fast heuristic logic minimizer', *Proc. IEEE Int. Conf. on CAD*, November 1984, pp. 206–8.
37. Bouhasin, G., 'EDA pushes towards logic synthesis', *User's Guide to Design Automation*, CMP Publications, Manhasset, NY, 1988, pp. 10–16.
38. Kernighan, B. S. and Lin, W., 'An efficient heuristic procedure for partitioning graphs', *Bell System Technical Journal*, Vol. 49, 1970, pp. 291–307.
39. Breur, M. A. (ed.), *Design and Automation of Digital Systems*, Prentice-Hall, NJ, 1972.
40. Schwartz, A. F., *Computer-Aided Design for Microelectronic Circuits and Systems*, Vol. 1 (736 pp.) and Vol. 2 (772 pp.), Academic Press, London, UK, 1987.
41. Breur, M. A., 'Min-cut placement', *J. Design Automation and Fault Tolerant Computing*, Vol. 1, 1977, pp. 343–62.
42. Lauther, U., 'A min-cut placement algorithm for general cell assemblies', *ibid.*, Vol. 4, 1980, pp. 21–34.
43. Sechen, C. and Lee, K.-W., 'An improved simulated annealing algorithm for row-based placement', *Proc. IEEE Int. Conf. Computer-Aided Design*, 1987, pp. 478–81.
44. Veechi, M. P. and Kirkpatrick, S., 'Global wiring by simulated annealing', *IEEE Trans. CAD*, Vol. CAD2, 1987, pp. 165–72.
45. Rutman, R. A., 'An algorithm for placement based upon minimum wire length', *Proc. AFIPS Conf. SJCC*, 1964, pp. 477–91.
46. Wilson, D. C. and Smith, R. J., 'An experimental comparison of force directed placement techniques', *Proc. IEEE Design Automation Workshop*, 1974, pp. 194–9.
47. Chyan, D.-Y. and Breur, M. A., 'A placement algorithm for array processors', *Proc. IEEE Design Automation Conf.*, 1983, pp. 182–8.
48. Tsay, R.-S., Kuh, E. S. and Hsu, C.-P., 'PROUD: a sea-of-gates placement algorithm', *IEEE Design and Test of Computers*, Vol. 5, December 1988, pp. 44–56.
49. Macaluso, E., 'Graphical floorplan design of cell-based ICs', in *User's Guide to Design Auomation*, CMC Publications Inc., Manhasset, NY, 1988.

50. Praes, B. T. and van Cleemput, W. M., 'Placement algorithms for arbitrary shaped blocks', *Proc. IEEE Design Automation Conf.*, 1979, pp. 474–80.
51. Kuh, E. S. and Marek-Sadowski, M., 'Global routing', in [14].
52. Massara, R. E. (ed.), *Design and Test Techniques for VLSI and WLSI*, IEE Peter Peregrinus, London, 1989.
53. van Cleemput, W. M., 'Mathematical models for the circuit layout problem', *Trans. IEEE*, Vol. CAS23, 1976, pp. 759–67.
54. Chen, N. P., 'New algorithms for Steiner trees', *Proc. IEEE ISCAS*, 1983, pp. 1217–19.
55. Garey, M. R. and Johnson, D. S., *Computers and Intractability: A guide to the theory of NP-completeness*, Freeman, CA, 1979.
56. Loberman, H. and Weinberger, A., 'Formal procedures for connecting terminals with a minimum total wire length', *J. ACM*, Vol. 4, 1957, pp. 428–37.
57. Kruska, J. B., 'On the shortest spanning subtree of a graph and the travelling salesman problem', *Proc. American Math. Society*, Vol. 7, 1956, pp. 48–50.
58. Abel, L. C., 'On the ordering of routes for automatic wiring routing', *IEEE Trans.*, Vol. C21, 1972, pp. 1227–33.
59. Lee, C. Y., 'An algorithm for path connection and its application', *IEEE Trans. Computers*, Vol. EC10, 1961, pp. 346–65.
60. Akers, S. B., 'A modification of Lee's path connection algorithm', *ibid.*, Vol. EC16, 1967, pp. 97, 98.
61. Hoel, J. H., 'Some variations on Lee's algorithm', *ibid.*, Vol. C25, 1976, pp. 19–24.
62. Hightower, D. W., 'A solution to line-routing problems on the continuous plane', *Proc. IEEE Design Automation Conf.*, 1969, pp. 1–24.
63. Heyns, W., Sansen, W. and Beke, H., 'A line expansion algorithm for the general routing problem, with a guaranteed solution', *ibid.*, 1980, pp. 243–9.
64. Hashimoto, A. and Stevens, J., 'Wire routing by optimising channel assignment with large apertures', *Proc. IEEE Design Automation Conf.*, 1971, pp. 155–69.
65. Kernigham, S., Schweikert, D. and Persky, G., 'An optimum channel routing algorithm for polycell layouts of integrated circuits', *ibid.*, 1973, pp. 50–9.
66. Deutsch, D. N., 'A dog-leg channel router', *ibid.*, 1976, pp. 425–33.
67. Yoshimura, T. and Kuh, E. S., 'Efficient algorithm for channel routing', *IEEE Trans.*, Vol. CAD1, 1982, pp. 25–35.
68. Rivest, R. L. and Fiduccia, C. M., 'A "greedy" channel router', *Proc. IEEE Design Automation Conf.*, 1982, pp. 418–24.
69. Burstein, M. and Pelavia, R., 'Hierarchical channel router', *ibid.*, 1983, pp. 591–7.
70. Burstein, M., 'Channel routing', in [14].
71. Hurst, S. L., *Custom Specific Integrated Circuits*, Marcel Dekker, NY, 1983.
72. Soukup, J., 'Circuit layout', *Proc. IEEE*, Vol. 69, 1981, pp. 1281–304.
73. Antognetti, P., Pederson, D. O. and de Man, H. (eds.), *Computer Design Aids for VLSI Circuits*, Martin Nijhoff, Boston, MA, 1984.
74. Rothermel, H.-J. and Mlynski, D. A., 'Routing method for vlsi design using irregular cells', *Proc. IEEE Design Automation Conf.*, 1983, pp. 257–62.
75. Damnjanović, M. S. and Litorski, V. B., 'A survey of routing algorithms in custom IC design', *J. Semicustom ICs*, Vol. 7, December 1989, pp. 10–19.
76. Yoshida, K., 'Layout verification', in [14].
77. Haskard, M. R. and May, I. C., *Analog VLSI Design: nMOS and CMOS*, Prentice-Hall, NJ, 1988.
78. Allan, P. and Holberg, D., *CMOS Analog Circuit Design*, Holt, Reinhart and Winston, NY, 1987.
79. Nordholt, E. H., *Design of High-Performance Negative Feedback Amplifiers*, Elsevier, Amsterdam, 1983.

80. Trontelj, J., Trontelj, L. and Shenton, G., *Analog Digital ASIC Design*, McGraw-Hill, NY, 1989.
81. Gregorian, R. and Temes, G. C., *Analog MOS Integrated Circuits for Signal Processing*, Wiley, NY, 1986.
82. Grebene, A. B., *Bipolar and MOS Analog Integrated Circuit Design*, Wiley, NY, 1984.
83. Gayakwad, R. A., *Op-Amps and Linear Integrated Circuits*, Prentice-Hall, NJ, 1988.
84. Gray, P. R. and Meyer, R. G., *Analysis and Design of Analog Integrated Circuits*, Wiley, NY, 1984.
85. Tsividis, Y. and Antognetti, Y., *Design of MOS VLSI Circuits for Telecommunications*, Prentice-Hall, NJ, 1985.
86. Bray, D. and Irissou, P., 'A new gridded bipolar linear semiconductor array family with CAD support', *J. Semicustom ICs*, Vol. 4, June 1986. pp. 13–20.
87. Crolla, P., 'A family of high-density, tile-based bipolar semicustom arrays for the implementation of analogue integrated circuits', *ibid.*, December 1987, pp. 23–9.
88. Allen, P. E., 'Computer-aided design of analogue integrated circuits', *ibid.*, December 1986, pp. 22–31.
89. Spence, R. and Soin, R. S., *Tolerance Design of Electronic Circuits*, Addison-Wesley, MA, 1988.
90. Trontelj, L., Trontelj, J. *et al*, 'Analogue silicon compiler for switched-capacitor circuits', *Proc. ICCAD*, 1987, pp. 506–9.
91. Assael, J., 'A switched-capacitor filter silicon compiler', *IEEE J. Solid-State Circuits*, Vol. 23, 1988, pp. 116–74.
92. Weder, U. and Möschwitzer, A., 'SCF, a gate-array switched-capacitor filter design tool', *J. Semicustom ICs*, Vol. 7, March 1990, pp. 15–21.
93. Davidse, J. and Nordholt, E. H., 'Basic considerations concerning the application of semicustom IC techniques for the processing of analogue electronic signals', *ibid.*, Vol. 5, December 1987, pp. 5–11.
94. Habekotté, E., Hoefflinger, B., Klein, H.-W. and Beunder, M. A., 'State of the art in analog CMOS circuit design', *Proc. IEEE*, Vol. 75, 1987. pp. 816–28.
95. Tong, C. S., 'A new MAX-EPLD architecture which provides logic density, speed and flexibility', *J. Semicustom ICs*, Vol. 7, September 1989, pp. 12–19.
96. Bolton, M. J. P., *Digital System Design with Programmable Logic*, Addison-Wesley, MA, 1990.
97. Bostock, G., *Programmable Logic Handbook*, Blackwell, Oxford, 1987.
98. Advanced Micro Devices, *Programmable Logic Handbook*, Advanced Micro Devices, CA, 1987.
99. Monolithic Memories, *PAL Programmable Logic Handbook*, Monolithic Memories Inc., CA, 1983.
100. Philips Components, *Programmable Logic Devices*, Philips Components, London, 1986.
101. Xilinx, *Programmable Gate Array Handbook*, Xilinx Corporation, CA, 1989.
102. Altera, *The MAXimalist Handbook*, Altera Corporation, CA, 1990.
103. Osann, R., 'Designer's guide to programmable logic', Parts 1, 2 and 3, EDN January/February, 1985, reprinted in [98].
104. Stump, H., 'A designer's guide to simulation models', *Computer Design*, Vol. 29, January 1990, pp. 91–8.
105. Meyer, E., 'Mixed-signal simulators take divergent paths', *ibid.*, January 1990, pp. 49–56.
106. The Open University, *System and Logic Simulation*, Microelectronics for Industry, Publication PT505SIM, The Open University, UK, 1988.

107. Russell, G. and Sayers, I. L., *Advanced Simulation and Test Methodologies for VLSI Design*, Van Nostrand Reinhold, UK, 1989.
108. *IEEE Design and Test*, Vol. 4, August 1987, special issue *Modeling and Switch Level Testing*.
109. Sedra, A. D. and Smith, K. C., *Microelectronic Circuits*, Holt, Reinhart and Winston, Toronto, 1987.
110. Hodges, D. A. and Jackson, H. G., *Analysis and Design of Digital Integrated Circuits*, McGraw-Hill, NY, 1983.
111. The Open University, *Circuit and Device Modelling*, Microelectronics for Industry, Publication PT505MOD, The Open University, UK, 1988.
112. Vladimirescu, A., Newton, A. R. and Pederson, D. O., 'SPICE version 2G.1 users' guide', University of California, Berkeley, Dept of EE and CS, 1980.
113. Di Giacomo, J. (ed.), *VLSI Handbook*, McGraw-Hill, NY, 1989.
114. Hitchcock, R. B., 'Timing verification and the timing analysis program', *Proc. IEEE Design Automation Conf.*, 1982, pp. 594–603.
115. Tamura, E., Ogawa, K. and Nakano, T., 'Path delays analysis for hierarchical building block layout system', *ibid.*, 1983, pp. 403–10.
116. Wei, Y-P., Lyons, C. and Hailey, S., 'Timing analysis of VLSI circuits', *VLSI System Design*, Vol. 8, August 1987, pp. 52–8.
117. Friedman, M., 'A swifter way to simulate analog-and-digital ICs', *Electronic Products*, Vol. 10, 15 September 1987, pp. 28–33.
118. Vucurevich, T., 'SPECTRUM: a new approach to event-driven analog/digital simulation', *Proc. IEEE Custom Integrated Circuits Conf.*, 1990, pp. 5.1.1–5.1.5.
119. Kurker, C. M., Paulos, J. J., Cohen, B. S. and Conley, E. S., 'Development of an analog hardware language', *ibid.*, pp. 5.4.1–5.4.6.
120. Moser, L., 'Behavioural analog circuit models for multiple simulation environments', *ibid.*, pp. 5.5.1–5.5.4.
121. Blank, T., 'A survey of hardware accelerators used in computer aided design', *IEEE Design and Test of Computers*, Vol. 1, No. 3, 1984, pp. 21–39.
122. Pfister, G. F., 'The Yorktown simulation engine: an introduction', *Proc. IEEE Design Automation Conf.*, 1982, pp. 51–4.
123. Frank, E. H., 'Exploiting parallelism in a switch-level simulation machine', *ibid.*, 1986, pp. 20–6.
124. Johnson, P., 'Software vs hardware models for system simulation', in [9].
125. Rossbach, P. C., Linderman, R. W. and Gallacheer, D. M., 'An optimising XROM silicon compiler', *Proc. IEEE Custom Integrated Circuits Conf.*, 1987, pp. 13–16.
126. Kang, S. and van Cleemput, W. M., 'Automatic PLA synthesis from a DDL-P description', *Proc. IEEE Design Automation Conf.*, 1981, pp. 391–7.
127. Leith, J. W., 'Crystal oscillator compiler', *Proc. IEEE Custom Integrated Circuits Conf.*, 1987, pp. 17–19.
128. Rabaey, J., Vanhoof, J., Goossens, G., Catthoor, F. and De Man, H., 'CATHEDRAL II computer aided synthesis of digital signal processing systems', *ibid.*, pp. 157–60.
129. Helms, W. J. and Byrkett, B. E., 'Compiler generation of A to D converters', *ibid.*, pp. 161–4.
130. Trimberger, S., 'Automating chip layout', *IEEE Spectrum*, Vol. 19, 1982, pp. 38–45.
131. Seattle Silicon, *CONCORDE Documentation*, Seattle Silicon Corporation, Bellevue, WA.
132. Silicon Computers, *Genesil Documentation*, Silicon Compiler Systems Corporation, San Jose, CA.

133. Gajski, D. D. (ed.), *Silicon Compilation*, Addison-Wesley, MA, 1987.
134. Parker, A. C. and Hayati, S., 'Automating the VLSI design process using expert systems and silicon compilation', *Proc. IEEE*, Vol. 75, 1987, pp. 777–85.
135. Gajski, D. D., Dutt, N. D. and Pangrle, B. M., 'Silicon compilation: a tutorial', *Proc. IEEE Custom Integrated Circuits Conf.*, 1986, pp. 453–9.
136. Evanczuk, S., 'Results of a silicon compiler design challenge', *VLSI Systems Design*, Vol. 6, July 1985, pp. 46–54.
137. Bosworth, M. F., 'The management of documentation and design databases', *J. Semicustom ICs*, Vol. 4, No. 1, September 1986, pp. 13–17.
138. Burgess, L., 'PCs for CAE – how powerful are they?', *ibid.*, Vol. 4, No. 2, December 1986, pp. 44–8.
139. Mokhoff, N., 'Differences blur as PCs take on engineering workstations', *Electronic Design*, Vol. 35, September 1987, pp. 15–31.
140. Editorial Staff, 'Multi-MIPS workstations under $15 K', *VLSI System Design*, Vol. 8, August 1987, pp. 78–81.
141. Watson, G. and Cunningham, D., 'FDDI and beyond', *IEE Review*, Vol. 36, 1990, pp. 131–4.
142. Andrews, W., 'Bridging today's buses to Futurebus', *Computer Design*, Vol. 29, No. 3, 1990, pp. 72–84.
143. Bresford, R., 'Circuit simulators at a glance', *VLSI Systems Design*, Vol. 8, August 1987, pp. 76, 77.
144. Personal CAD Systems, booklet *Answers to the Most Commonly Asked Questions on ICB CAD*, Personal CAD Systems, Inc., San Jose, CA, 1989.
145. Kirkpatrick, J. M., *Electronic Drafting and Printed Circuit Board Design*, Chapman and Hall, NY, 1989.
146. Clark, R. H., *Printed Circuit Engineering*, Chapman and Hall, NY, 1989.
147. Sloan, J. L., Design and Packaging of Electronic Equipment, Chapman and Hall, NY, 1985.
148. O'Reilly, W. P., *Computer Aided Electronic Engineering*, Chapman and Hall, NY, 1986.
149. Browne, J., Harhen, J. and Shivnan, J., *Product Management Systems: a CIM perspective*, Addison Wesley, Reading, MA, 1988.
150. Geber, S.I.C., *Gerber Format*, Gerber Scientific Instrument Company, Document No. 40101-S00-066A, 1983.
151. Factron, *CADDIF Version 2.0 Engineering Specifications*, Schlumberger-Factron Corporation, 1985.
152. Calma, *GDS II Stream Format*, Calma Corporation, 1984.
153. Parker, K. P., *Integrated Design and Test: Using CAE tools for ATE programming*, IEEE Computer Society Press, Washington DC, 1987.
154. Peiper, C., 'Stimulus Data Interchange Format (SDIF)', *VLSI System Design*, Vol. 7, July 1986, pp. 76–81 and August 1986, pp. 56–60.
155. Newton, A. R., 'Electronic Design Interchange Format: Introduction to EDIF 2.0.0', *Proc. IEEE Custom Integrated Circuits Conf.*, May 1987, pp. 571–5.
156. Etherington, E., 'Interfacing design to text using the Electronic Design Interchange Format (EDIF)', *Proc. IEEE Int. Test Conf.*, September 1987, pp. 378–83.
157. The Open University, *Microelectronic Decision*, Microelectronics for Industry, Publication PT505MED, The Open University, UK, 1988.
158. VLSI System Design, *Survey of Commercial CAD Systems*, usually published annually in the June edition.
159. McClean, W. J. (ed.), *ASIC Outlook 1990*, Integrated Circuit Engineering Corporation, Scotsdale AZ, 1990.

6 Test-pattern generation and design-for-testability

6.1 General introduction

The need to test custom microelectronic circuits has been noted previously, but, due to its significance, has been left until now in order to consider the whole subject area within one chapter. As has been seen, for example in Figure 5.3, aspects of testing may be involved in the CAD resources used during the design phase of a custom IC, but, in any case, these requirements must always be considered as early as possible in the design of any new circuit.

The testing of any product involving one or more ICs usually involves some verification of the individual ICs before product assembly, plus some final test of the product before despatch. The only exception to this may be in very cheap 'give-away' goods, where the cost of individual testing is not justified. Hence, normal testing procedures involve both the OEM and the vendor, the former doing the final product test, and the vendor, and possibly the OEM, doing the individual IC tests.

Figure 6.1 shows the usual prototype and production test procedures. The total testing of the USIC involves the following:

- tests to ensure that all the fabrication steps have been implemented correctly during routine wafer manufacture (fabrication checks);
- tests to ensure that the prototype ICs are functional-satisfactory in all respects to meet the product requirements (design checks); and
- tests to ensure that subsequent production ICs finally used in the product are defect-free (production checks).

The first of these three testing categories is the province of the vendor, and does not involve the OEM in any way. The vendor will ensure that the wafer containing the custom ICs has 'drop-ins' located at scattered points on the wafer, these being small circuits or structures from which the correct resistivity and other parameters of the wafer fabrication can be checked before any more comprehensive tests are performed on the surrounding circuits. This is illustrated in Figure 6.2. In addition, every mask used in the wafer fabrication will have some identification symbol on

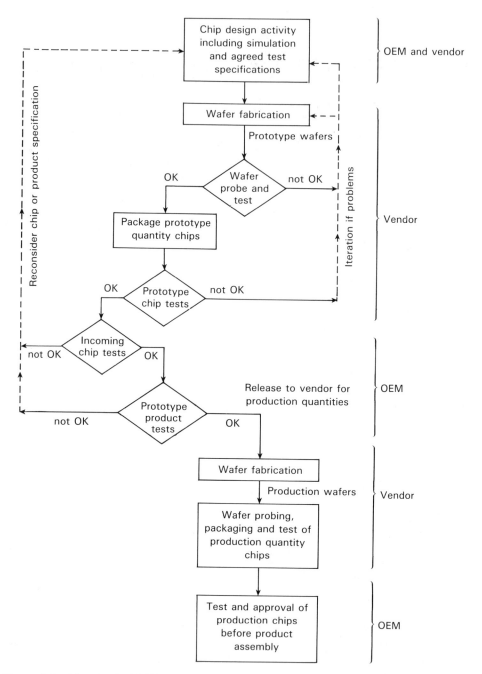

Figure 6.1 The vendor/OEM test procedures for a custom IC before assembly into the final production equipment. Ideally, the OEM testing should be a 100% fully functional test in a working equipment or equivalent test rig, although this may not always be possible.

General introduction 273

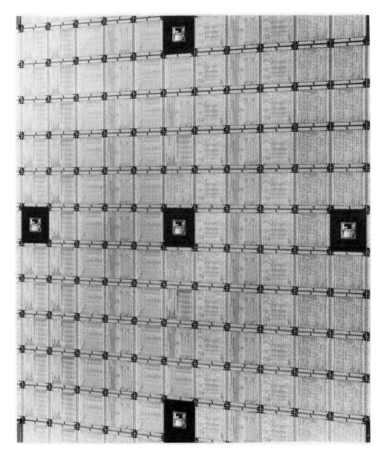

Figure 6.2 The vendor's means to check on correct wafer fabrication, using 'drop-in' special test circuits at selected points on the wafer. These 'drop-in' circuits may alternatively be known as 'process evaluation devices' (PEDs), 'process device monitors' (PDMs), 'process monitoring circuits' (PMCs), or similar terms by different authorities. (Courtesy of Micro Circuit Engineering Ltd., UK.)

it, which enables the vendor to confirm that all masks required have been correctly used.

The remaining approval procedures on the individual ICs are joint vendor/OEM activities, as shown in Figure 6.1. However, these activities involve the following considerations:

- the size and complexity of the custom circuit and how it will be tested;
- which tests are to be undertaken by the vendor and which by the OEM;
- how much the OEM is prepared to pay the vendor for testing;

- how many tests the vendor is willing to apply to each production USIC before shipping to the OEM;
- what resources the OEM has for both IC and product testing.

All these factors must be considered at the design stage and cannot be left until after the design has been completed.

In the case of very small USICs containing, say, only a few hundred digital logic gates or a very small number of analogue circuits, the problem is not very acute. With such small circuits the OEM can usually undertake a fully functional check by plugging each IC into a test rig or even a final product held for test purposes, and checking the correct operation of the complete product. However, if production quantities increase, then there may be pressure to do a restricted series of tests on each IC, which will not exhaustively test its correct functionality. Which restricted tests to do are a matter for discussion, since there is always a possibility of some faulty circuits being despatched to final users.

Production testing of the IC is necessary because it is impossible to guarantee that the fabrication is completely free of defects. If the wafer manufacturing processes and the subsequent chip bonding and packaging procedures were perfect, then there would be no need to do any testing − every circuit would be fully working, assuming that the original design had been fully verified. Such perfection cannot be achieved with such complex fabrication processes, and thus testing of all production ICs is necessary.

However, unless a fully comprehensive, fully exhaustive test of an IC is undertaken there always remains the possibility that a circuit will pass the given tests but will still not be completely fault-free. The lower the yield of the production process and the less exhaustive the testing, then the greater will be the probability of not detecting faulty circuits during test.

This probability may be mathematically analysed. Suppose the production yield of fault-free circuits is Y, where Y has some value between 0 (all circuits faulty) and 1 (all circuits fault-free). Suppose also that the tests applied to the circuit have a fault coverage (test efficiency) of FC, where FC also has some value between 0 (the tests do not detect any possible faults) and 1 (the tests detect all possible faults). Then the percentage of circuits which will pass the tests but will still be faulty circuits (the 'defect level' after test, DL) is given by

$$DL = \{1 - Y^{(1-FC)}\} \times 100\%$$

This equation is shown in Figure 6.3.

The significance of this very important probability relationship is as follows. Suppose the production yield was 25% ($Y = 0.25$). Then if no testing at all was done ($FC = 0$), 75% of the ICs would be faulty when they come to be used. If testing is now done and the efficiency of the tests is 90% ($FC = 0.9$), then the percentage of faulty circuits when they come to be used has dropped to about 15% (85% fault-free, 15% faulty). This is still about one IC in seven faulty, which is far too high

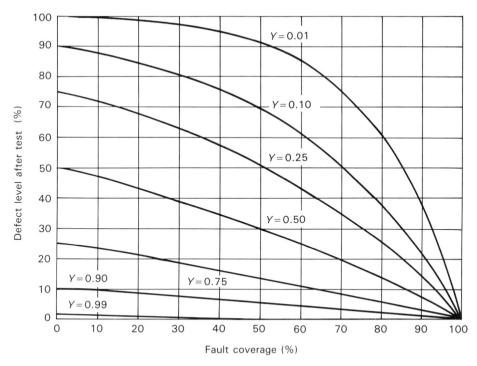

Figure 6.3 The defect level (*DL*) of a circuit after test, with different manufacturing yields (*Y*) and fault detection coverage (*FC*). Note, only $FC = 100\%$ or $Y = 100\%$ will guarantee 100% functional circuits (zero defects) after test.

for most manufacturers. Hence, to ensure a very high percentage of fault-free circuits after test, either the manufacturing yield *Y* must be high, or the fault coverage *FC* under test must be high, or both.

Unfortunately *Y* will always be low due to the complexities of the fabrication processes. Therefore, to ensure a very low percentage of faulty ICs after test, the testing efficiency *FC* must be high.

This, therefore, is the dilemma of testing complex integrated circuits. If the testing efficiency is low, faulty ICs will not be detected. But achievement of 100% fault coverage ($FC = 1.0$) may require such extensive testing as to be non-cost effective.

To revert to the problem of acceptance testing of the first prototype USIC: during the design phase, simulation of the circuit will have been undertaken and approved before any fabrication was commenced. This procedure invariably involves the vendor's CAD resources for the final simulation check, and from this simulation a set of tests for the chip may be automatically generated which can be down-loaded into the vendor's test equipment (see Figure 6.4). The vendor's tests on prototype circuits may therefore be based on this pre-manufacture simulation

276 Test-pattern generation and design-for-testability

Figure 6.4 A comprehensive general-purpose VLSI test system as may be used by USIC vendors. The wafers under test are carried on test jigs on the carousel-type structure, with the magnetic tape unit containing the specific test data for the circuits on the wafer. (Courtesy of Avantest Corporation, Japan.)

data, and if the circuits pass, this means that they conform to the data approved by the OEM.

Unfortunately, prototype custom ICs which pass the vendor's tests often fail to satisfy the OEM's tests when tested under working-product conditions. This is not to say that the ICs are faulty, but rather that they were not designed in the first place to provide the exact functionality or performance required in the final product. The original system design or the IC specification was somehow incomplete or faulty, perhaps in a very minor way, such as a logic 0 input condition being specified instead of a logic 1, or perhaps due to last-minute product changes not being incorporated in the IC design specification. Historically, this has been the main reason why custom ICs fail to meet OEM requirements, and it thus demands close managerial control during the product design phase to ensure final compatibility of product and custom circuit.

In production testing the vendor is primarily interested in being able to test the production USICs as rapidly as possibly with the minimum number of test input vectors, using expensive in-house general-purpose test equipment. For these tests a minimum set of test vectors is usually sought, this minimum test set being the subject of discussion and agreement between the OEM and the vendor. It may, for example, be a set of input conditions suggested by the OEM which checks that the IC performs a certain series of key functions correctly, or it may be a minimum test set based upon fault models (see Section 6.3.2) which will detect all specific faults within the circuit. However, problems may still arise, such as the following:

- The vendor's general-purpose, computer-controlled test equipment may not be capable of applying some of the input conditions met in the final product, particularly analogue input signals.
- Similarly, some of the output conditions which the custom circuit provides may not be precisely monitored by the vendor's test equipment.
- The vendor's test equipment may not have the capability to check the custom circuit at the operating speeds of the final product.

All these aspects become more serious as the size and performance of the custom IC increases and/or when mixed analogue/digital designs are involved.

In the following sections we will consider the fundamentals of testing in more detail, and how design-for-test becomes increasingly necessary for large circuits. However, in all cases detailed discussions between OEM and vendor are always necessary to ensure acceptable testing details. Further general details on testing may be found in [1]-[8].

6.2 Basic testing concepts

6.2.1 *Digital circuit test*

Before continuing with a discussion of digital test methods, it may be appropriate to clarify the following three terms:

1. Input test vector (or input vector or test vector): this is a combination of logic 0 and 1 signals applied in parallel to the accessible ('primary') inputs of the circuit under test. For example, if eight primary inputs are present, then one test vector may be 01101110. A test vector is the same as a word, but the latter term is rarely used in connection with test.
2. Test pattern: a test pattern is the same as a test vector but with the addition of the fault-free output response of the circuit to the input test vector. For example, if there are four primary outputs, then with the above input test vector the four expected outputs may be 0, 0, 0, 1.
3. Test set: a test set is a set of test patterns which in total should determine whether the given circuit under test is fault-free or faulty. A test set may be fully exhaustive, reduced or minimum, see Section 6.3 following. Continuing the above example, a test set may begin as shown in Table 6.1. Unfortunately, the terms 'vectors', 'patterns', and 'sets' are sometimes loosely used, and thus care is necessary in reading literature from different sources.

Consider the test of a simple network containing combinational logic only (no latches or other bistable circuits), as shown in Figure 6.5. With n binary inputs 2^n

Table 6.1. An example test set for a combinational network with eight inputs and four outputs

	Test vectors $x_1\ x_2\ x_3\ x_4\ x_5\ x_6\ x_7\ x_8$	Test response $y_1\ y_2\ y_3\ y_4$
First test pattern	0 1 1 0 1 1 1 0	0 0 0 1
Next test pattern	0 1 1 0 1 1 1 1	0 0 1 1
Next test pattern	1 0 0 1 1 1 1 1	1 0 0 1
•	•	•
•	•	•
•	•	•

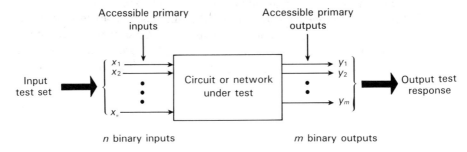

Figure 6.5 The test of a simple combinatorial circuit containing logic gates only.

input test vectors are required to test the circuit exhaustively – an exhaustive input test set – with the output response being checked on each of these 2^n input vectors. If $n = 20$, then it is necessary to apply and check the response to approximately one million test vectors, which, in effect, means checking through a one-million-line truthtable. (For CMOS logic gates using pairs of p-channel and n-channel transistors, there is also a question of certain open-circuit faults which are difficult to detect and which may require more than 2^n input vectors for a complete test. This will be referred to again in Section 6.3.2.)

The output response check may be undertaken in two distinct ways. It may be possible to check the response of the circuit under test by comparing its output(s) with that of a known good circuit, sometimes termed a 'gold unit', as shown in Figure 6.6(a). This procedure may be used for the production line testing of standard off-the-shelf SSI and MSI circuits, but for more complex circuits and for custom ICs it is more appropriate to use a general-purpose, computer-controlled test resource which can be programmed to apply the required test set, monitoring the output response(s) on every input test vector. This is illustrated in Figure 6.6(b).

Because every input test vector and the associated healthy circuit response has to be programmed into a general-purpose tester, this can involve the storage and recall of very large amounts of test data, particularly when sequential circuits (see below) are also involved. Hence there is usually pressure to formulate some reduced

Basic testing concepts 279

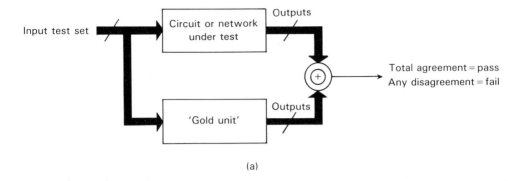

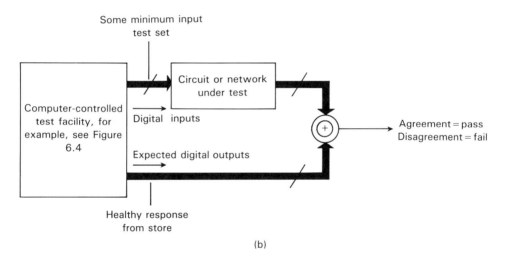

Figure 6.6 Output response methods: (a) testing against a known good circuit; the input test set is usually an exhaustive test; (b) testing against the stored healthy response to the circuit held in the computer memory; the test may be exhaustive or more usually some reduced test set in order to save testing time.

or minimum test set which requires less storage and can be applied more quickly, but which will still ensure an acceptable test of the circuit.

By definition, sequential circuits involve bistable (storage) circuits ('latches' or 'flip-flops') to remember present states and to influence the next state of the network. For a network containing s bistable circuits it is theoretically necessary to test all 2^s possible combinations of states in order to provide an exhaustive test. This is clearly necessary if all s circuits form a single counter, but is less obvious if these circuits are scattered throughout the network in smaller groups. Nevertheless, there may be bridging faults between logic gates which are common to two or more groups, and thus a test of all possible combinations of states may still be needed.

However, real-life circuits invariably consist of both combinational logic and bistable circuits as shown in Figure 6.7, and both need to be tested together unless means are provided to separate them under test conditions (see Section 6.6.2 following). To test a combined combinatorial/sequential network with n primary inputs and s internal latches under all possible input conditions and internal data storage, now requires $2^n \times 2^s$ input test vectors for fully exhaustive testing. This requirement has been considered on p. 209, where it was shown that a simple 16-bit accumulator circuit theoretically required $2^{35} = 34\,359\,738\,368$ input test vectors to test the circuit exhaustively through all its possible input conditions and internal stored states. The correct 16-bit output response on every one of these 2^{35} input test vectors would also need to be checked, which is a very tedious and time-consuming procedure for such a simple basic circuit.

There is, however, one very common sequential circuit which can be successfully tested with other than 2^s clock pulses to generate 2^s states. This is the shift register configuration, see Figure 6.8. Since the action of a healthy shift register is that the state of a flip-flop is always passed on to its neighbour on receipt

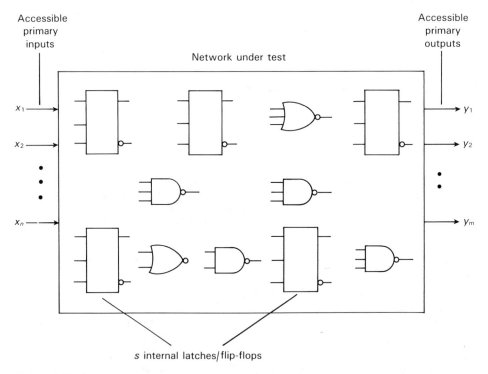

Figure 6.7 A network containing both combinatorial logic and latches/flip-flops. Under test, only the primary inputs and outputs are accessible, with the internal connections to and from the storage circuits (the secondary inputs and outputs, see Figure 6.26) not directly available for test purposes.

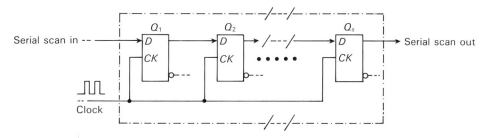

Figure 6.8 A s-stage shift register assembly which may be tested by scanning in a serial data stream at the left and checking that the same data stream emerges s clock pulses later at the output.

of a clock pulse, it is only necessary to test that logic 0s and 1s can be passed through the register from beginning to end, and that each stage can hold a logic 0 or a logic 1 when its immediate neighbours are in the opposite logic state. A sequence of tests may thus be as follows:

1. Clear all the stages in the shift register to logic 0, and then shift through a single 1 with $s + 1$ clock pulses, thus ensuring that 1 is possible surrounded by near-neighbour 0s.
2. Set all stages in the register to logic 1, and then shift through a single 0 with $s + 1$ clock pulses, thus ensuring that 0 is possible surrounded by near-neighbour 1s.
3. Shift through a pattern of 00110011..., this sequence exercising each stage through the remaining combinations of self- and near-neighbour states.

For long shift registers it is possible to combine these three types of test by shifting through a sequence such as 0001011100010..., which combines the self- and near-neighbour conditions tested by (a), (b) and (c). Such tests are sometimes known as 'flush tests'.

Hence it is possible to test a shift register configuration with considerably less than the 2^s clock pulses which are required by other s-stage networks. This is a very significant factor. As a result the shift register is the most powerful circuit configuration available when testing considerations are involved; it is used in many 'easily testable' and 'designed-for-testability' (DFT) IC designs, and will be referred to many times in later sections.

6.2.2 Analogue circuit test

The testing of analogue (linear) custom ICs is usually less complex than digital IC testing because, in general, analogue ICs contain far fewer primitives (amplifiers, etc.) than the very large number of gates and other macros involved in digital LSI or VLSI networks. As a result the input/output behaviour can be specified more readily, although the parameters involved may be more complex.

The following may be among these parameters:

- voltage amplification (gain),
- bandwidth,
- signal-to-noise ratio,
- common-mode rejection (CMR),
- offset voltage,

and others. A check on all parameters may be essential at the prototype stage, but, provided the performance of the prototype circuits is found to be satisfactory, then production testing may possibly be relaxed to a sub-set of the full set, for example gain and offset measurements, provided that the fabrication processes continue to be monitored by means of the drop-in test circuits.

The actual test of analogue USICs involves standard test instrumentation such as waveform generators, signal analysers, voltmeters, etc., as used in the testing of any type of analogue system. This instrumentation may be an assembly of 19-inch rack-mounted instruments all under the control of a dedicated microcomputer, such an assembly sometimes being termed a 'rack-and-stack' resource [1], [5]. An alternative to such standard instrumentation may, however, be a test circuit specially designed to perform the required tests on the custom design.

It is essential for the OEM to discuss the testing requirements of an analogue custom IC very closely with the vendor. In the case of very complex analogue networks, both the design and the subsequent testing may be beyond the capabilities of an OEM just starting to use custom ICs, in which case the expertise of a specialized design house or vendor will be necessary.

6.2.3 Mixed analogue/digital circuit test

The test of the analogue part and the test of the digital part of a combined custom IC each require their own distinct forms of test, and hence it is usually necessary to have the interface between the two brought out to observable test points so that analogue testing and digital testing may be performed separately. To some extent, this mirrors the problem of mixed analogue/digital simulation considered in Chapter 5, where it was seen to be difficult to combine these two parts within one simulation resource due to the dissimilar signal characteristics involved.

In the case of a relatively simple mixed analogue/digital IC containing, say, an input A-to-D converter, some internal digital signal processing and an output D-to-A converter, it may be possible to define a test schedule without requiring access to the internal digital/analogue interfaces. All such cases must be individually considered, and no hard and fast rules can be subscribed. There are, however, on-going research activities which seek to combine both analogue and

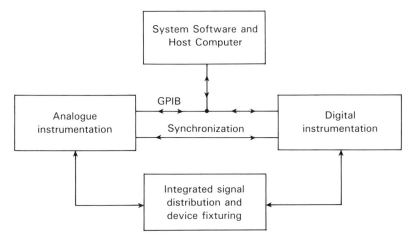

Figure 6.9 A mixed analogue/digital test resource, with the analogue and digital test halves synchronized under host computer control. (Courtesy of Integrated Measurement Systems Inc., OR.)

digital testing within one test resource, using, for example, multiple discrete voltage levels to represent analogue signals, or digital signals as voltage-limited analogue signals, but no commercial instrumentation using these concepts is currently available.

However, certain mixed-signal test systems, such as those illustrated in Figure 6.9, are currently available. Here, separate digital and analogue resources are present, linked by a G.P. Instrumentation Bus, the whole being under the control of a dedicated host computer. If such resources are used, then it is essential to consider their capability during the design phase of a mixed analogue/digital custom IC, in order that unforeseen difficulties are not encountered at the prototype testing stage.

6.3 Digital test-pattern generation

To summarize the principal difficulties of digital logic testing, the two fundamental problems in testing a VLSI digital circuit are as follows:

(a) the restricted access to the circuit since only the primary input and output pins of the IC are available for test purposes (it is possible for a vendor to probe within an IC if it is necessary to locate some obscure trouble, but this invariable causes scar damage, making the IC no longer usable – some newer techniques using scanning electron microscopy (SEM) may also be used to monitor internal logic voltages without causing any damage [9]);
(b) the time and cost to test is prohibitive if a fully exhaustive test is attempted.

Hence, some reduced test set or other means of reducing the test time is needed for large digital networks. As we will discuss later, test-pattern generation invariable involves the determination of some non-exhaustive set of input test vectors which will test the circuit to an acceptable confidence level.

6.3.1 Controllability and observability

Two terms need to be defined before discussing certain aspects of testing: 'controllability' and 'observability'. The broad concept of controllability is simple: it is a measure of how easily a given node within a circuit can be set to logic 0 or logic 1 by signals applied to the accessible primary inputs. Similarly, the concept of observability is simple: this is a measure of how easily the state of a given node (logic 0 or logic 1) within a circuit can be determined from the signals available at the accessible primary outputs. These two terms are central factors when discussing the testability of digital networks.

Consider the small circuit shown in Figure 6.10(a). Nodes 1, 2 and 3 are immediately controllable, since they connect directly to the primary inputs. Node 7, on the other hand, is clearly not so readily controllable; it requires either node 5 to be switched between logic 0 to 1 ('toggled') with node 6 held at 0, or vice versa, which in turn requires, say, F held at 1, G held at 1 and E toggled. Hence the input test vectors to control node 7 can be

Input vector . . E F G . .	Node 7
. . 0 1 1 . .	0
. . 1 1 1 . .	1

Looking at the requirements of observability, consider the small circuit shown in Figure 6.10(b), and suppose it is desired to observe the value of internal node 1. In order to propagate this nodal value to the output, i.e. to make the observable output voltage of 0 or 1 depend upon this node only, it is clear that nodes 2 and 4 must be set to 1 and node 6 to 0. (This is sometimes termed 'sensitizing' or 'forward-propagating' the path from the required node to an observable output.) Hence the input signals must be such as to give these logic conditions on nodes 2, 4 and 6 in order that the output is dependent upon the value of node 1.

The general features of controllability and observability are shown in Figure 6.11. The further away a node is from a primary input or output, then the ease of controlling or observing this node diminishes. However, provided there are no redundant nodes in a network, that is, all paths are necessary to produce fault-free outputs, then it is always possible to determine two (or more) input vectors that will check whether a given node switches from 0 to 1 and back again correctly. The complexity of determining the smallest set of such vectors for every internal node of a circuit is extremely high, way beyond the bounds of hand computation for all except very small circuits. If sequential circuits are also present, then there will be

Digital test-pattern generation 285

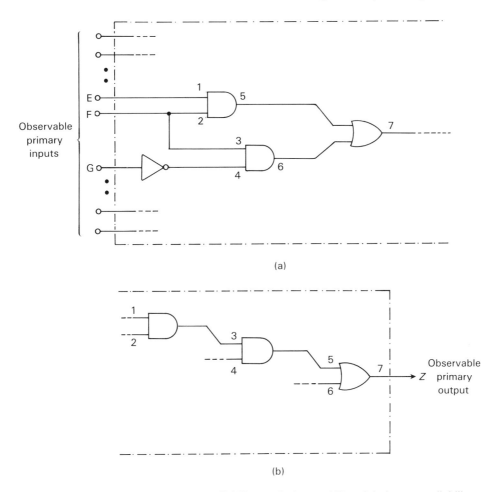

Figure 6.10 The concepts of controllability and observability: (a) the controllability of node 7 (see text); (b) the observability of node 1 (see text).

the additional complexity of driving the latches/flip-flops to specific states to give the required nodal observability, which may require a large number of clock pulses to achieve the required states.

With internal circuit redundancy (deliberate or accidental), it will not be possible to control certain internal nodes without incorporating additional circuit details. Consider the monitoring circuit shown in Figure 6.12(a), which has been included to check that the outputs of macros A and B always agree. With a fault-free circuit, no means exist to influence the output of the monitoring circuit; an addition such as that shown in Figure 6.12(b) is necessary if controllability and observability of this monitoring circuit is required. This is an example of the well-known feature of building in redundancy in any type of system, electrical,

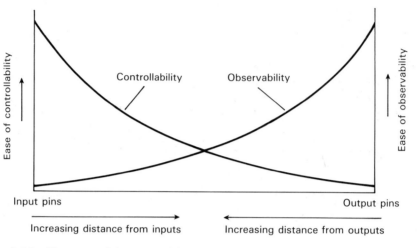

Figure 6.11   The general features of both controllability and observability.

electronic or mechanical: unless means are provided for disconnecting or otherwise overriding the redundancy there is no way in which the correct functioning of this part of the system can be fully checked. On the bonus side, however, if it is found to be impossible to control and observe a node within a logic circuit, then this must be a redundant part of the circuit which the designer has perhaps overlooked.

Many attempts have been made to quantify the two parameters, controllability and observability, for digital networks. They include the following:

1. controllability defined as

$$1 - \frac{N(0) - N(1)}{N(0) - N(1)}$$

where $N(0)$ is the number of different input vectors which establish logic 0 at the node in question and $N(1)$ the number of different ways of establishing logic 1; and observability defined as

$$\frac{N(P)}{N(P) + N(NP)}$$

where $N(P)$ is defined as the number of input vectors which allow the logic value on the node to propagate to the output and $N(NP)$ the number of input vectors which do not allow this propagation;

2. a count of the number of nodes which have to be set to produce a 0 or 1 at the node in question, and a count of subsequent nodes which have to be set in order to propagate this signal to the output.

These and other approaches require considerable complexity to accommodate

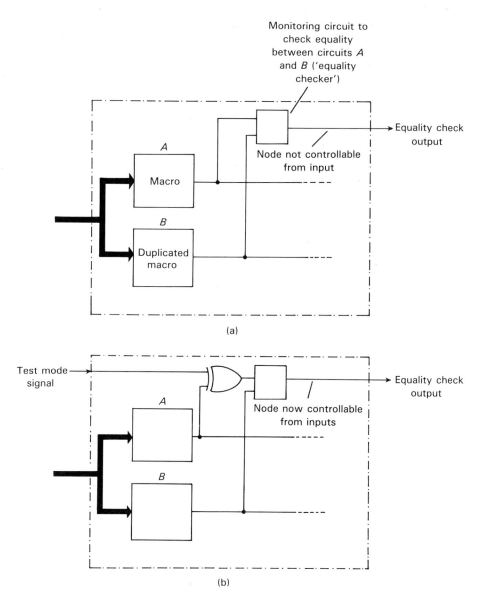

Figure 6.12 A circuit with an internal node not directly controllable from the primary inputs: (a) basic circuit with duplicated macros *A* and *B*; (b) addition required in order to give controllability of the equality checking circuit.

feedback loops, re-convergent fan-out and storage elements. Finally, an average or other amalgamation of the individual controllability and observability values for all the nodes must be made to give some overall measure of the circuit testability.

Testability analysis was widely pursued in the late 1970s–early 1980s. In particular two commercial software packages may be mentioned, namely HI-TAP (formerly CAMELOT) from GenRad Inc., based upon (1) above, and SCOAP (Scandia Inc. Controllability/Observability Analysis Program), based upon (2) above. Other commercial packages include TMEAS, COMET, TESTSCREEN, VICTOR and other proprietary software from CAD vendors [10]. Such tools normally require only a circuit description file from which to run. However, the limitations of testability analysis include the following:

- The analyses do not always give an accurate measure of the ease or otherwise of testing.
- They are applied after the circuit has been designed, and do not give any help at the initial design stage or guidance on how to improve the testability.
- They do not give any help in formulating a minimum set of test vectors for the circuit test.

Hence, although the concepts of controllability and observability are key concepts in the consideration of testability, quantification of these parameters is not particularly valuable. In particular, testability analysis should only be regarded as an aid to identify possibly difficult-to-test parts of a circuit as a post-design exercise. We will return to these concepts again in Section 6.3.3.

6.3.2 Fault-effect models

The most common way of determining some minimum set of input vectors to test a digital network involves a consideration of what faults are likely in the circuit. A particular input test vector may then be determined which would give, say, an observable logic 0 output when the circuit is fault-free, but logic 1 if this fault is present on a given node. This procedure, known as fault modelling, can be used to determine a minimum set of test patterns which will detect the presence or absence of the given type of fault on every internal node of the circuit.

The most commonly used fault model is the 'stuck-at' model, in which it is assumed that any physical defect in a digital circuit causes the input or output of a gate to be permanently at logic 0 or logic 1. Thus stuck-at faults may be stuck-at-0 (s-a-0) or stuck-at-1 (s-a-1).

For a 3-input NAND gate, for example, there are eight possible stuck-at faults, as shown in Figure 6.13. However, it will be seen that four of these are indistinguishable at the gate output node, all giving the stuck-at-1 output condition. A total of four test inputs applied to the gate will detect (but not identify) the presence of these stuck-at faults, as shown in Table 6.2.

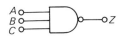

Inputs			Fault-free	Output Z							
A	B	C		A stuck-at-0	A stuck-at-1	B stuck-at-0	B stuck-at-1	C stuck-at-0	C stuck-at-1	Z stuck-at-0	Z stuck-at-1
0	0	0	1	1	1	1	1	1	1	0	1
0	0	1	1	1	1	1	1	1	1	0	1
0	1	0	1	1	1	1	1	1	1	0	1
0	1	1	1	1	0	1	1	1	1	0	1
1	0	0	1	1	1	1	1	1	1	0	1
1	0	1	1	1	1	1	0	1	1	0	1
1	1	0	1	1	1	1	1	1	0	0	1
1	1	1	0	1	0	1	0	1	0	0	1

Figure 6.13 Stuck-at faults on a 3-input NAND gate.

Table 6.2. The minimum test set to detect all stuck-at faults in a 3-input NAND gate

Input A B C	Expected output	Wrong output	Faults detected
0 1 1	1	0	A s-a-1 or Z s-a-0
1 0 1	1	0	B s-a-1 or Z s-a-0
1 1 0	1	0	C s-a-1 or Z s-a-0
1 1 1	0	1	A or B or C s-a-0 or Z s-a-1

Automatic test-pattern generation programs to determine test sets based upon this stuck-at model are widely available. These ATPG programs (see Section 6.3.3) assume that only single stuck-at faults are present, and determine the minimum set to detect any single node stuck-at-0 or stuck-at-1. The test vector applied to the circuit must be such as to apply particular signals to the node being tested (the controllability requirement) and also to propagate the signal on this node to the output (the observability requirement) in order to check for the particular stuck-at-condition. For example, in the simple circuit of Figure 6.14(a), if it is required to check that the output of the Invertor gate I_1 is not stuck at 0, the input test vector shown must be applied. If this node is s-a-0 the observable output will be 0 instead of the healthy value of 1. Note that this s-a-0 fault is indistinguishable from the Invertor input s-a-1, or the other input of NAND gate N_1 s-a-0, or its output s-a-1. Hence, in total, there is a great deal of commonality in the set of test vectors checking for individual nodes s-a-0 and s-a-1, from which a minimum test set can be compiled. Figure 6.14(b) gives a more comprehensive example, showing that fifteen input vectors will detect all the s-a-0 and s-a-1 faults.

The possibility of multiple stuck-at-faults in a network clearly exists, but the number of possibilities ($3^\eta - 2\eta - 1$ compared with 2η single stuck-at faults, where η = number of nets in the circuit) becomes too great to consider. However, it has been found in practice that the single stuck-at-fault model will normally cover multiple stuck-at faults also – in other words it is extremely unlikely that one stuck-

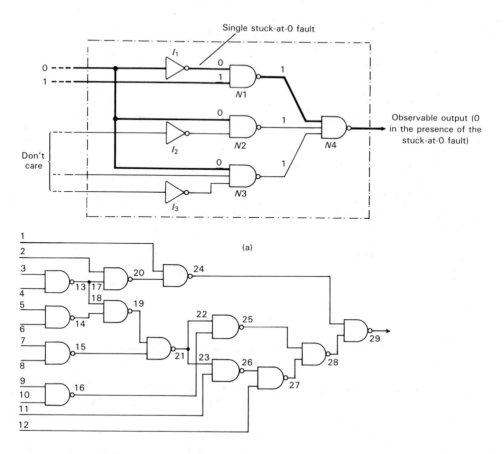

Figure 6.14 The propagation of stuck-at faults to an observable output: (a) simple example showing the test to propagate the output of gate I_1 stuck-at-0 to the output, with the three lower primary inputs being don't care; (b) a more comprehensive example, showing that 15 input test vectors will detect all the s-a-0 and s-a-1 faults in the circuit. (Courtesy of the University of Oxford, UK.)

at fault will mask another and give a fault-free output from the minimized input test set.

Bridging faults are electrical connections between two (or more) points in a circuit which should be logically independent. The theoretical number of bridging faults possible in a digital network, if any interconnection line (net) is assumed to be bridged to any other net, is extremely high, and thus only 'near-neighbour' bridging faults are normally considered in the bridging-fault model.

However, considerable difficulties are associated with bridging faults, such as the following:

- It is necessary to know the topology of the circuit in order to identify all the 'near-neighbour' connections.
- Does the bridging fault result in the logical AND (wired-AND) or logical OR (wired-OR) of the logic signals on the two lines?
- Does the bridging fault result in the formation of a latch by establishing an erroneous feedback path?

The possibility of forming a latch is illustrated in Figure 6.15, although the exact result of such a bridging fault depends upon the sink or source impedance of the logic signals involved.

Bridging faults within a logic gate, particularly in complex CMOS gates, can raise further difficulties since the logic function may be modified. For example, in Figure 6.16 the presence of the bridging fault, shown dotted, will pull the gate

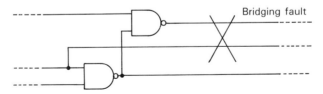

(a)   Resultant equivalent circuit of two cross-coupled NAND gates:

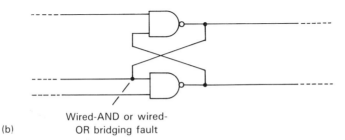

(b)   Wired-AND or wired-OR bridging fault

Figure 6.15   Bridging fault conditions which may produce a feedback path and hence a bistable (latch) configuration: (a) the circuit and the bridging fault; (b) the equivalent circuit with the fault present.

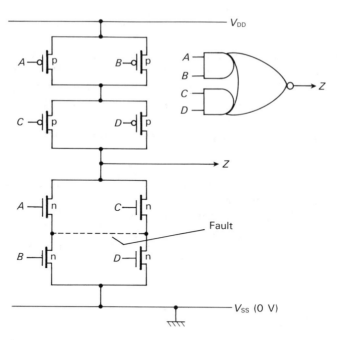

Figure 6.16 A possible bridging fault shown dotted within a complex CMOS logic gate, the fault-free gate output being $Z = \overline{(AB + CD)}$. (Note that the equivalent Boolean gate representation cannot model this fault.)

output down towards logic 0 under the input conditions $AB + CD + AD + CB$ instead of the normal conditions of $AB + CD$, although if the p-channel transistors remain fault-free, considerable power dissipation will result when $AD + CB$ is present. (Such faults in CMOS circuits are usually monitored by measuring the current taken from the V_{DD} supply line rather than by functional tests. Excessive dissipation is taken to be some short-circuit condition somewhere inside the circuit.)

In general, bridging-fault models are not widely used for the test-pattern generation of random layout ICs. They are particularly relevant, however, for PLAs and other regularly structured layouts; this will be considered separately in Section 6.4.

Finally, one may consider open-circuit faults. Unfortunately, this is one fault which is particularly awkward to detect in CMOS circuits, although less difficult in other technologies where it is usually indistinguishable from a stuck-at fault as far as the observable outputs are concerned.

Consider the problem which can arise in a 2-input CMOS NAND gate, see Figure 6.17(a). In the fault-free circuit the four transistors behave as shown in Table 6.3. Suppose now that the p-channel transistor $T1$ becomes open-circuit so that it is effectively always 'off'. With inputs 00, 10 and 11 the gate output will be unaffected since there is always on 'on' transistor from V_{DD} or V_{SS} to the gate output. With input 01, however, there will be no 'on' path through $T1$, and hence no path to the output from either supply rail. The output will thus be floating.

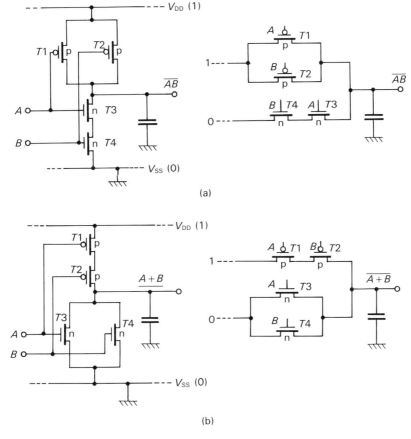

Figure 6.17 The problem of CMOS open-circuit faults: (a) a CMOS 2-input NAND gate drawn conventionally and in an alternative way which emphasizes the paths from V_{DD} and V_{SS} to the gate output; the capacitor represents the lumped capacitance of all the circuitry connected to the output; (b) CMOS 2-input NOR gate, again drawn conventionally and in an alternative form.

Table 6.3. The fault-free action of a 2-input CMOS NAND gate

Inputs		State of the transistors				Output
A	B	T1	T2	T3	T4	Z
0	0	on	on	off	off	1
0	1	on	off	off	on	1
1	0	off	on	on	off	1
1	1	off	off	on	on	0

If under test conditions the inputs are applied in the order shown above, input vector 00 will establish a logic 1 at the gate output through working transistor $T2$. When the input changes to 01 there will be no immediate change of gate output, since the circuit capacitance will maintain the output voltage at the logic 1 value. Should input 01 be held for a long period (possibly minutes), then the gate output may decay sufficiently to no longer be an effective 1 signal, but if (as is invariably the case) the test input vector is changed to 10, then no fault will have been detected. The output has 'remembered' the correct logic value and the fault is said to be a 'memory fault'.

However, if in practice the input condition 01 had been preceded by input 11, then the gate output would not have been pre-charged to logic 1, and on the input 01 it would remain at logic 0. Hence, to detect this particular open-circuit CMOS fault it is necessary for the test vector sequence to be 11 followed by 01.

A similar problem can arise with n-channel transistors in parallel, see Figure 6.7(b), where a lack of discharge path to V_{SS} (logic 0) may not be detected if the output is already at logic 0. It can be shown that a fully exhaustive test set for any n-input combination CMOS network requires $n2^n + 1$ input vectors to detect all possible NAND/NOR failings, rather than the 2^n vectors which is normally considered to be the exhaustive set. The fully exhaustive set for all 2-input CMOS gates is as follows:

```
00
01
11
10
00
10
11
10
00
```

Details of how such exhaustive sequences may be generated can be found in [11], [12]. Each overlapping pair of vectors may be regarded as an initialization vector followed by a test vector to check for correct pull-up or pull-down should any change of output from 0 to 1, or vice versa, be required. If the precise transistor configuration of all the gates in the network under test is known, then fewer than $n2^n + 1$ vectors may be possible.

The problems with CMOS testing can thus be seen to be inherently related to the fact that every CMOS gate consists of two dual parts: the p-channel and the n-channel logic. A logical fault in one half and no fault in the other half, which is perhaps the most likely occurrence in practice, gives rise to problems of testing which are not present in bipolar and nMOS logic. The question of open-circuits in CMOS transmission gates and flip-flop circuits such as illustrated in Figure 5.22, does not seem to have been specifically considered, or at least has not been published. CMOS must therefore be regarded as the most difficult digital technology to test thoroughly.

Currently, most custom IC vendors do not consider open-circuit faults in their recommended test procedures, in spite of CMOS being the most dominant technology for custom applications. Instead, the stuck-at fault model is by far the most widely used. However, the following points should be appreciated:

- The stuck-at model and the actual circuit failure present may have no relationship with each other; stuck-at tests do not give any reliable diagnostic information as to what is wrong.
- The collective set of stuck-at tests will not exercise the circuit under test through all its logic states.

Nevertheless, it has been found in practice that if single stuck-at faults are considered at every internal node of a circuit, then these tests collectively cover most of the other possible circuit failures. Hence the stuck-at fault model is widely used throughout industry to derive digital test patterns for both printed-circuit boards and ICs. PLDs (see Section 6.4) may, however, have a different basis for their test vector generation. For further details of faults and fault modelling see [13], [14], and the comprehensive references contained therein.

6.3.3 Test-pattern generation

Test-pattern generation is the design process of generating the test patterns required to test a given digital system, whether it is a printed-circuit board (PCB) assembly or a custom IC. As previously noted, exhaustive testing is usually prohibitively long and hence some reduced test set is normally sought. However, as will be seen later, if partitioning of a large circuit into smaller sections is done for test purposes it may be possible to test each small partition exhaustively, in which case the test-pattern-generation problem does not arise.

The generation of a reduced set of test patterns acceptable to both the OEM and the vendor may be obtained in any of the following three ways:

(a) manual generation;
(b) algorithmic generation; and
(c) pseudo-random generation.

The formulation of a reduced test set normally requires the internal details of the circuit under test to be known, perhaps from the netlist generated from the circuit schematic. It may therefore be referred to as 'structural testing' as distinct from fully exhaustive testing ('functional testing'), which does not require knowledge of the precise internal details.

Manual test-pattern generation is invariably undertaken by the OEM, who knows what the circuit is intended to do, rather than by the vendor. The test patterns may be compiled by either of the following two methods:

(a) by considering a comprehensive range of working conditions and listing the input vectors and the expected output responses involved in these operations; or
(b) by listing the input vectors which cause every gate in the circuit to toggle from one state to the other.

If a working breadboard identical to the custom IC has been made, this can assist the OEM in the test-pattern-generation procedure.

On receipt the vendor will usually compute the acceptability of a manually generated test set by checking whether this test set gives 100% toggling of all nodes in the circuit netlist. If it does not, then the vendor will usually ask the OEM for additional tests, or discuss with him the acceptability of the given testing details. (It should be noted that a 100% toggle check only checks that all nodes in the circuit switch from 0 to 1, and vice versa, under test conditions. This does not necessarily confirm the correct functionality of the complete circuit under all conditions.)

Manually generated test sets usually become far too time-consuming to produce when more than a few hundred gates and macros are present in a custom IC. Hence, for larger or more complex circuits, algorithmic or pseudo-random test-pattern generation is necessary, possibly with formal design-for-testability (DFT) techniques built in at the design stage, as will be considered later.

Algorithmic test-pattern generation, or automatic test-pattern generation (ATPG), usually uses a gate-level representation of the circuit, with all gate input and output nodes identified. A fault model is used which assumes that a given type of fault is present on each node of the circuit, the test pattern to detect this fault being determined for every node of the circuit.

The fault model used almost exclusively is the stuck-at-fault model introduced above. The following four features may be identified in most ATPG programs:

(a) listing the necessary gate inputs which will generate a gate output different from that occurring when a stuck-at fault is present on the gate;
(b) determining the primary inputs necessary to establish these gate input conditions ('fault sensitizing') and propagating the gate output to a primary output ('path sensitizing');
(c) repeating this procedure until all single stuck-at faults have been covered; and
(d) combining and sorting this test pattern listing into a minimum test set, utilizing the fact that a single test vector using all the primary inputs and primary outputs may simultaneously test a large number of stuck-at faults in the circuit.

The most difficult part in the above is the path sensitizing. The most commonly used methods are based upon Roth's D-algorithm [15], which uses the following five logic values:

0 = logic 0
1 = logic 1
D = fault-sensitive value, D = 1 for fault-free conditions and D = 0 for faulty conditions
$\bar{D}$ = fault-sensitive value, D = 0 for fault-free conditions, and D = 1 for faulty conditions
X = unassigned (don't care) value which can take any value 0, 1, D or $\bar{D}$

From these five logic values the 'primitive D-cubes of failure' for logic gates may be defined. See Table 6.4 for the case of 3-input gates.

The requirement to propagate any D-value through the following (fault-free) gates, or to propagate a stuck-at input of a gate to the gate output, the 'propagation D-cubes', is shown in Table 6.5. For a stuck-at gate input we have, for example, the D-notation D 1 1 $\bar{D}$, see the NAND gate below detailed in Table 6.5.

The full 5-valued logic relationships between inputs and output for any logic gate can also be defined; for example for 2-input logic gates it is as shown in Figure 6.18; the propagation D-cubes above are merely particular entries of the full 5-valued input–output relationships. An example of this notation applied to a simple circuit is shown in Figure 6.19, from which it will be observed that the fault source on the output of $G3$ is driven to the observable output by the appropriate signals sensitizing the path from $G3$ to the output Z. The problem that this poses in a large

Table 6.4. The primitive D-cubes of failure for 3-input logic gates

	Inputs			Output	D-notation			
	A	B	C	Z stuck-at	A	B	C	Z
AND gate:	1	1	1	s-a-0	1	1	1	D
	0	X	X	s-a-1	0	X	X	$\bar{D}$
	X	0	X	s-a-1	X	0	X	$\bar{D}$
	X	X	0	s-a-1	X	X	0	$\bar{D}$
NAND gate:	1	1	1	s-a-1	1	1	1	$\bar{D}$
	0	X	X	s-a-0	0	X	X	D
	X	0	X	s-a-0	X	0	X	D
	X	X	0	s-a-0	X	X	0	D
OR gate:	0	0	0	s-a-1	0	0	0	$\bar{D}$
	1	X	X	s-a-0	1	X	X	D
	X	1	X	s-a-0	X	1	X	D
	X	X	1	s-a-0	X	X	1	D
NOR gate:	0	0	0	s-a-0	0	0	0	D
	1	X	X	s-a-1	0	0	0	$\bar{D}$
	X	1	X	s-a-1	X	1	X	$\bar{D}$
	X	X	1	s-a-1	X	X	1	$\bar{D}$

Table 6.5. The propagation of D values through healthy 3-input logic gates

	Inputs			Output
AND gate:	1	1	D	D
	1	D	1	D
	D	1	1	D
NAND gate:	1	1	D	$\bar{D}$
	1	D	1	$\bar{D}$
	D	1	1	$\bar{D}$
OR gate:	0	0	D	D
	0	D	0	D
	D	0	0	D
NOR gate:	0	0	D	$\bar{D}$
	0	D	0	$\bar{D}$
	D	0	0	$\bar{D}$

circuit is of course how to establish all the required gate input conditions on the forward path to an output without encountering any conflictions, bearing in mind that only the primary inputs are available for controlling the state of all internal nodes.

Roth's algorithm for test-pattern generation thus consists of the following three main operations, namely:

(a) listing of the primitive D-cubes of failure for each gate and stuck-at condition;
(b) the forward-drive operation to specify the signals required to propagate a faulty D condition to an observable output; and
(c) a backward-trace (or consistency) operation to establish the required logic 0 and 1 conditions on all gates and primary inputs in order to forward-propagate (b) above.

The algorithm guarantees to find a test for any fault that can be modelled if such a test exists (see p. 301), but considerable complexity can arise in the backward-trace operations particularly by (i) re-convergent fan-out, that is, when a fan-out signal from a logic gate travels through distinct paths but re-converges at some further gate before reaching a primary output, and (ii) by Exclusive-OR gates. For further details of Roth's algorithm, see, in particular [16] plus other general sources of information [13], [14], [17].

More recent developments, based on the 5-valued fundamentals of the D-algorithm, provide faster and more efficient automatic test-pattern generation than Roth's original search mechanisms. In particular, RAPS (Random Path Sensitizing), PODEM-X (Path-Oriented Decision Making) which incorporates RAPS, and FAN (Fan-out oriented test generation) may be mentioned [18]–[21].

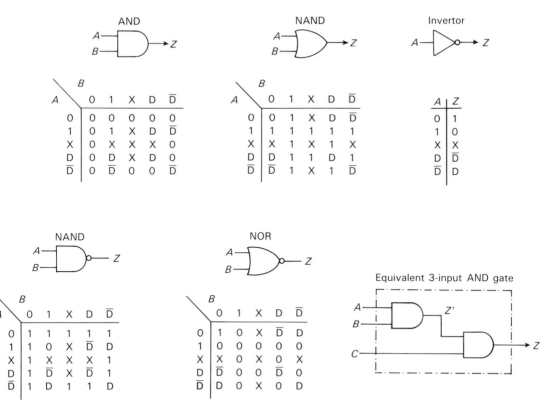

Figure 6.18 Roth's 5-valued logic relationship (D-notation) applied to 2-input logic gates. The relationships for a 3- (or more) input gate can be derived by considering a cascade of 2-input gates, since commutative and associative relationships still of course hold.

PODEM-X is an ATPG system in which several distinct programs may be used during the test-pattern generation procedure. In particular it encompasses the following:

1. An initial global procedure using RAPS, which first sets all nodes in the circuit to X and then arbitrarily selects paths back from each primary output to the primary inputs, determining what fault cover this provides by simulation after each path has been completed. [13], [19].
2. Following this procedure (if used), directed tests for each of the nodes which have not been covered by this global search are carried out in turn by the main PODEM algorithm, which again uses a backward-trace procedure followed by simulation to check how many faults have been covered by each test vector.

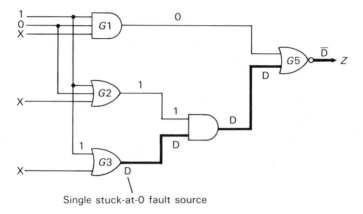

Figure 6.19 An example of Roth's D-notation, where a fault is assumed to be present on the output of gate to 3 denoted by the symbol D.

The fault simulator used in the RAPS and PODEM searches is FFSIM (Fast Fault Simulator), and is a 5-valued zero-delay simulator using the five logic values previously defined [20]. Additionally, PODEM-X includes a controllability analysis of the circuit to define the 'distance' of each node from a primary output; this information is used to guide the search procedures in the choice of a shortest path, which is in contrast to Roth's D-algorithm which propagates information along all possible paths during a trace procedure. Finally, a compaction program in PODEM attempts to merge as many test patterns as possible in order to produce a final minimized test set.

Worked examples of the RAPS and PODEM algorithms may be found in [10], and of the PODEM algorithm only in [14], together with flowcharts showing the search procedures.

FAN is similar to, but is reputedly a more efficient ATPG than PODEM. It considers the circuit topology in greater detail, in particular the fan-out points in a network, in order to reduce the amount of computer time wasted in re-trying multiple choice paths. The heuristics incorporated in FAN are designed to minimize the number of arbitrary decisions made, which can result in considerable backtracking if conflicting signals are subsequently encountered, by keeping account, at each phase of the test-generation algorithm, of the logic values which are still free to be assigned to each node in the circuit. The concept of 'headlines' is also incorporated, a headline being a net in the circuit up to which all gates from the primary inputs have single fan-outs. The advantage of identifying headlines is that once a backtrace procedure has reached a headline, then no further backtracking to the primary inputs is necessary since the inputs can always be set to give the required logic values at headline points. This is in contrast with PODEM, which continues to trace backwards until it reaches the primary inputs.

Like PODEM-X, the FAN ATPG algorithm can be incorporated into a wider

ATPG system, namely FUTURE [22]. This incorporates a global test generator which can be run before FAN is used to complete the test set, but, unlike the RAPS procedure used in PODEM-X, the global test generator used in FUTURE applies a set of pseudo-random test vectors, the fault coverage of which is determined by a fault simulator. This is a very simple and fast technique, with the user being able to specify when the pseudo-random tests are to be terminated. It is a characteristic of fault coverage that the first few input test vectors applied to a circuit will normally detect a high percentage of the total number of possible stuck-at faults, but the scoring rate for subsequent test vectors drops off as the percentage increases, being asymptotic to 100% fault coverage. Hence, somewhere between 10 and 100 pseudo-random test vectors are normally applied before FAN is brought in to complete the test-pattern generation. Even then, it may take FAN (and the previous PODEM) a considerable (unacceptable?) amount of computer time to achieve 100% fault coverage with very large and complex circuits.

Further details of FUTURE and FAN may be found in [13], [14], [21], [22]. Further details of fault simulators may be found in [14], [23], whilst an overall discussion of fault models and their problems may be found in [24].

All the above ATPG techniques guarantee to find a test pattern for any fault modelled in a circuit, if such a test exists. However, the following should be appreciated:

1. If a stuck-at-0 fault cannot be detected at a primary output, then this node does not uniquely contribute a logic 1 to the output path under any input combination.
2. If a stuck-at-1 fault cannot be detected at a primary output, then this node does not uniquely contribute a logic 0 to the output path under any input combination.
3. If neither (1) or (2) can be found for a given node, then this particular part of the circuit is redundant; all non-redundant nodes must at some time uniquely control the 0 or 1 value of a primary (observable) output.

In connection with this last point, there are many published examples of circuits which purport to show why a stuck-at-0 or a stuck-at-1 fault cannot be detected by fault modelling. Figure 6.20 shows one example. What is not always made clear is that these may be examples of redundancy in the circuit, which may or may not be appreciated. In the circuit shown in Figure 6.20, input C is completely redundant, together with the second 2-input AND gate. The 2-input NOR gate may also be replaced by an Inverter.

The final technique which may be used for test-pattern generation is pseudo-random-pattern generation. This has already been mentioned as the method used for global test generation in FUTURE to precede the FAN ATPG procedure. Theoretically, if a sufficiently long pseudo-random test set is applied, then all possible faults in a circuit will be detected, but such a length is in effect a fully exhaustive test of the network rather than some minimized test set.

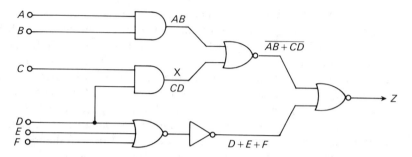

Figure 6.20 An example circuit where a check for stuck-at-0 faults will show that node X stuck at 0 is undetectable. The circuit illustrated is realizing the function $Z = \overline{(\overline{(AB + CD)} + (D + E + F))}$ but this may easily be shown to minimize to $Z = AB\overline{D}\overline{E}\overline{F}$.

The great advantage of pseudo-random test-pattern generation is that it completely avoids the difficulties involved in deriving such minimum ATPG programs as have been considered previously. However, its serious disadvantage is that if the test set length is short then the fault coverage produced, which may be determined by fault simulation, will be inadequate.

Pseudo-random TPG is used extensively in Built-In Self-Test (BIST), a particularly powerful form of design-for-testability which we will consider in detail in Section 6.6.3. There we will consider not only the generation of pseudo-random test patterns, but also how pseudo-random generators may be used to capture the test output responses and provide an output 'signature' of the circuit under test.

The preceding discussions of ATPG have all considered networks which have been flattened to the gate level, with (normally) stuck-at conditions being considered on all nodes. Because the number of individual gates in a complex VLSI circuit is high, consideration has been given to the use of higher levels of functionality in test-pattern generation procedures. However, so far this has proved a difficult problem, since although there are fewer macros to consider than gates, the difficulty of defining the forward-propagation of test signals and the backward-tracking to establish primary input conditions is considerable. Currently, the effort necessary to tackle this is greater than that involved for ATPG at the gate level, and hence, in view of the increasing emphasis upon design-for-testability which minimizes the test-pattern-generation problem, there seems little reason to pursue high-level ATPG procedures unless it is an integral part of a high-level HDL design environment. Details of work in this area may be found in [13], [14], [25]–[29].

Finally, in these discussions on test-pattern generation, it will be appreciated that all the preceding details have implicitly assumed combinatorial logic networks only, and have not addressed sequential networks specifically. It is exceedingly difficult to formulate test patterns for sequential networks because it is necessary to set or reset or clock the circuits into known states before nodal information can be driven to observable outputs. This may involve either a master reset on the circuit or the running of some 'homing sequence' which will drive the circuits into a known state. The latter technique may not be possible, and in any case a large number of clock cycles may be required before any test can begin.

Therefore, in spite of some R and D work [16], [30], it is generally considered to be impossible to derive effective test patterns using ATPG techniques for circuits containing latches/flip-flops which are deeply buried inside a circuit, that is, where there is poor controllability and observability of these circuit elements. Instead, it is increasingly essential as circuit complexity passes the stage where functional testing becomes impractical, for the circuit to be designed in the first place to enable testing of the combinatorial part of the circuit to be done separately from the testing of the sequential elements; the former can then be tested using ATPG test sets or other means, with the sequential elements tested by their own unique mode of test, ideally using a small number of clock pulses for this purpose. This is where partitioning and design-for-testability (DFT) techniques come into play, as will be covered in Section 6.6 following.

6.4 Test-pattern generation for memory and programmable logic devices

In the widest sense, programmable logic devices include both RAM and ROM as well as programmable logic arrays (PLAs) and their variants. All are characterized by internal architectures consisting of regular structures, and hence the most likely internal faults can be defined.

The testing of standard off-the-shelf RAM and ROM ICs has received considerable attention, and has resulted in established techniques for the test of such circuits (see Sections 6.4.1 and 6.4.2 below). Should RAM and ROM macros be included within standard-cell custom IC, then the same test techniques are usually applied, the design being such as to give access to the memory I/Os. Should this not be done, there arises the very difficult problem of testing the embedded memory through the surrounding logic; this problem has been studied and analysed mathematically [31]–[33], but it is a situation which should be avoided if at all possible – see Figure 6.23.

6.4.1 Random-access memories

The fundamental problem with RAM testing is that every possible memory location in the RAM may be required to store a logic 0 or a logic 1 during normal system operation, but the stored patterns of 0s and 1s are completely fluid during normal operation. The principal test problem is therefore to derive a series of tests, based on the device structure, which is not excessively long but which covers the type of failures most likely to occur. Among these failures are the following:

- one or more bits in the memory array stuck at 0 or 1;
- coupling between cells such that a transition from 0 to 1, or vice versa, in one cell erroneously causes a change of state in another (usually adjacent) cell – a 'pattern sensitive' fault;
- bridging between adjacent cells causing the logical AND or OR of their stored outputs;
- faults in a decoder which causes it to address additional rows (columns) of cells or entirely wrong rows (columns);
- other faults such as driver circuit faults between bit lines.

Additionally, there may be parametric faults such as a DRAM cell failing to hold its charge state for the specified time. These and other possible failings have been extensively studied in [3], [8], [13], [14], [25], [34]–[36].

The following algorithms have been developed to test RAM circuits; each involves writing a 1 in each cell and performing a readout, and/or writing a 0 in each cell and performing a readout, in some sequence of tests. The tests are therefore functional tests, and not tests based upon fault models:

1. *Marching patterns*: in the 'marching-one' pattern test the memory array is first filled with 0s and read out. A single 1 is then written into the first address and this location is checked.

   The first and second addresses are then written to 1 and checked. This procedure of progressively filling up the array with 1s continues until all addresses are full. The converse, the 'marching-zero' pattern test, is then performed, progressively setting the array back to all 0s and checking the 0-valued cells at each step.

   This algorithm and its variants are known as MARCH test procedures.

2. *Walking and galloping patterns*: the memory array is first filled with 0s and read out. A single 1 is then written into the first address and all locations except this one are read out to check that they are still at 0. This is then repeated for all other locations individually set to 1 against a background of all 0s, and finally the whole procedure is repeated for a single 0 at each location against a background of 1s.

   Variations on the above involve the reading of the single 1 address between each reading of the 0 values, and the reading of the single 0 address between each reading of the 1 values, or other variants, which therefore breaks up the long sequences of the same output value that would otherwise be present.

   These tests are variously known as WALKPAT or GALPAT tests.

3. *Diagonal patterns*: here a single diagonal across the memory array is filled with 1s against a background of 0s, and every cell is checked. This diagonal is then moved progressively across the array until all cells have at some time been set to 1, checking every cell output at each stage. The converse, namely a diagonal of 0s against a background of 1s, may also be used.

4. *Nearest-neighbour patterns*: since in the WALKPAT and GALPAT tests large

blocks of 0s (or 1s) remote from the single 1 (or 0) are repeatedly checked, it is much more economical to identify the cells in the immediate surround of the single 1 (or 0) cell, and verify the fault-free nature of this cluster of cells at each step. However, this requires the identification of such cells as part of the test algorithm, but it will provide a considerably reduced total number of read operations in the test procedure.

In connection with the latter consideration, it will be appreciated that the walking and galloping pattern tests require $O(N^2)$ read operations, where N is the total number of cells in the memory array. This is clearly a completely impractical number when memory size exceeds a few thousand bits, even if the tests are performed at MHz clock rates, and hence there is considerable pressure to use tests which do not require an order of N^2 operations. In the custom microelectronics area the vendor must always advise the OEM on the way any memory macros are to be tested, which may reflect back to the most likely fault mechanisms occurring in his particular technology and fabrication process.

Recent R and D work has, however, been undertaken in order to provide some measure of self-test for memory circuits, usually by building in redundancy and incorporating some means of eliminating the effect of faulty cells or rows of cells [37]. These considerations are beyond our immediate concern here, but may become increasingly significant for very large memory macros built into standard-cell custom designs.

6.4.2 *Fixed memories (ROM and PROM)*

Unlike RAMs, read-only memories are of course permanently dedicated to give known input–output relationships when in service. In practice, it is not sufficient to check that the outputs are at logic 1 on the appropriate input vectors, but also that they are not at 1 under any other input conditions. This implies that the full 2^n truthtable of the programmed device, where n is the number of inputs, should be tested.

If programmable read-only memory circuits are in use, then the programmer which dedicates the PROM to a particular requirement will invariably also have test facilities which will check the dedicated device exhaustively and functionally. No test-set generation or other complexities are therefore involved. In the case of ROM macros built into a standard-cell custom IC, then controllability and observability of the ROM I/Os should be provided in order that an exhaustive test can be carried out.

The vendor's resources for undertaking both RAM and ROM testing may be very sophisticated, involving test equipment costing several millions of dollars and operating at clock speeds up to several hundred MHz [38]. The tester illustrated in Figure 6.4, for example, has a built-in memory test capability enabling it to test VLSI memory circuits as well as ultra-high speed custom circuits.

6.4.3 Programmable logic arrays

Programmable logic arrays are being used increasingly in custom IC applications, since they provide a ready means of realizing random combinatorial logic.

Off-the-shelf PLAs of small to moderate complexity pose no problems for test, since exhaustive testing can be performed. The programming units for such devices will include a 100% functional test. However, for large PLAs there is an increasing need to find some minimum test set and derive ATPG programs for this purpose; a 30-input PLA, for example, would require over 10^9 input vectors for an exhaustive test at minterm level, but internally there may only be, say, 50–100 product terms generated by the AND matrix. Also, the 'marching' and 'galloping' procedures used for RAM testing are not appropriate for testing PLAs since the internal architecture and action of the two is completely dissimilar (see Figures 2.31 and 3.11).

The particular architecture of a PLA does, however, lead to certain fundamental properties which can ease the testing problem. Basically, the logic structure is a two-level AND–OR structure, as shown in Figure 6.21(a), and does not naturally have a re-convergent fan-out topology (see Figure 6.21(b)), which ATPG programs find difficult to handle. However, the suitability of using random or pseudo-random test patterns for PLA testing is poor since to sensitize a particular product term from the AND array requires application of the exact input variable values of the product term and the chance of random input vectors achieving this is correspondingly poor.

The PLA outputs are dependent upon the presence or absence of the connections at the crosspoints in the AND and OR matrices. A crosspoint fault model can therefore be proposed which represents the presence or absence of these connections. Smith [39] has classified PLA crosspoint faults as follows:

- *Growth fault*: a missing contact in the AND array causing a product term to lose a literal and hence have twice as many minterms as it should.
- *Shrinkage fault*: an erroneous contact in the AND array causing a product term to have an unwanted literal and hence lose half of its original minterms, or have none at all if the product term is a single literal and the complement is erroneously connected. (If the erroneous contact is the complement of an existing literal, then this is sometimes referred to as a 'vanish fault', since this input variable disappears from the resulting product [40].)
- *Disappearance fault*: a missing contact in the OR array causing a product term to be lost in an output.
- *Appearance fault*: an erroneous contact in the OR array causing an additional product term to be added to an output.

These various crosspoint faults are illustrated in Figure 6.22. Note that all of them either add or subtract minterms from the fault-free output function, but do not cause a complete shift of the minterms to another part of complete input n-space.

A crosspoint fault is termed a functional fault if it alters at least one output

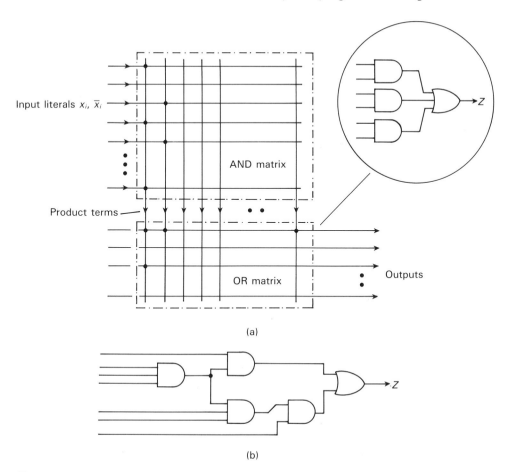

Figure 6.21 The PLA logic architecture: (a) the 2-level AND/OR realization produced by the input AND matrix and the output OR matrix (note: in practice it may be a NAND/NAND or NOR/NOR structure, but this is logically equivalent); (b) a reconvergent fan-out topology which is never present in a PLA dedication.

of the PLA. This qualification is necessary since an appearance fault, for example, may merely introduce an additional but redundant product term into an output. However, the majority of crosspoint faults will result in functional faults.

The process of test-pattern generation for PLAs essentially considers single possible crosspoint failures and determines appropriate input test vectors to show their existence, if present. One of the earliest detailed considerations was by Smith [39], followed later by Ostapka and Hong [41] who developed the PLA test-pattern-generation algorithm TPLA. Subsequent variations have been published, for example Eichelberger and Lindbloom with their PLA/TG algorithm [42], and others. Reference to these and further works may be found in [13] and in Zhu and Breur [43] who consider and compare a number of published algorithms. Multiple crosspoint faults have been considered by Agrawal [44], who has indicated that

308 Test-pattern generation and design-for-testability

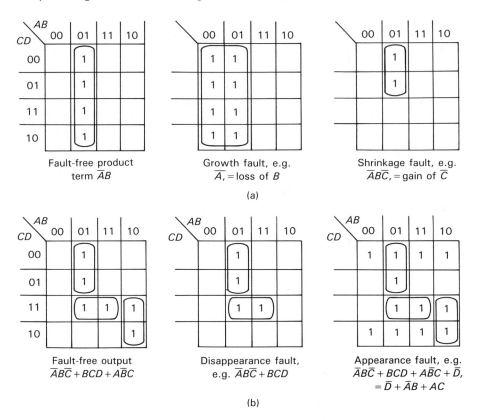

Figure 6.22 Crosspoint faults in a PLA: (a) growth and shrinkage faults in the AND matrix; (b) disappearance and appearance faults in the OR matrix.

inclusion of all single crosspoint faults will cover most multiple crosspoint faults (cf. the single stuck-at model considered earlier). However, because of the increasing significance of the PLA as a means of logic design, theoretical discussions continue as to how to guarantee the detection of all possible multiple crosspoint faults, for example see [45] and the further references contained therein.

In the context of custom ICs, the testing requirement for PLA macros should be part of the vendor's responsibility, and form part of the CAD support. The OEM should thus have no need to become involved in manual generation of any test vectors for a PLA macro which may be part of his custom circuit.

However, the PLA test-pattern-generation procedures considered above are applied to standard PLAs which do not have any additional built-in circuits for test purposes. Also, the test patterns are entirely dependent upon the functions being realized by the PLA, and are not independent of dedication, which would be attractive. Hence considerable research effort has been undertaken to augment the standard PLA architecture in some way so as to produce more easily testable or self-checking circuits. All involve some circuit additions to the basic AND/OR structure [13], [14], [43], with the minimization of this additional circuitry being a prominent requirement. We will consider testable PLA designs in Section 6.7 when we deal more specifically with design-for-testability (DFT) techniques.

6.5 Microprocessor testing

The microprocessor is the most complex circuit to test, whether it is a standard off-the-shelf unit or a macro for incorporation into a standard-cell custom USIC. The OEM can do very little innovation in the test of such circuits due to the following:

- The vendor does not usually disclose the full internal circuit details, and certainly not the gate-level equivalent circuit.
- General controllability and observability of the circuit is limited to the address and data busses.
- The key failure modes of the circuit are not usually revealed.

In any case it is not feasible to test at the gate level, and hence the only appropriate way to test is to perform a series of functional tests designed to verify the major components of the architecture, rather than to apply any form of fault modelling [3].

The evolution and present status of microprocessor testing may be found in [46]–[49]. Until recently, the most widespread practice has been to undertake such functional checks as the following:

- program counter test;
- scratchpad memory test;
- stack pointer and index register tests;
- arithmetic logic unit (ALU) and associated register tests;
- further tests on control lines and other peripheral parts.

The most difficult of these to test is the ALU since it is completely impractical to verify exhaustively that it can handle all possible arithmetic operations for all data patterns. Instead, it is for the vendor to propose a minimum series of tests to check that the ALU operates correctly and transfers data for a range of typical operating conditions.

Some theoretical studies of microprocessor testing have been pursued, for example modelling the microprocessor as a system graph (the 'S-graph'), wherein each register in the microprocessor is represented by a node in the graph [46]. Main memory and I/O ports are represented by additional nodes. Another technique has been functional-level fault modelling for bit-sliced microprocessors [48]–[50], but as bus widths and internal complexity have increased, so the requirement for some form of built-in test and/or design-for-testability has escalated. Hence, the latest microprocessor designs have built-in means, usually some form of scan test (see Sections 6.6 and 6.8), to ease this difficult test problem [51].

The OEM is very largely in the hands of the vendor when it comes to testing microprocessor macros, and full discussions should be held between parties to clarify the test requirements in any custom circuit environment.

6.6 Design-for-testability (DFT) techniques

Because of the difficulties of formulating acceptable tests as digital ICs become larger, some features must be built in at the design stage of the IC, and not as an afterthought once the placement and routing has been completed.

DFT techniques normally fall into the following three principal categories:

(a) *ad-hoc* design methods;
(b) structured design methods; and
(c) self-test.

The first two of these require the use of some form of comprehensive test facility, usually a very expensive one for the vendor's tests, for example see Figure 6.4, but the third minimizes, to some extent, the external test resources required.

6.6.1 Ad-hoc *design methods*

Ad-hoc design methods consist largely of a number of recommended or desirable practices which a designer should use when undertaking an IC design. One of the most obvious tasks is to partition a large circuit into functionally identifiable blocks, so that testing of individual blocks can be considered. It is generally accepted that test costs are proportional to the square of the number of logic gates in a circuit, so that halving the size of a circuit block to be tested reduces the testing problems by a factor of four.

During the design procedure, therefore, *ad-hoc* methods should be adopted with a view to providing increased controllability and observability. A common method is to multiplex appropriate input and output connections (see Figure 6.23). Additional I/O pins will be required, which is the penalty to be paid for providing this accessibility. With very large standard cells in a standard-cell custom chip design, such as a PLA, memory, or microprocessor, the vendor will previously have worked out how his large macros should be tested. The OEM designer will therefore have no need to worry about how to test such standard designs, but the chip layout must provide the appropriate test access to them.

Long counter assemblies require a large number of input clock pulses to test them through all their states. This can be extremely time consuming, especially if

Design-for-testability (DFT) techniques

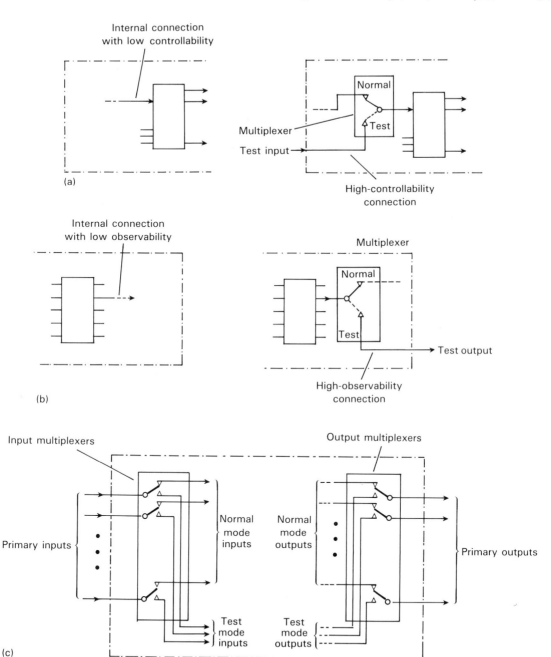

Figure 6.23 The use of multiplexers to increase internal controllability and observability. The multiplexer paths are switched from normal to test under the control of appropriate normal mode/test mode control signal(s) (not shown).

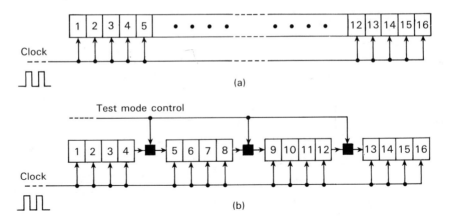

Figure 6.24 The requirements of dividing up long counters during test so as to reduce the testing time. 65536 clock pulses would be required to test the 16-bit counter (a), but only 16 to test each of the four partitions in (b).

a given signal from a counter is required in order to be able to test a further part of the circuit. Long counters must therefore be divided into smaller sections which can be individually tested in order to avoid application of an extremely large number of clock pulses (see Figure 6.24).

Other partitioning which must be considered is the segregation of analogue and digital parts of a circuit test, so that one does not have to be accessed via the other.

Partitioning, therefore, can be an *ad-hoc* process based upon the designer's experience, but it can also be part of a structured design methodology (see Section 6.6.2).

Apart from general consideration of the internal controllability and observability of the design being undertaken, the designer should also keep the following features in mind in order to facilitate the final test procedures:

- Make a list of key points during the design phase of the chip, and check that they can be easily accessed.
- Never build in logical redundancy unless it is specifically required; by definition such redundancy can never be checked from the I/O pins.
- At all costs avoid asynchronous operating circuit assemblies; make all sequential circuits operate synchronously under rigid clock control.
- Use level-sensitive storage circuits see p. 318, rather than edge-triggered circuits if possible.
- At all costs avoid on-chip monostable circuits, these must be off-chip if really necessary; on-chip use some form of synchronous counter-timer.
- Ensure that all storage circuits, not only in counters but in PLAs and memory, can be readily initialized before test; this also applies to any internal cross-coupled NANDs and NORs which act as local latches.

- Provide facilities to break all feedback connections from one partition of the chip to another ('global' feedback), so that each partition can be tested independently.
- Keep analogue and digital circuits on a chip physically as far apart as possible, with completely separate d.c. supply connections.
- Avoid designing 'clever' circuits which perform different duties under different circumstances.
- Limit the fan-out from individual gates as much as possible so as not to degrade performance, and to ease the task of manual or automatic test-pattern generation.
- It may be useful to consider putting in a separate clock control line for testing purposes, see Figure 6.25, in order that all or some part(s) of the circuit may be tested more easily, perhaps by one clock pulse at a time.
- Some commercial testers may not cater for unusual input and output signals, such as very slow rise or fall times or non-standard 0 and 1 voltages; the vendor should be consulted if any unusual logic signals are involved.
- Remember that a vendor's 100% toggle test on internal nodes is not a guarantee of correct functionality.
- Controllability/observability/testability numbers produced by analysis programs such as SCOAP, HI-TAP, etc., do not give any indication as to how to design a circuit for ease of test; they only highlight hard-to-test nodes which may require further design thought.
- Statistical analysis programs which give an estimate for the number of possible faults covered by a particular test sequence should be treated with care; they may work reasonably well for random combinational logic, but not so accurately for sequential networks or bus-structured networks.
- Never waste any spare I/Os on the custom IC and its packaging – use them if possible for additional test points for internal controllability or observability!

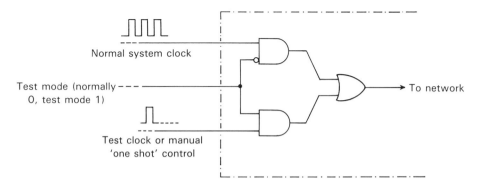

Figure 6.25 Provision of a separate test clock for test purposes, which may be useful to step through sequential parts of a circuit separately from other parts or at different rates.

Having produced the ASIC design, and perhaps having applied a controllability and observability analysis if this resource is available, it remains necessary to formulate the test vectors with which the circuit will be tested. If the circuit is sufficiently small to be tested exhaustively, there is no problem, but for larger circuits there has to be formulation of an appropriate reduced test set. Test-pattern generation may be manual or algorithmic (see Section 6.3.3); it is unlikely to be pseudo-random, since pseudo-random test-pattern generation is far more commonly associated with self-test methods of DFT (see Section 6.6.3), rather than with *ad-hoc* methods.

Thus these *ad-hoc* design rules can be summarized by the following four broad statements:

1. Maximize the controllability and observability of all parts of the circuit by partitioning or other means.
2. Do not use asynchronous logic or redundancy unless absolutely essential.
3. Provide means whereby all internal latches and flip-flops can be initialized for test purposes.
4. Consult the vendor at all stages of the design so that mutually acceptable test procedures can be formulated.

6.6.2 *Structured DFT methods*

The above good design practices should still be followed, even when using more formal design-for-testability techniques, which we will now consider.

Structured DFT methods are formal concepts built in at the design stage. They are principally concerned with sequential logic, and involve re-configuration of the sequential parts of the circuit when in test mode to give controllability and observability not only of the storage elements themselves but also of the blocks of combinational logic associated with the storage elements.

Digital logic systems which contain both combinational logic (gates) and sequential logic (latches or flip-flops) can always be re-drawn as shown in Figure 6.26(a). This is known as the Huffman model [52]. In most integrated circuit layouts it may be difficult to distinguish the two parts in this representation since they are inextricably mixed, but the two will always be present. On the other hand, certain programmable logic devices which have a separate row or block of latches or flip-flops may show this division very clearly. (Some uncommitted array USICs may also show this segregation, their floorplan being rows of gates interspersed with rows of latches.)

Scan path testing, sometimes known as 'scan test', is a DFT method which builds upon this general model. It usually provides the following three basic facilities when the circuit is switched from its normal operating mode to its test mode:

Design-for-testability (DFT) techniques 315

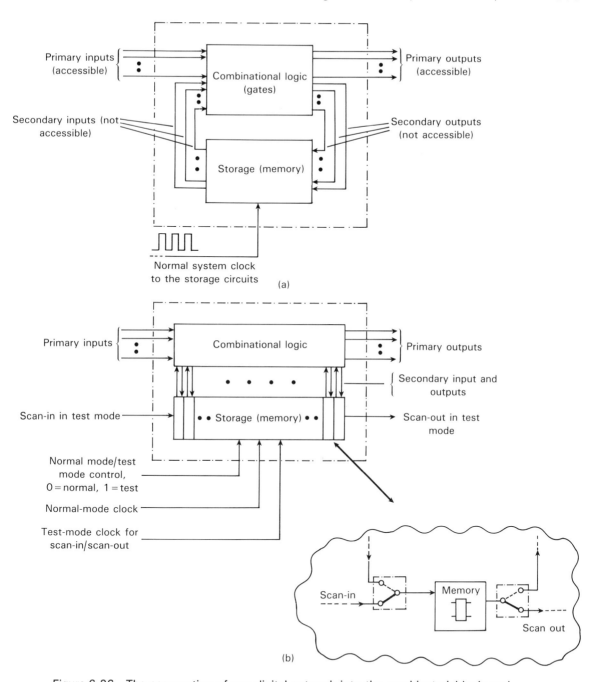

Figure 6.26 The segregation of any digital network into the combinatorial logic and the sequential (memory) logic: (a) the theoretical model; (b) adaptation of this model to provide the shift-register scan-path means of test.

(a) re-configuration of all the sequential storage elements into a continuous shift register configuration, known as the 'scan path';
(b) a 'scan-in' procedure, whereby any pattern of 0s and 1s may be fed into the shift register serially; and
(c) a 'scan-out' procedure, whereby any pattern of 0s and 1s held in the shift register may be read out serially.

This is shown in Figure 6.26(b). Procedures (b) and (c) above may take place simultaneously if required.

The test procedure for a complete circuit generally proceeds as follows:

1. The circuit is switched from normal mode to scan mode, which converts the storage elements into the scan path.
2. The correct action of this shift register is then checked by clocking through a pattern of 0s and 1s under the control of the test clock, the normal system clock being inoperative. This 'flush test' (see p. 281) confirms that all the storage elements are functional.
3. A chosen test pattern of 0s and 1s is then loaded into the shift register under the control of the test clock, to act as secondary inputs to the combinational logic.
4. The circuit is switched back to normal mode, and the normal system clock operated once so as to latch the resultant secondary outputs from the combinational logic into the shift register.
5. With the system switched back again to test mode, the test clock is operated so as to sequence the latched data to the scan-out terminal for verification purposes.

Steps (3), (4) and (5) are then repeated as many times as necessary in order to test all the combinational logic fully.

It will be appreciated that step (3) is applying an input test vector to the combinational logic which, with the primary (accessible) inputs, will exercise the combinational logic. Step (4), with the primary (accessible) outputs, is the response of the combinational logic to the test input data.

The principle behind the scan path testing is thus simple: it provides controllability and observability of internal nodes of a circuit when in test mode, utilizing the facility of a shift register to apply and monitor this internal data without the expense of having a large number of additional I/O terminals for this purpose. It also capitalizes on the fact that test vectors for combinational logic networks with no feedback can be automatically generated by ATPG programs, whereas test-program generation for sequential networks is extremely difficult — the need for the latter has of course been entirely eliminated by the re-configuration of the storage elements into a shift register configuration, isolated from the rest of the internal logic gates. It finally provides a means of partitioning the circuit into smaller blocks of combinatorial logic which may be more easily tested than one

Design-for-testability (DFT) techniques 317

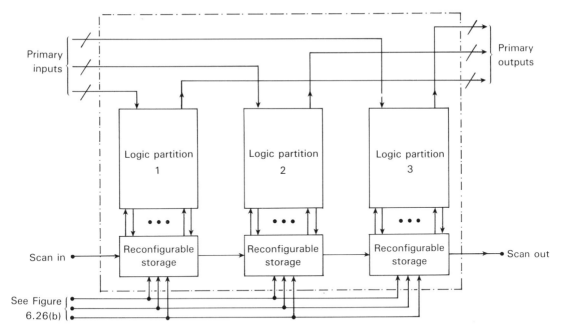

Figure 6.27 The partitioning of a very large circuit into smaller blocks for test purposes, using scan paths to give controllability and observability of each partition.

large block (see Figure 6.27), although, as will be seen in Figure 6.33, this form of partitioning is frequently associated with formal built-in, self-test methods. Notice that in the circuit shown in Figure 6.27 it is still necessary to scan-in and scan-out a lengthy bit stream of data even though the individual blocks may be more easily tested. Separate scan paths for each block would reduce the bit stream length but at the expense of addition scan-in/scan-out I/Os. The penalties to be paid for this method of testing must, however, be appreciated. They include the following:

- Each of the individual storage elements now becomes a much more complicated circuit in order that it may be re-configured, see below.
- The maximum speed performance is lower due to the extra gates in the storage elements, possibly a loss of 5 ns per flip-flop.
- Chip size will be larger, not only due to larger storage elements but also to the routing of the additional interconnections.
- Its adoption imposes a set of rigorous rules on the chip designer in what he must do, including careful consideration of the timing of clock pulses.
- A considerable amount of data has to be handled by the external test equipment in order to perform the large number of individual load and test operations, and hence the total time to test a circuit may be long.

Several variations of scan path testing have been used by different companies, differing mainly in the circuit configurations used in the re-configurable storage circuits and whether one, two or more clocks are required. Different company names will also be found. The first and most widely known one is the Level-Sensitive Scan Design (LSSD) system introduced by IBM, which uses the shift-register-latch (SRL) circuit shown in Figure 6.28 [53]. The terminology 'level-sensitive' refers to the action of the latches in the LSSD system: a latch (flip-flop) is said to be level-sensitive if all changes to the state of the circuit are controlled by the *voltage level* of a clock input signal, and not by its rise and fall times, and also the steady-state response to a change of value on the data inputs is independent of the gate and interconnection delays within the circuit. What this means is that the action of the circuit is raceless, not being dependent upon its transient (a.c.) performance which may vary or degrade, but only on the clock signals, whose timings are the dominating factor. (The opposite of level-sensitive is edge-triggered: here the switching action of a latch depends upon the propagation times of pairs of gates to give the latching action [53].)

Looking more closely at the circuit of Figure 6.28(b), in normal mode the scan shift and the scan transfer clocks $C2$ and $C3$ are inoperative (logic 0), and only the system clock $C1$ is in use. When $C1$ is high (logic 1) system data is free to enter latch $L1$; when the clock returns low (logic 0) this data is latched. Latch $L2$ is inoperative during this normal mode. Scan data is latched into $L1$ in place of system data by the scan shift clock $C2$, the system clock $C1$ now being inoperative, and, finally, either the system data or the scan data held in $L1$ is latched into $L2$ and fed out by means of the scan transfer clock $C3$. Hence the pattern of the clock pulses applied to $C1$, $C2$ and $C3$, which come from primary inputs and which must be non-overlapping, provides the following three operating conditions:

(a) normal operation, output from $L1$;
(b) latch in scan data into $L1$ instead of system data;
(c) latch in normal data or scan data from $L1$ into $L2$ and feed out to the scan path.

A variant on this mode of operation is to use both $L1$ and $L2$ in the normal operation of the circuit, clocking the system data first into $L1$ under the control of clock $C1$ and then into $L2$ under the control of clock $C2$. Both the system data and the scan-path data is then taken from the output of $L2$. This operating arrangement is termed the 'double-latch' LSSD system as distinct from the 'single-latch' system

---

Figure 6.28 The IBM LSSD scan-path shift-register-latch (SRL). Note the number of additional I/O pins to give internal controllability and observability is only four. Variants on this SRL circuit are possible, see text: (a) general schematic arrangement with one system clock and two test mode clocks per circuit; (b) NAND implementation; (c) typical systems use, with one scan path through all the $L1$, $L2$ latches.

Design-for-testability (DFT) techniques 319

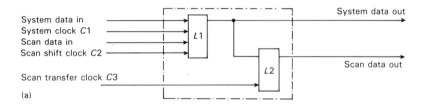
(a)

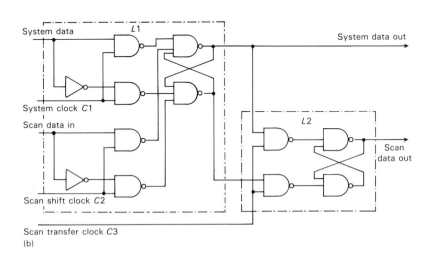
(b)

(c)

where the system data is taken directly from $L1$. Notice that in this double-latch arrangement, where both $L1$ and $L2$ participate in the normal circuit operation, in any feedback loop around the combinatorial network two latches and two clocks are always operating, and hence no feedback races can ever occur; to ensure this feature in the previous single-latch configuration two separate interleaved system clocks $C1(A)$ and $C1(B)$ are necessary rather than one common clock $C1$ to all stages, the combinatorial logic being partitioned such that where any feedback loop is present, then both clocks $C1(A)$ and $C1(B)$ are involved in the loop. Further details of the clock requirements for LSSD may be found in [10], [13], [14], [17], [54], [55].

These LSSD arrangements involve high circuit overheads — they have two to three times the amount of logic per shift-register-latch compared with a basic D-type circuit. An improvement can be achieved by providing both $L1$ and $L2$ with a system data input and a system clock, thus allowing the output from $L1$ *and* $L2$ to be used during normal system operation. This arrangement is known as the $L1/L2^*$ shift-register-latch [10], [14], and clearly reduces the overhead in comparison with the previous LSSD arrangements. These overhead requirements have been detailed by Williams [14], but care should be taken in reading these and other figures to make sure of what is being compared, particularly whether the non-LSSD basis of comparison is a master–slave circuit which by definition contains two latches, the input master and the output slave, and which is thus very nearly as complex as the LSSD circuit with latches $L1$ and $L2$ shown in Figure 6.28.

Alternative scan testing methods for VLSI circuits have been proposed, of which the three other most publicized are as follows:

(a) the scan path technique of NEC [56];
(b) the scan set technique of Sperry-Univac [57]; and
(c) the random-access scan technique of Fujitsu [58].

The scan path approach has a number of similarities to LSSD. It employs what was termed a 'raceless master–slave D-type flip-flop with Scan Path', the circuit of which is shown in Figure 6.29(a). As will be seen, the circuit closely resembles a conventional level-sensitive master–slave D-type circuit, but with the addition of a scan-in port and a second clock $C2$ to control the circuit when in test mode. The action of the circuit to load test inputs into the combinatorial logic is indicated in Figure 6.29(b), and the scan-out of the resulting logic signals is illustrated in Figure 6.29(c).

An essential difference between LSSD and scan path is that the latter only uses one clock for normal mode and one clock for test mode. This means that data passes through the master–slave circuit on one clock cycle, and therefore there is not the absolute security against race conditions in feedback loops which there is in the LSSD arrangement with two non-overlapping clock pulses for all data transfer. Further details, including the testing strategy, may be found in [10] and [14].

Design-for-testability (DFT) techniques 321

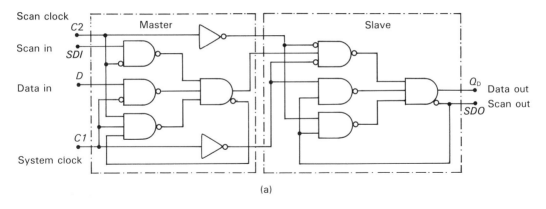
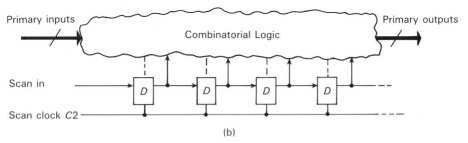
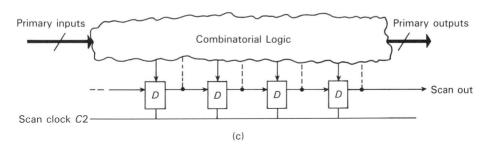

Figure 6.29 The scan-path method of NEC which employs raceless master–slave D-type flip-flops.

The scan set test technique is somewhat different from LSSD and scan path in that it does not employ any form of re-configurable shift-register-latch in the working circuits; instead it leaves the working mode storage circuits unchanged and adds an entirely separate shift register chain to give controllability and observability of the internal parts of the circuit. This is shown in Figure 6.30.

The advantages of scan set include the following:

- It may be applied to any type of digital network, and does not impose the strict design rules necessary in the LSSD and similar cases.

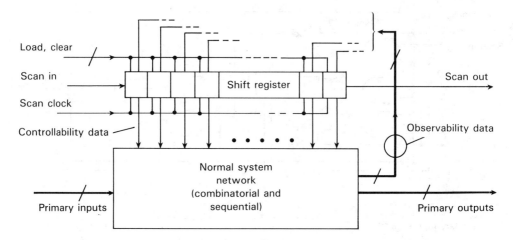

Figure 6.30 The scan-set technique, which employs an entirely separate shift register for test mode operations.

- It does not require more complex latches/flip-flops in the working circuit.
- The working system may be monitored under normal running conditions by taking a 'snapshot' of the data being fed into the shift register.
- Being separate from the working system the scan path testing will involve no additional race hazards and no partitioning problems of the working system.

Against these advantages must be weighed the following disadvantages:

- The additional cost of the entirely separate shift register and the possibly difficult-to-route interconnections within the working system may be high.
- It may be difficult to formulate a good test strategy and decide which nodes of the working system shall be controlled and observed when in test mode, bearing in mind that the normal system latches are not automatically disengaged when in test mode.

However, from the point of view of the system designer, scan set may be more immediately acceptable than the rules necessary to incorporate LSSD or scan path.

Finally, random-access scan: unlike all three preceding methods, random-access scan does not employ a shift register configuration to give internal controllability and observability. Instead, it employs a separate addressing mechanism which allows each internal latch to be selected individually so that it can be either controlled or observed. This mechanism is very similar to a random-access memory (RAM), consisting of a matrix of addressable latches which may be set either by the test input data or by the normal system latches, the outputs of which may be used to control the system latches or be read out for observation. Figure 6.31 gives the general schematic arrangement.

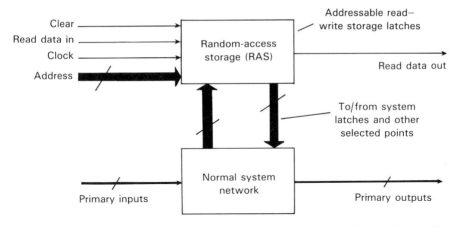

Figure 6.31 The random-access scan (RAS) technique, which employs a form of random-access memory for controllability and observability. An (optional) shift register may be used to read in data and record the data out.

The advantage of the RAS method is its flexibility, since it allows any latch or any node in a circuit to be addressed. Its disadvantages, however, include a high overhead to provide the RAS matrix and all the system interconnections, and a great deal of time is needed to set input values into the circuit and to observe the output responses.

Further details of all these techniques may be found in the previously cited references.

Whilst scan test methodologies and their variants give internal controllability and observability with the minimum number of additional I/O pins, and make automatic test-pattern generation feasible for the combinatorial blocks of logic, strict adherence to a full scan-based design methodology requires CAD resources which implement the required design requirements from the basic schematic capture or functional level data. This requirement, together with the considerable silicon area overheads and the often long testing times involved in scan test, has meant that such techniques have largely been confined to major project manufacturers such as IBM where testing requirements dominate. In an attempt to make the basic principles of scan test more acceptable to a wider range of OEMs, attempts have been made to define a partial scan test strategy, the objectives being to avoid having to make all the latches in the system convertible into scan path circuits when in test mode, thus saving silicon area and reducing the very long length of shift registers which can otherwise accrue.

The problem with a partial scan approach, however, is to decide which latches to make convertible, and which to leave as normal bistable circuit elements. One approach, by Agrawal et al. [59], requires the OEM to supply a set of test vectors which caters for as many circuit faults as possible – this is not an unusual requirement (see p. 296) – following which the faults which are not covered are identified and the latches required to be put into a scan mode to access these nodes are determined. However, it would appear that a high percentage (over 50%) of the latches may need to be in scan path mode to achieve the required controllability and observability of these 'difficult-to-test' nodes, and therefore it is not obvious that the silicon overhead and test-time savings achieved are justified. A further technique, RISP (reduced intrusion scan path), has been proposed by HHB Systems, [60], and it is likely that further developments along these lines will be pursued. However, one fundamental limitation will always remain with scan test, that is, the need to load in and verify test vectors one at a time, thus making it impossible to excercise the circuit under test at its normal operating speed [10], [14], [55].

We will return to a further form of scan test later in Section 6.8 when considering I/O testing; this is known as 'boundary scan' for obvious reasons, and has particular significance in giving controllability and observability of the pins of ICs after they have been assembled into their final PCB or other assembly.

6.6.3 Self-test DFT methods

Self-test, or, as it is more usually termed for IC design applications, 'built-in self test' (BIST), becomes increasingly significant as IC size increases to greater than, say, 5000 gates per circuit. As will be seen BIST has the following three considerable advantages:

1. It eliminates the need to generate any test patterns for the circuit under test.
2. It eliminates the need to have to use very expensive test equipment to apply test patterns and monitor the results, and expensive jigs ('test probe fixtures') to connect a circuit under test to the test equipment (see Figure 7.18).
3. The tests can usually be performed at the full operating speed of the circuit.

But against these advantages must be weighed the fact that the tests are not fully exhaustive functional tests under normal working-mode conditions, and hence there is always a small probability of a faulty circuit being passed as fault-free. However, this is an accepted risk with VLSI complexity circuits, since 100% functional testing can never be afforded.

The general hierarchy of all built-in self-test procedures is shown in Figure 6.32. On-line testing techniques are specialized and are particularly applicable to vital communications and control systems [17], [61]–[63]. Off-line testing, particularly with built-in hardware for test purposes, is currently more relevant for custom microelectronic circuits, although we will return to discuss certain on-line redundancy concepts in Section 6.9.

The off-line testing of computer systems with test programs is standard computer practice and will not concern us here. Instead, we will concentrate on the extreme right-hand path of Figure 6.32, where additional hardware and/or

Design-for-testability (DFT) techniques 325

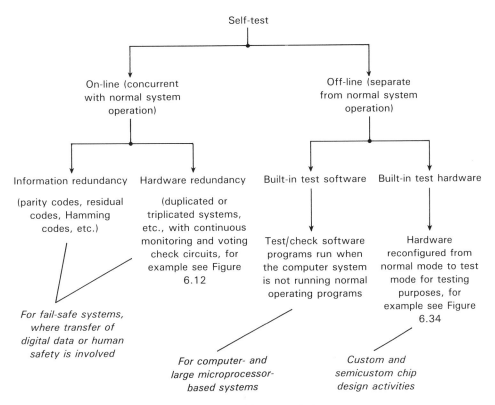

Figure 6.32 The hierarchy of possible self-test techniques.

re-configurable circuits are employed to provide the *in situ* test procedures. As in all our immediately preceding discussions, we are dealing with digital-only networks, and considering some set of test vectors which will provide an acceptable test set for the circuit under consideration.

The most widely known BIST technique is built-in logic block observation, or 'BILBO'. Its general arrangement is shown in Figure 6.33(a). Each partition of the complete circuit is a block of combinational logic, separated by a set of storage (register) circuits known as BILBOs, which in normal mode form part of the working circuit. In test mode, two sets of BILBO registers, one preceding and one following each block of logic, are used to test the block, the first to provide the input test vectors and the second to provide a test output signature. This signature can be read out in a serial scan-out mode for checking purposes. The first BILBO register can act as the output register for the last block of logic, as shown in Figure 6.33(b).

The BILBO registers, however are complex circuits, as shown in Figure 6.34. They operate not only in normal circuit mode and as a scan-in/scan-out shift

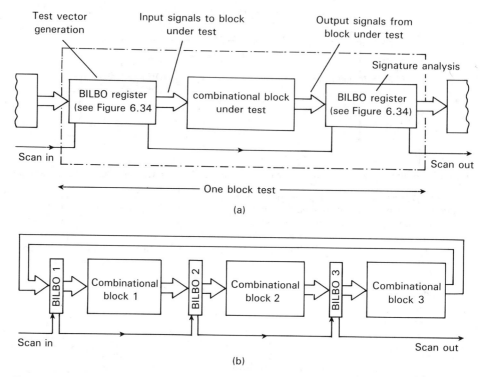

Figure 6.33 The partitioning of a digital circuit by BILBO registers; the control signals and clocks have been omitted for clarity: (a) the basic concept; (b) example circuit under test partitioned into three blocks when in test mode, each block being individually tested.

register, but also as a pseudo-random sequence generator for input test generation and output data capture. The control signals $C1$ and $C2$ operate as follows:

- $C1 = 0$, $C2 = 1$: reset mode, to set all the storage circuits to reset state on receipt of a clock pulse;
- $C1 = 1$, $C2 = 1$: normal operating mode with each storage circuit operating independently (see Figure 6.34(b))
- $C1 = 0$, $C2 = 0$: scan path mode with the storage circuits in a shift register configuration (see Figure 6.34(c)): notice there is an inversion between stages, but this does not cause problems; and finally,
- $C1 = 1$, $C2 = 0$: linear-feedback shift register (LFSR) mode, to provide either a pseudo-random test input sequence to the following block under test, or to give a signature analysis (see Figures 6.34(d) and (e)).

Each BILBO is controlled by its clock and control signals $C1$ and $C2$. The scan-in and scan-out leads may be connected all in series, or kept separate for each block.

'Flush-testing' of all the BILBO registers themselves can be done in scan-in/scan-out mode, as with any scan test system. Loading of individual input test vectors serially into the input BILBO to act as an input test set for each combinational block can then be done, and the resultant output response from the combinational block can be scanned out using the output BILBO in scan-out mode.

However, the principal reason for BILBO is to eliminate the need to formulate any set of input test vectors and apply them individually. As previously mentioned, this is a time-consuming process. Instead, the input BILBO is normally used in LFSR mode so as to apply a pseudo-random set of input test vectors; the output BILBO is simultaneously run in LFSR mode, but the output test data is mixed in the shift register so as to modify the counting sequence. Notice from Figure 6.34(e) that each combinational output is capable of modifying the sequence and hence the 'test signature' which is finally scanned out.

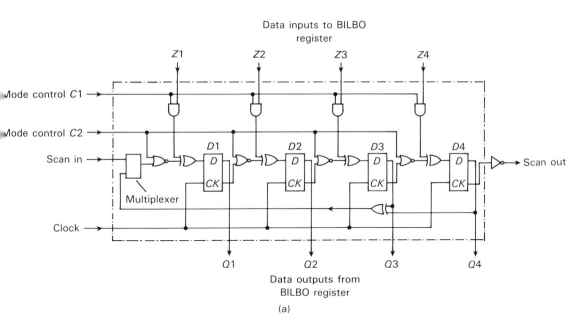

Figure 6.34 The functional attributes of a BILBO register, showing four stages only for clarity: (a) the overall circuit arrangement; (b) normal mode, with $C1 = 1$, $C2 = 1$; (c) scan-path mode with $C1 = 0$, $C2 = 0$; (d) pseudo-random input test vector generator (see Figure 6.35), $C1 = 1$, $C2 = 0$; (e) output signature capture (see Figure 6.36) $C1 = 1$, $C2 = 0$. (Note, $C1 = 0$, $C2 = 1$ (not shown) resets all stages to zero. The clock signal is applied in all modes of operation. Detailed variations on this circuit arrangement may also be found.)

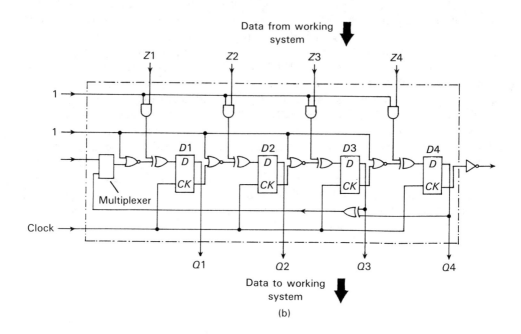

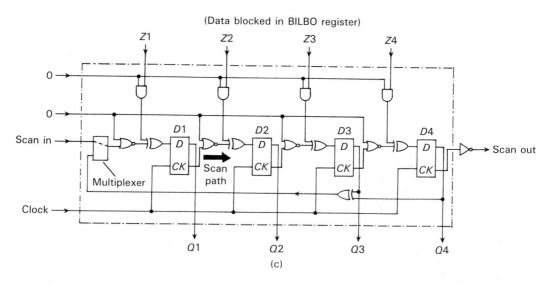

Figure 6.34 (Continued)

Design-for-testability (DFT) techniques 329

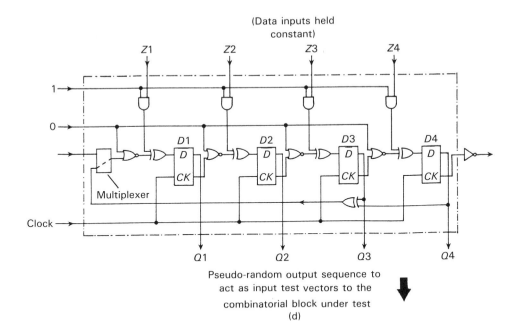

Pseudo-random output sequence to act as input test vectors to the combinatorial block under test

(d)

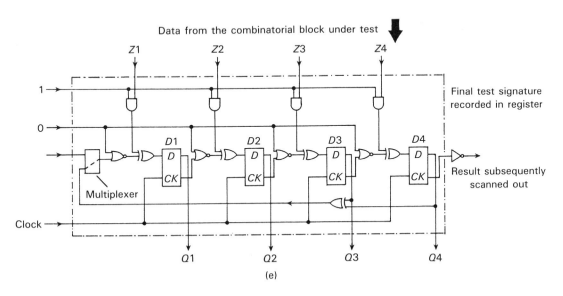

(e)

Figure 6.34 (*Continued*)

The principle of a linear-feedback shift register (LFSR) is shown in Figure 6.35. A LFSR is a normal shift register configuration which when clocked progresses its stored pattern of 0s and 1s from the first to the last stage, but which has feedback Exclusive-ORed from selected stages in the register to form the serial data input to the first stage. When the feedback taps are appropriately chosen an n-stage LFSR will count in a pseudo-random manner through $2^n - 1$ states; this is known as a maximum-length pseudo-random sequence, and contains all possible states of the register except for one 'forbidden' state. When the shift register is assembled by cascading the Q output of each stage directly into the D input of the following stage, this forbidden state is the all-zero $(00\ldots00)$ state; in the case of the configuration shown in Figure 6.34(d) the use of $\bar{Q}$ followed by the NOR inversion gives the same result. The all-zero case may be provided by resetting the shift register, but then a seed $1\ldots$, is necessary in order that a pseudo-random sequence may be started.

The taps which are necessary on an LFSR to produce a maximum length sequence may be found in many publications, for example [31]. For up to 300 stages it has been shown that more than four taps are never required, one of which must always be the last stage in the register. Lengths of multiples of eight stages always require a minimum of four taps, but other lengths require only two or one (never three) taps. Alternative taps are always possible. Thus the LFSR is an extremely simple test-pattern generator for generating a set of test vectors which contains all input patterns except one; the major limitation is that it is not possible to order the test vectors in other than an inherent pseudo-random sequence.

If additional 0 and 1 logic signals are fed into an LFSR then, clearly, some alternative sequence will result. This is the principal of the test 'signature', such as that introduced in Figure 6.34(e). The earliest application of this concept was for the testing of printed-circuit boards rather than ICs [64]. Here a hand-held probe is used to sample the logic signals at selected nodes of a circuit under test, as shown in Figure 6.36(a). Unless this probe signal is a constant 0, it will clearly influence

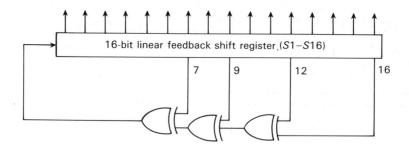

Figure 6.35 The linear-feedback shift register consisting of n D-type flip-flops in cascade with feedback so arranged as to generate a maximum-length pseudo-random sequence when clocked, this length being 65535 $(2^{16} - 1)$ in the case of a 16-stage assembly. The clock signal is omitted here for clarity.

Design-for-testability (DFT) techniques 331

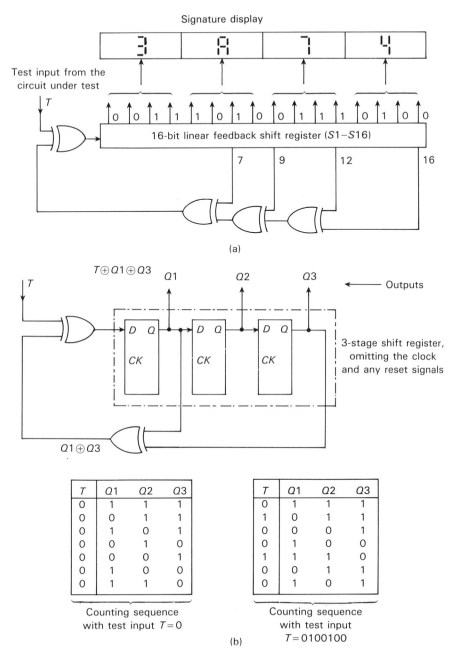

Figure 6.36 The principle of signature analysis using LFSR signature generation: (a) the addition of a test input signal to a maximum-length LFSR; (b) example LFSR sequences (signatures) with different test input bit streams, a three-stage circuit only being shown for convenience.

the sequence which the LFSR follows, and result in a different number being registered after a given number of clock pulses have been applied. Under test, therefore, the LFSR is reset to some known state, usually 11...11, and the circuit under test is likewise reset to a known state, following which a given number of clock cycles is applied to the circuit and the LFSR. The number registered after this test, the test signature, is then inspected; if it agrees with some pre-determined number no fault is considered to be present; if the signature is not as pre-determined for this test the circuit node is faulty. This is illustrated in Figure 6.36(b). By probing a number of nodes this analysis confirms the functionality of the whole board or identifies part(s) which are faulty.

There is still a possibility that a faulty bit stream from a node under test will produce the same signature as a fault-free bit stream. This possibility is known 'error masking' or 'aliasing'. However, provided the number of bits monitored is considerably higher than n (the number of stages in the LFSR), then the probability of generating a healthy signature from a faulty bit stream approaches the limit of 1 in 2^n [65].

To revert to the BILBO test philosophy of Figures 6.33 and 6.34: from Figure 6.34(e) it will be seen that a test signature from the block under test is generated, but, unlike the signature analysis method of Figure 6.36, multiple test inputs to the LFSR are now present. (Because of these multiple test inputs the circuit is sometimes referred to as a 'multiple-input signature register' or 'MISR'.) The test procedure for each BILBO block of combinatorial logic is therefore as follows:

1. Initialize the combinational block under test and the preceding and following BILBO registers to known starting states.
2. Apply a given number of clock cycles to both BILBO registers so as to generate the pseudo-random input test set and record the output signature.
3. Scan out the signature for approval.

This procedure is repeated for each of the logic partitions.

As with the single-input test signature there is always a possibility that a faulty circuit may give the same signature as a fault-free one. However, if the number of stages in the LFSR is reasonably large, then the error-masking probability will theoretically be

$$P(M) = \frac{2^{L-1} - 1}{2^{L+n-1} - 1}$$

$$\approx \frac{1}{2^n}$$

where

n = number of bits wide in the data input, = the number of stages in the LFSR and

L = length of the test sequence
  = $2^n - 1$ if maximum length

This compares with the theoretical error-masking probability of the single-input signature analyser of

$$P(M) = \frac{2^{L-n} - 1}{2^L - 1}$$

which reduces to the same value of $1/2^n$ provided $L, n \gg 1$ and $L > n$ [13], [17], [31], [65]. These analyses are based upon the assumption of a uniform probability of errors in the test data, an assumption which need not necessarily be true [14]. However, it is generally accepted that the error-masking probability of LFSR signatures, whether single-input or multiple-input, may be taken as $1/2^n$, which gives an acceptably small probability of error when n is, say, 16 or more.

BILBO will therefore be seen to offer an attractive method for the test of large digital networks at normal system speed, which would otherwise require complex testing resources. An n-bit maximum-length pseudo-random test set is effectively a fully exhaustive test for an n-input combinatorial block, the missing all-zero input vector being possible as an initial or final test if needed. The principal disadvantages of BILBO, however, are as follows:

- the overhead cost in silicon area to incorporate the BILBO registers — perhaps as much as 20–30% in die area;
- the inability to apply the test vectors in other than the fixed pseudo-random sequence;
- the difficulty in deciding upon the partitioning of the complete system into the optimum number of BILBO partitions;
- the difficulty of determining the correct fault-free signature from each BILBO partition — the 'good machine' signatures.

If a known-good circuit is available, then the test signatures can be obtained by running the BILBOs, but this is a dangerous procedure to attempt with prototype circuits. On the other hand, full simulation to determine the healthy signatures may be a very lengthy process, and thus it is possible that some judicious mix of simulation and fault modelling of a restricted number of nodes may be used.

Variations on the BILBO circuit arrangement shown in Figure 6.34 have been published [65]–[72], but the essential principles covered above remain. An alternative circuit to the LFSR for the pattern generation and signature analysis has also been pursued and will be mentioned in Section 6.9.7.

6.7 PLA design-for-testability techniques

It is generally accepted in industry that PLAs with an upper limit of eighteen to twenty inputs can be exhaustively tested; a 10 MHz tester clock rate will test such a circuit in about 100 ms. However, for greater than twenty inputs there is an increasing need for alternative test procedures, but in all cases appropriate controllability and observability of the I/Os is necessary if the PLA is a buried macro in a VLSI layout.

As covered in Section 6.4.3, the simple stuck-at fault model is not fully appropriate to PLA testing, and hence the test-pattern-generation programs which have been developed are based upon more than the simple stuck-at-fault model [39]–[45].

In this section we will introduce some of the concepts which have been proposed for making PLA architectures more easily testable; all involve some addition(s) to the basic AND/OR matrix, and hence the minimization of this additional circuit complexity is of paramount practical importance.

Figure 6.37 shows the span of PLA test methodologies, with the initial division being between on-line (concurrent) test and off-line techniques. Our interests here will not be with all these theoretical divisions, but principally with those which involve some circuit enhancements to achieve improved testability. As will be seen, the overall picture includes adoption of signature analysis and BILBO structures [14], [31], [73], as well as code checkers in the case of concurrent testability [13], [74]–[76]. The signature analysis and BILBO techniques may involve the replacement of the normal input decoding circuits which provide x_i and $\bar{x}_i$ from each primary input x_i (see Figures 3.10–3.13) with input latches which, when in test mode, are re-configured into an LFSR so as to provide the input test vectors. However, in general, the addition of signature analysis of BILBO adds an unacceptable silicon overhead to the basic array architecture *unless it is necessary* because of restricted test access to the PLA I/Os.

Concurrent checking is also an area which will not be considered in any depth here, but we will look at some fundamentals in Section 6.9.6. However, it is worth noting that complete duplication of PLAs provides an extremely powerful on-line method of checking (see Figure 6.12), particularly if:

- the primary inputs are arranged dissimilarly between A and B, each PLA AND array being programmed accordingly;
- the primary outputs are likewise dissimilarly allocated;
- different product terms are used in A and B to generate each required output function, or if B is designed always to generate the complement of the A outputs.

However, there remains the problem of verification of the equality (or complementarity) of the two outputs by means of some fail-safe checker, and even if this is done there still remains a situation that the primary input data from

PLA design-for-testability techniques

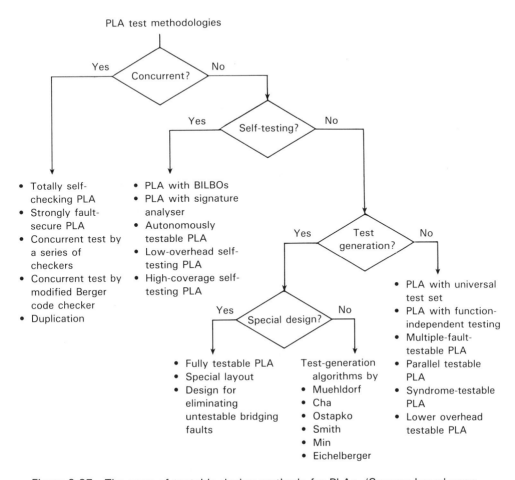

Figure 6.37 The span of testable design methods for PLAs. (Source: based upon [43].)

preceding circuits may be faulty and/or subsequent circuits fed by the PLA outputs may be faulty. (Ideally, the output from any form of checking circuit, electrical, electronic or mechanical, should not normally be a static condition, e.g. steady logic 0 or logic 1, but rather some dynamic (continuously switching/oscillating) condition. It is left to the reader to consider the implications of such a fail-safe philosophy.) Hence, excessive checking of just one part of a circuit should always be considered in the context of the complete system or product.

To revert to the other PLA DFT techniques: recall from Figure 3.11 that a PLA has n primary inputs, $x_1, \ldots, x_n$, p internal product lines, and m primary outputs $f_1(X), \ldots, f_m(X)$. Although referred to as an AND/OR array architecture for convenience, the actual arrays may be NAND or NOR structures, but this does not

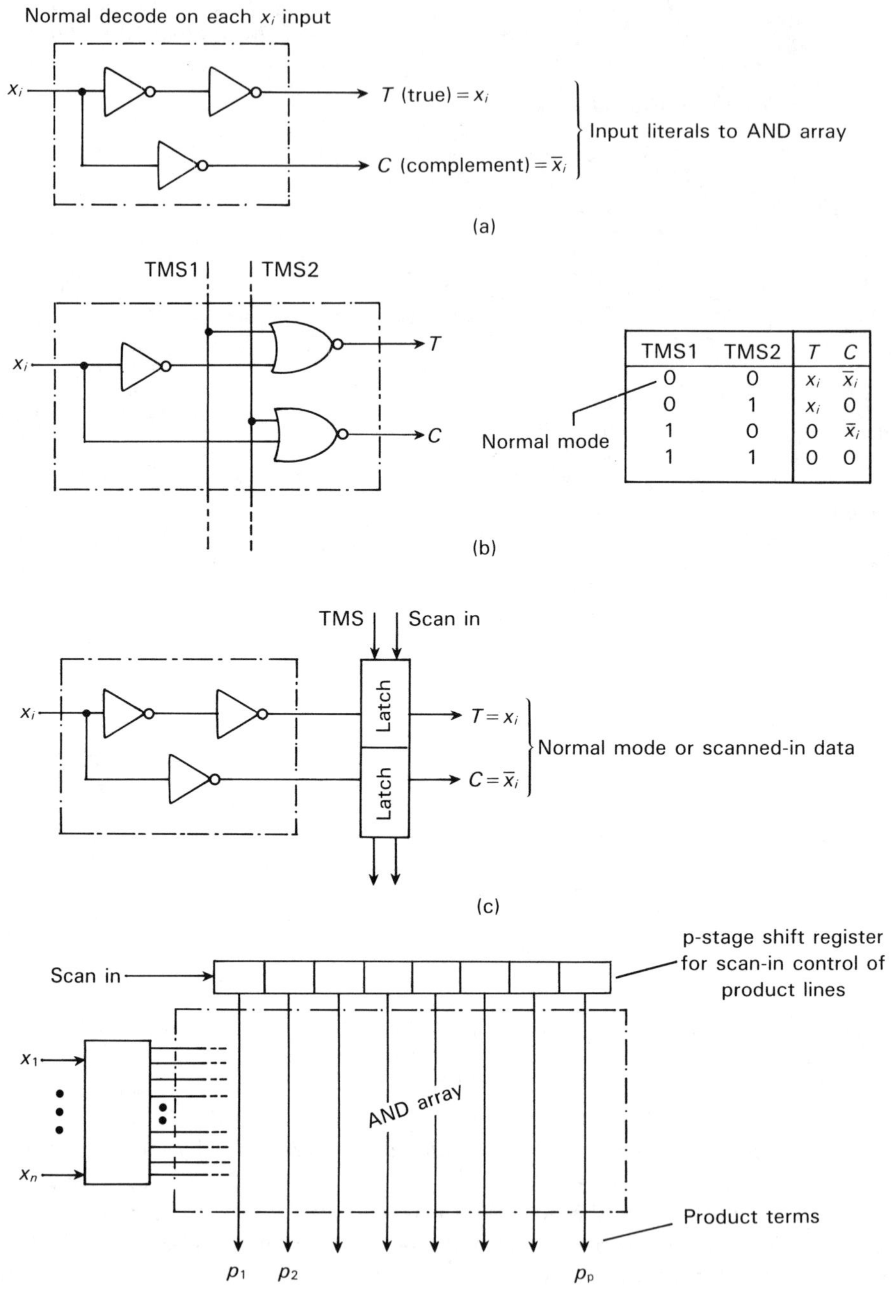

Normal decode on each x_i input

T (true) = x_i
C (complement) = $\bar{x}_i$
} Input literals to AND array

(a)

TMS1 | TMS2

→ T
→ C

Normal mode

TMS1	TMS2	T	C
0	0	x_i	$\bar{x}_i$
0	1	x_i	0
1	0	0	$\bar{x}_i$
1	1	0	0

(b)

TMS   Scan in

Latch
Latch

$T = x_i$
$C = \bar{x}_i$
} Normal mode or scanned-in data

(c)

Scan in → p-stage shift register for scan-in control of product lines

x_1
⋮
x_n

AND array

p_1 p_2 p_p

Product terms

(d)

Figure 6.38 DFT techniques in programmable logic array architectures. Combinations and variations on these may be found: (a) the normal primary input decoding; (b) modification to (a) to modify the x_i or $\bar{x}_i$ inputs or both; (c) scan-path addition to override the normal x_i inputs — TMS = 'test model signal'; (d) scan-path addition to allow removal of product terms from the AND array; (e) an alternative to

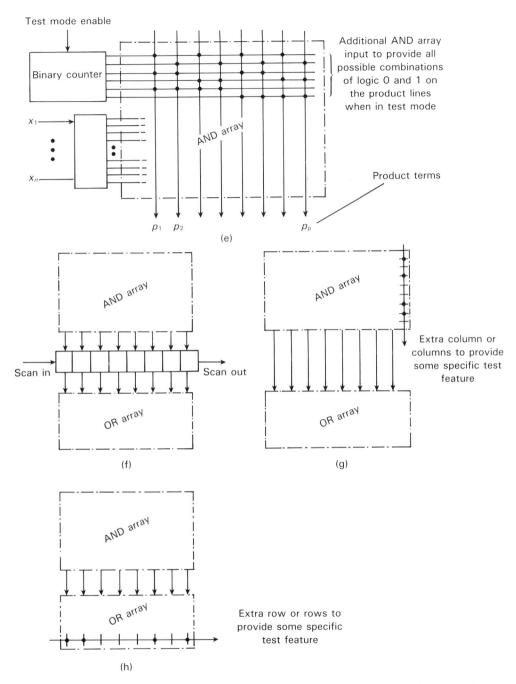

(d), where the binary counter should have $\log_2 P$ stages so as to be able to supply all possible signal combinations to the P product lines; (f) scan-path addition to provide observability and controllability between the AND and OR arrays; (g) and (h) additional column or row lines in the AND and OR arrays, respectively, so as to give, say, some parity or other check feature.

affect our general consideration of the normal PLA mode of operation. It does, however, need to be taken into consideration when looking at actual test mode signals, in order to ensure that logic 1s and logic 0s are appropriately chosen.

The principal needs when considering PLA DFT techniques are as follows:

- controllability of the 2^n x_i and $\bar{x}_i$ input literals, either individually or collectively;
- controllability and perhaps observability of each of the p product lines (the AND array outputs); and
- controllability of each of the m OR array output lines.

Typical ways in which these requirements may be approached are shown in Figure 6.38. As will be seen there are three principal ideas:

1. Alter the normal x_i input decoders so as to be able to modify the x_i and $\bar{x}_i$ AND array inputs.
2. add scan path shift registers so as to be able to shift in specific 0s and 1s to the AND or OR arrays, or to scan-out product line data for test purposes.
3. Add additional product columns to the AND array and/or additional rows to the OR array for particular tests.

However, there is no one accepted 'best' method for realizing and combining these DFT features.

Examples of DFT techniques which illustrate the above basic concept are those of Fujiwara *et al.* [77], [78], Hong and Ostapko [79] and Khakbaz [80]. Figure 6.39 gives the FK testing technique of Fujiwara [77], from which it will be seen that the architectural modifications include the following:

(a) modification of the input decoders to give outputs $x_i + Y_1$ and $\bar{x}_i + Y_2$, thus allowing the true and the complementary input literals collectively to be set to a steady logic value;
(b) addition of a shift register to the AND array to enable the selective choice of any one product line at a time; and if the AND array is really a NOR matrix then this requires a logic 1 on all product lines except the selected one. A NAND matrix would require the opposite logic; and
(c) an extra column in the AND array and an extra row in the OR array, which can be programmed such that the total number of crosspoint connections in each AND row and in each OR column is odd.

The odd parity check is made by cascaded Exclusive-ORs on the product lines, giving test output Z1, and on the sum lines giving test output Z2.

Details of the test patterns required to test for all single faults in the PLA and the additional DFT circuitry may be found in [77]. These tests are independent of the functions provided by the PLA. An extension of this FK scheme is the later FKS

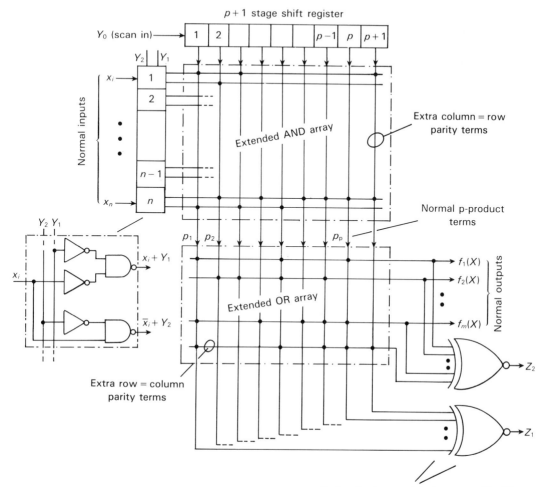

Figure 6.39 The basic DFT PLA architecture of Fujiwara and Kinoshita, which provides function-independent tests covering all single cross-point and stuck-at faults. Note, the $(p+1)$-input Exclusive-OR gate shown here for output $Z1$ would be composed of a cascade of p 2-input XOR gates, and similarly for output $Z2$ there would be a cascade of m 2-input XOR gates.

proposal [78], which adds a shift register on the AND array input lines so that any row in the AND array may be individually selected. This addition gives complete multiple-fault coverage, but requires a higher number of input test patterns [13], [14], [78].

The FITPLA test scheme of Hong and Ostapko [79] is similar to the FK and FKS schemes in that a shift register is employed to select individual product lines, with parity checking included in the AND and OR arrays, but the individual input line decoding and the parity checking is more complex. Like the FK and FKS proposals, the test set is independent of the functions programmed in the PLA; more additional circuitry is involved in the FITPLA method, but the number of test patterns required is somewhat less [14]. All these proposals, however, require lengthy cascades of Exclusive-OR gates for parity checking, which impose a timing constraint on the test procedures and require a considerable silicon area to implement.

The K-scheme of Khakbaz [80] does not employ any Exclusive-OR parity checking and thus does not suffer from these silicon area and timing penalties. However, the test patterns are function-dependent, and hence the test set has to be determined for each PLA dedication. All that is added in the K-scheme DFT method is a shift register on the AND array to select individual product lines, plus one additional row in the OR array which is the OR of *all* the product terms. Knowing the dedication of the PLA, a test set can be formulated which tests (i) the correct operation of the shift register, (ii) the correct presence of each individual product term, and (iii) the correct outputs from the OR matrix. All single and multiple stuck-at faults and all crosspoint faults which give rise to incorrect outputs can be detected.

These and many other PLA testing strategies have been proposed. Attempts have also been made to compare the advantages and disadvantages of many proposals [43], but the fact that research continues in this area indicates that no single strategy has received universal acceptance. We will come back to some further PLA testing ideas at the research level in our discussions in Section 6.9.

As far as PLAs as buried macros in custom VLSI circuits are concerned, vendors' views on:

- on-line concurrent checking,
- built-in self-test,
- enhanced DFT designs with function-independent test sets,
- enhanced DFT designs with function-dependent test sets, or
- no DFT enhancements at all,

will influence what the OEM may have to consider in future custom designs.

Further details of and discussion on PLA testing may be found in [13], [14], [43]–[45] and the references contained therein.

6.8 I/O testing and boundary scan

The individual I/O circuits at the perimeter of an IC are frequently extremely complex circuits, since they may be designed to be inputs or outputs as required, and have to interface between various combinations of on-chip and off-chip circuit requirements. Protection against static voltage damage may also be built in. Further

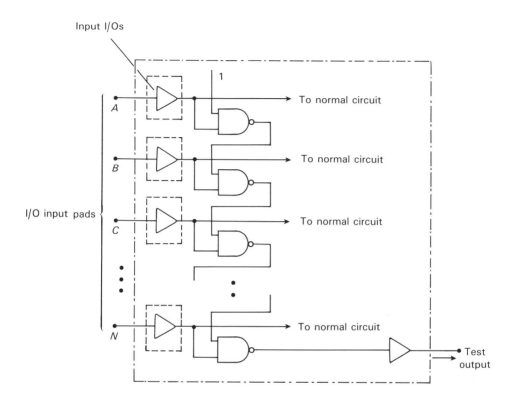

Figure 6.40 Vendor's testing of input I/Os, with the test input vectors being applied at the voltage threshold limits of V_{IL} and V_{IH}.

consideration usually has to be given to the fact that when packaged ICs are finally assembled on a printed-circuit board (PCB), final testing of the complete board or product normally has to be undertaken.

6.8.1 *I/O testing*

It is particularly important that the input signals provided by the I/Os are correct to drive the subsequent on-chip internal logic; the output signals provided by the I/Os can be monitored by external test instrumentation. Hence, custom IC vendors may measure (i) the output voltage of each output pin under some defined worst-case loading or supply conditions, and (ii) the satisfactory performance of each input pin under low voltage input conditions (V_{IL}) and high voltage input conditions (V_{IH}) – the 'threshold' design limits of the input circuits. Testing over a range of temperatures may also be performed, particularly for military or avionic applications.

The I/O input tests require additional on-chip logic to check that the input circuits do switch the logic correctly. Figure 6.40 illustrates one internal test circuit whereby each input drives one of a cascade of 2-input NAND gates. To check that each input responds correctly to low and high input signals, ($N+1$) tests at V_{IL} (low) followed by the same ($N+1$) tests at V_{IH} (high) are applied, the resultant switching signal at the single test output pin confirming the action of each input being monitored [81].

If the I/Os are bidirectional and have to operate as either inputs or outputs on demand, then an additional test-enable pin has to be provided to disable the output portion of the bidirectional circuits when in test mode. The same V_{IL} and V_{IH} patterns may then be applied and response checked through the NAND cascade.

Tristate-ing of I/Os is another feature provided by many vendors. When switched by an appropriate control signal all I/Os take their tri- (third) state, becoming a high impedance not allowing current to flow in either direction. By this means the I/C on an assembled PCB can be effectively disconnected, allowing tests to be made on the rest of the board or product as though the IC was not present.

However, the most powerful technique for handling the testing problems associated with ICs assembled on PCBs is a scan path technique, as covered in the following section. Once again the principle is simple: it is to provide controllability and observability of I/O signals, this time within a possibly large and complex PCB with difficult test accessibility, so that either the assembled IC can be separately tested, or the board surrounding the IC can be tested, or both.

6.8.2 *Boundary scan and board testing*

The development of boundary scan originated from design engineers in Europe who were involved in testing considerations for very complex PCBs, with increasing use of surface-mount ICs which could not easily be removed after assembly, and decreasing spacings between conductors and components, making probing more difficult. As a result, the international Joint Test Action Group (JTAG) was established, the efforts of which have now resulted in the IEEE Standard 1149.1 for boundary scan architectures and protocols.

The essence of boundary scan is to include a scan latch in every I/O circuit, as shown in Figure 6.41. During normal-mode operation, system data passes freely through all the boundary scan cells, but in test mode this transparency is interrupted and replaced by a number of possible test mode conditions, providing the following seven facilities in total:

(a) normal mode, transparent on-chip operation;
(b) test mode, scan in test data to the input boundary cells to act as a test vector;
(c) test mode, use data (b) to exercise the on-chip circuits, and capture the resultant response in the output boundary cells;
(d) test mode, scan out test results (c);
(e) test mode, scan in test data to the output boundary cells;
(f) test mode, use data (e) as a test vector to exercise the circuits on the PCB surrounding the particular IC, and capture the resultant PCB response in input boundary cells;
(g) test mode, scan-out test results (f).

it will be seen that tests (b)–(d) above are the usual load-and-test methods associated with scan path IC testing, but the provision of a boundary scan cell on every I/O now enables the same load-and-test technique to be applied to the circuitry outside the IC, the particular IC effectively being isolated from the rest of the system. This is the power of boundary scan. Notice that modes (b), (d), (e) and (f) above are identical scan modes with the boundary cells in cascade, and hence in total only two signals are necessary for boundary scan control purposes, these two signals being decoded on-chip to give four test conditions (see Figure 6.41(b)).

Multiple ICs on an assembled PCB are all under the control of the JTAG 4-wire test bus with its four lines TDI, TDO, TMS and TCK. Figure 6.42(a) indicates how each IC has its own boundary scan which enables it to be scan tested independently, or to receive or transmit test data to other circuits or components. Still further facilities are now being commercially pursued, with the availability of off-the-shelf chip sets from IC vendors providing such facilities as the following:

- test bus selector circuits, which can select certain scan paths on a complex PCB layout and short-circuit the remainder;
- test bus controller circuits for overall management of the scan path data; and
- scan test circuits for capturing data on busses, or loading test data on to busses in place of normal data (see Figure 6.42(b)).

Considerable housekeeping may accrue in developing and keeping track of all the

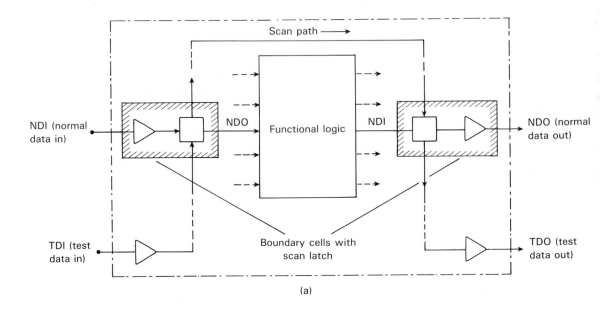

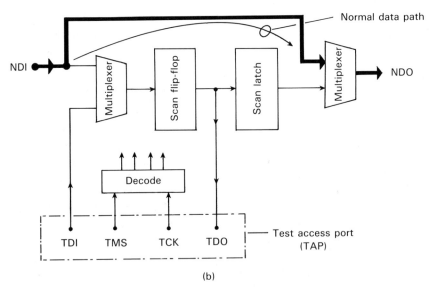

Figure 6.41 The principles of boundary scan: (a) provision of boundary scan on each I/O so as to give a complete scan path; (b) simplified details of a boundary cell; exact details may be proprietary.

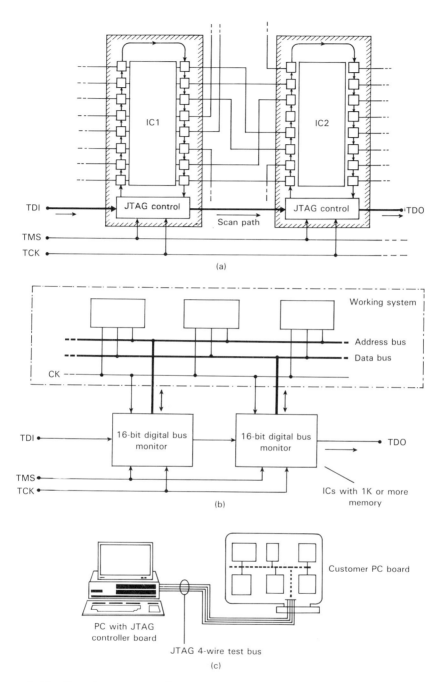

Figure 6.42 Boundary scan applications: (a) the scan path through several on-board ICs; (b) scan-path principles applied to bus monitoring, with the monitor ICs able to capture system data or load in test data; (c) CAD resources to assist in generation of test data formats and general housekeeping. (Source: based upon Texas Instruments publicity data.)

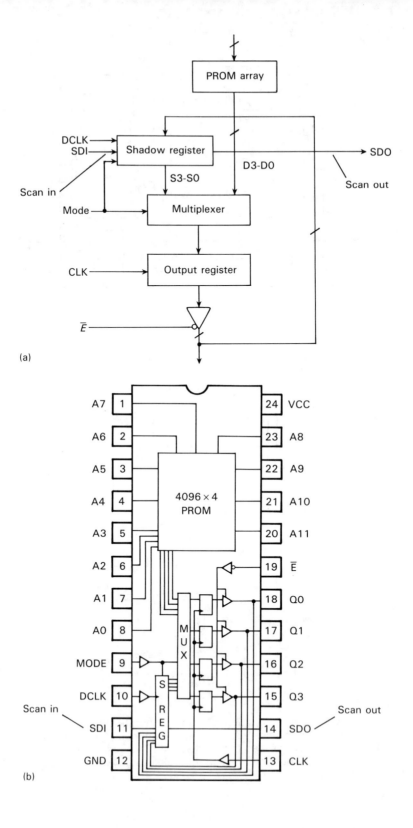

test mode data in a complex system, and therefore a further commercial development is that of CAD resources designed specifically for the protocols and system test-data generation. This is illustrated in Figure 6.42(c).

All these developments have evolved from the acceptance of the four-wire JTAG test bus and the IEEE Standard 1149.1, and further evolution will undoubtedly follow, including the use of pseudo-random test-pattern generation for the scan-in test data. It is evident that the increasing complexity of both individual ICs and the complete systems in which they are used have fostered these DFT developments, with scan path being the key element in providing increased controllability and observability without requiring an unacceptable number of additional test pins.

Further details of boundary scan and its considerable complexity of standardization may be found in IEEE 1149.1 and elsewhere [71], [82]–[89]. Finally, it may be noted that a simple form of scan testing was jointly introduced by Advanced Micro Devices Inc. and Monolithic Memories Inc. in the early 1980s, which therefore pre-dated the present boundary scan circuits by several years. This was the Serial Shadow Register (SSR^{TM}) architecture for MSI ROM and PROM circuits. From Figure 6.43 it will be seen that a 'shadow register' was added to the normal memory circuits. This register could function as a serial-in/serial-out shift register, or as a parallel-in/serial-in/parallel-out register. By this means, system data could be captured and serially scanned out for test purposes, or test data serial scanned in and loaded into the working register to replace system data. The scan-in/scan-out principle, however, resulted in the minimum number of required test pins, as it does for all scan test methods. Further details of these earlier developments may be found in [8], [10], [90].

6.9 Further testing concepts

Research into very many aspects of test for digital networks has been and continues to be undertaken, but this may not yet be reflected in common practice. We will, however, review and reference some of this work herewith in order to give a more complete overview of testing concepts.

As was seen in Section 6.1, one of the main problems with digital network testing is the volume of simulation data which has to be entered and its responses checked (see Figure 6.6). One ideal test procedure would be as shown in Figure 6.44(a), where:

- one additional input pin is added to the network under test so as to switch the network from normal mode to test mode,

Figure 6.43 The principles of early scan-path test applied to individual memory ICs, possibly now superseded by the more general boundary scan DFT: (a) the scan-path 'shadow register' in a programmable read-only memory; (b) 4 K × 4-bit PROM with shadow register diagnostics.

348   Test-pattern generation and design-for-testability

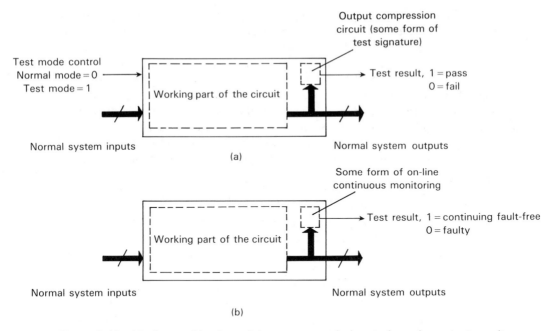

Figure 6.44 Ideal test objectives: (a) one test mode input pin and one test result output pin; (b) continuous (on-line) testing, with one test result output pin.

- one additional output pin is added to show the final result of testing, this being a simple 'pass' (fault-free) or 'fail' (faulty) indication, and where
- no failure mode is possible in the pass/fail check circuit(s),

all this being done with the minimum additional circuit complexity! Unfortunately, this ideal situation cannot be realistically achieved. Another ideal test procedure would be continuous checking, as shown in Figure 6.44(b), where one output test pin only is required, again with the minimum additional circuit complexity.

Most research work in digital test will be seen to be concerned almost exclusively with combinatorial logic only. This is because no realistic means apart from some form of functional test, can be formulated to test sequential networks with their possibly very long counting sequences and their inherent feedback connections, and hence the basic concepts of

- partitioning very long counters into manageable proportions (see Figure 6.24),
- separating the sequential logic elements from the combinatorial logic (see Figure 6.26(a)), and
- the use of shift register configurations as the most easily testable sequential architecture,

naturally arise. Beyond these basic concepts very few more useful ideas can be envisaged apart from some forms of self-checking, see Section 6.9.6. Test-pattern generation for sequential networks has very little significance.

On the combinatorial side, however, considerable R and D work may be found, including the following areas of investigation. On-going work is reported regularly in *IEEE Design and Test of Computers*, in the *Journal of Electronic Testing: Theory and Applications*, and in other professional journals, as well as in annual conferences such as the *IEEE International Symposium on Fault Tolerant Computing*, the *IEEE Design Automation Conference*, and others.

6.9.1 *Input compaction*

A number of authors have observed that although the total number of binary inputs in a system may be large, typical combinatorial outputs often depend upon a restricted sub-set of those inputs. Thus, although there may be, say, n inputs in total, where n is large, requiring 2^n input vectors for exhaustive global testing, individual outputs may be dependent upon $i, j, \ldots$ inputs, $i, j, \ldots, < n$, giving $2^i, 2^j, \ldots$ input vectors for local exhaustive test purposes.

It may also be observed that in many cases individual or local network outputs depend upon a disjoint set of input variables. Partitioning of the overall system at the design stage to achieve this may also be relevant. Hence, although, due to maximum test time considerations, one may be limited to an input test set of, say,

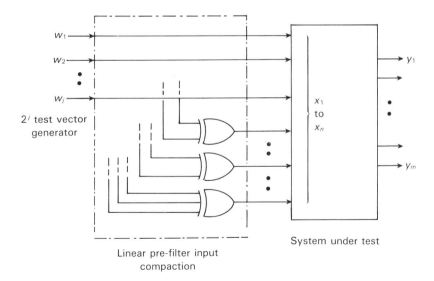

Figure 6.45 The use of a linear pre-filter to generate n binary inputs from the j-bit test vector generator; the system under test is the combinatorial elements only (see text).

Table 6.6. The seven possible input vectors available from three primary inputs by linear summation. Note the choice of any three of the seven vectors will give an exhaustive input test set provided all the subscripts 1, 2 and 3 appear (but not all an even number of times) in the three chosen vectors

Primary inputs						
w_1 w_2 w_3	$w_1 \oplus w_2$	$w_1 \oplus w_3$	$w_2 \oplus w_3$	$w_1 \oplus w_2 \oplus w_3$
0 0 0	0	0	0	0
1 0 0	1	1	0	1
0 1 0	1	0	1	1
1 1 0	0	1	1	0
0 0 1	0	1	1	1
1 0 1	1	0	1	0
0 1 1	1	1	0	0
1 1 1	0	0	0	1

2^j vectors, by judicious sharing and combining of input signals when in the test mode, exhaustive testing of several network outputs can be simultaneously undertaken.

A significant approach to the generation of reduced input test sets for exhaustive local testing is that of Akers [91], which continues earlier developments [92], [93]. The system total of n binary inputs is generated from j independent binary test signals, $j < n$, by linear (Exclusive-OR) relationships, as shown in Figure 6.45. The value of j is determined by consideration of the maximum number of x_i input variables involved in any output function y_i, $i = 1, ..., m$, and the input independency between the m output functions. In the case of a simple 3-bit test-vector generator, a total of seven bit streams may be generated (see Table 6.6), from which 28 exhaustive 3-bit data streams can be selected. The total number of bit streams possible from a j-bit test vector generator is $2^j - 1$.

Published details cover the derivation of appropriate exhaustive input test sets for given multiple-output networks, with the emphasis being upon minimizing the width of the test-vector generator w_j, and hence the test time.

6.9.2 Output compaction

The several discrete local or system outputs y_i, $i = 1, ..., m$, may also be compacted into fewer test output lines z_i, $i = 1, ..., l, l < m$, by the use of Exclusive-OR relationships (see Figure 6.46). This compaction may be termed 'space compression', since it does not modify the time for the exhaustive testing of any individual output [94], [95].

However, whilst exhaustive energization of the network under test is still a part of this approach, the principal thrust of output compaction has been to eliminate the need for specific evaluation of the resulting z_i output functions when in test mode. Instead, the aim has largely been to consider optional combinations of the

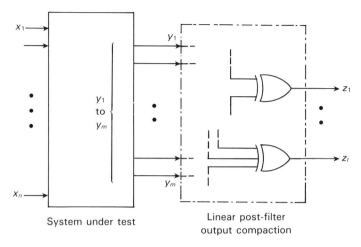

Figure 6.46 The use of a linear post-filter to generate l binary outputs from the m system output points.

y_i outputs such that maximum fault cover will be given by the reduced number of z_i test outputs.

If, for example, the final fault detection mechanism is a simple counting procedure which simply counts the number of logic 1s (or logic 0s) appearing on an output line when under test (this is otherwise known as the syndrome count, and will be specifically considered in the following Section 6.9.3), then maximum fault coverage (minimum fault masking) is obtained when this count tends to either the maximum of 2^j, where j is the number of binary inputs in the exhaustive input test sequence, or to zero [96]. The objective of output compaction and other forms of output modification involving Exclusive-OR combinations is therefore to maximize (or minimize) this value. A consideration of the choice of functions to combine in the output compaction method of Figure 6.46 is a non-trivial problem, since there are in total $2^n - 1$ possible combinations of the y_i outputs [94]–[96], but a further output compression proposal is to combine *all* the y_i outputs under test with a further function y_{m+1}, y_{m+1} being chosen such that the final output signature is all 1s or all 0s or some other precise signature. This is illustrated in Figure 6.47.

As in most forms of compression, a correct output signature does not absolutely guarantee the correct functionality of the circuit under test. For example, depending upon exactly how the circuit was designed, the test circuit of Figure 6.47 may not be able to detect a stuck-at fault in a primary input line which causes outputs $y_1, ..., y_{m+1}$ to lose this input variable, or an internal fault which affects two (or any multiple of two) network outputs identically. The more output compression involved, i.e. reduction in m outputs to the minimum of one, the more care needs to be taken in considering whether the test compression is acceptable – for example, a compressed output signature of a single 0 or a single 1 in the output

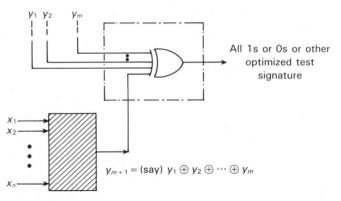

Figure 6.47 Linear output compaction of all y_i outputs, $i = 1, \ldots, m$, with a further test function y_{m+1} so as to give, say, all 1s in the compressed test output.

vector, rather than all 1s or all 0s, may be more secure against stuck-at input faults. Further details of input and output compression may be found in the reference list, but such techniques remain largely *ad-hoc* techniques rather than formalized methods to ease the testing problem.

6.9.3 *Syndrome and transition counting*

Mention of syndrome count has been made in the preceding section. The syndrome of a combinatorial function or network is formally defined as

$$S = \frac{k}{2^n}$$

where k is the number of 1-valued minterms in the function output and n is the number of independent primary input variables. With this definition S ranges between 0 (no 1-valued minterms) and 1 (all minterms one-valued). However, integer values are preferred in practice, and hence the normalizing factor of $1/2^n$ is usually omitted, leaving the syndrome S numerically equal to the number of 1s in the output of a given function.

Syndrome testing was first proposed by Savir, and much research work was undertaken to defining 'syndrome-testable' networks [12], [13], [31], [97]–[101]. The basic objective was to establish a unique syndrome signature from a network under test, usually with an exhaustive input test set, thus compressing the network output response into a simple 1s-count, as shown in Figure 6.48.

However, it may readily be shown by example that this very simple output test is not foolproof: certainly if the test count differs from the healthy count the circuit is faulty, but there may be a number of possible faults which produce an incorrect output function with the same number of logic 1s (or 0s) in the output bit stream.

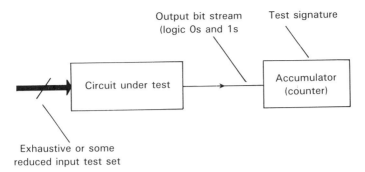

Figure 6.48 Syndrome counting, where the output accumulator is a simple counter counting the number of 1s in the output bit stream (cf. the more complex and secure output count used in LFSR signature analysis, see Figure 6.36).

A trivial example is an Exclusive-OR function, say $x_1 \oplus x_2$, which has an equal number of logic 0s and 1s in its output truth table. If x_2 is missing the output becomes just x_1, which still has an equal number of 0s and 1s. Hence a simple logic 1 count is not a practical proposition for general-purpose testing.

Further work on syndrome testing includes the following:

(a) checking to see whether a syndrome count for a particular network will detect all given internal faults such as stuck-at faults;
(b) modification of the network under test by the addition of test inputs and some internal logic which effectively adds minterms when under test so as to give a secure syndrome count for all possible circuit failures;
(c) constrained syndrome testing, which involves more than one syndrome test, certain inputs being held constant during each individual syndrome count procedure; and
(d) weighted syndrome testing for multiple-output networks, where the individual outputs are given weighting factors, the final test count being $\sum_{i=1}^{m} w_i S_i$ where w_i is the weight given to the syndrome count S_i from output y_i.

Further details of all these proposals may be found [99]–[101]; we will refer to (d) above again in Section 6.9.4 but, in practice, syndrome testing has not been pursued generally.

An alternative to the syndrome count is a count of the number of transitions from 0 to 1 (or vice versa) in the output bit stream. Like the syndrome test, this transition count test normally considers a fully exhaustive input test set. However, unlike the syndrome test where the count is independent of the order of application of the input test vectors, a transition count *is dependent upon the order of application*. Thus the exhaustive input test set may be ordered such that the number of transitions in the healthy output response is minimized, which in the extreme is a single 0-to-1 (or 1-to-0) transition [12], [102]–[105].

An ordering of the exhaustive input test vectors such that there is only one transition in the output bit stream (i.e. initially all 0s followed by all 1s, or vice versa) is not in itself a 100% secure output test. The danger here is if a block of logic 0s at the end of the healthy logic 0 bit stream becomes all 1s, or if a block of 1s at the transition point becomes all 0s, then this mode of failure will not be detected by a single transition count detection. However, this shortcoming may be very elegantly overcome by repeating the initial test vector of the block of 0s and the initial vector of the block of 1s at the end of their respective blocks, thus giving a total input test sequence of $2^n + 2$ vectors with a healthy transition count of one [105].

Yet another output count variant has been proposed by Saxena and Robinson based upon the syndrome count; this is the 'accumulator syndrome', and is effectively the area under the syndrome count over the exhaustive sequence of 2^n input test vectors [106]. For example, if the output bit stream begins

0 0 1 1 0 0 1 0 ...

then the accumulator syndrome count would increase as follows:

0 0 1 3 5 7 10 13 ...

Note that this count, like the transition count, is dependent upon the order of application of the input test vectors.

However, as shown by Table 6.7, the accumulator syndrome count is not inherently secure against all possible functional faults. Fault masking (aliasing) can occur in all these forms of output compression, except where the input test sequence has been specifically matched to the function under test so as to give some extreme signature count, e.g. the single transition count noted above.

Further details and analysis of the fault-masking properties of these test compression concepts may be found [12], [106]–[108]. A combination of transition counting and syndrome counting has also been proposed [12], but the fault coverage is difficult to determine if random input test-vector sequences are assumed. On the other hand, the use of functionally dependent input sequences to

Table 6.7. A simple example showing the failure of the three forms of output compression to provide either individually, or collectively, a unique fault-free signature. Specific re-ordering of the input test vector sequence can improve the coverage of the transition and accumulator syndrome signatures

	Output response on test	Syndrome count	Transition count	Accumulator syndrome count
(i) Healthy	0 1 1 0 0 1 0 0	3	2	16
(ii) Faulty	0 1 0 1 1 0 0 0	3	2	16
(iii) Faulty	0 1 1 0 0 0 1 0	3	2	15
(iv) Faulty	0 1 1 0 1 0 0 0	3	2	17
(v) Faulty	0 0 1 1 0 1 1 0	4	2	16
(vi) Faulty	1 0 0 1 0 1 0 0	3	3	16

give ideally a 100% secure output count raises the practical problem of generating such sequences, which cannot be done by such simple means as on-chip LFSRs or binary counters, etc.

We will return to further output count parameters in Section 6.9.5, from which it will be seen that the syndrome count is a particular case of a wider range of numerical parameters which can be suggested for test purposes. The transition and accumulator syndrome concepts, however, will not be mentioned further.

6.9.4 *Further PLA and BIST testing concepts*

PLA structures continue to be a prominent field of DFT research, with the objective of finding a means of test which involves the minimum addition to the basic PLA architecture.

Syndrome testing, and in particular weighted syndrome testing, has been widely considered as a possible means of PLA test. The most definitive work in weighted syndrome testing with a single compressed output count signature is that of Serra and Muzio [109]. Following earlier work by Yamada [110], which considered the individual output syndromes of PLAs under fault modelling, and that of Bazilai *et al.*, who first introduced the concept of a weighted syndrome summation for multiple-output networks [100], Serra and Muzio show that the weighted syndrome sum of all output syndromes is a sufficient single test for all types of single fault within PLA structures. The fault models considered are as follows:

1. Stuck-at-faults: one row or column section stuck at either logical 0 or 1.
2. Bridging faults: two adjacent rows or columns shorted together taking the logical value of either the AND or the OR of the signals.
3. Crosspoint faults: the loss or spurious presence of a contact in the internal matrix of connections.

The conditions are considered for all input and internal lines in any PLA implementation.

The analysis shows that any single PLA fault cannot cause 1-valued minterms to be both removed and added in the sum-of-products outputs, due to the structure of the PLA itself. Thus this fault effect is always unidirectional in a syndrome count, either increasing or decreasing its value. The only exception to this is where two internal array lines are stuck, one at 0 and the other at 1. Here it is conceivable for 1-valued minterms to be both added and removed from two (or more) outputs, resulting in an unchanged total syndrome count across all outputs.

A simple analysis of the latter possibility can be made at the system design stage. If present, then the final weighting given to the individual syndromes to provide the single weighted-syndrome-summation (WSS) test signature may be chosen such that masking of this fault does not occur. Alternatively, it may be appropriate in some cases to use an unused output line connected to an already existing product term to achieve a similar result. This is, however, a once-and-for-all design consideration.

The weighting suggested by Serra and Muzio is to select powers of 2 for the weights, so that the overall summation can be achieved by the bit position to which each output is connected in a single WSS accumulator. This is indicated in Figure 6.49. The single-fault models used in this analysis do not specifically cover multiple faults, but the problem has been addressed in [109]. This concludes that the probability of multiple faults not being detected by the WSS signature is small, this probability decreasing as the multiplicity of faults or the number of paths through the PLA increase.

Other PLA testing developments include concurrent testing using checker codes (see Section 6.9.6), pseudo-exhaustive testing wherein the primary inputs and the product lines are partitioned into sub-sets with exhaustive testing of each sub-set [111], and built-in self-test proposals which exploit the feature that simple non-linear feedback shift registers can generate marching 0s and 1s appropriate for testing NOR array structures [112]. Other PLA built-in self-test proposals include those of Hassan and McClusky [113] and Treur, Fujiwara and Agarwal [114]. More details and additional references may be found in [12], [13], [14].

Further BIST research has largely concentrated upon reducing the silicon overheads for incorporating this form of self-test. The basic principles covered in Section 6.6.3 remains unchanged.

However, there is increasing interest in the possibilities of merging the internal BIST-type structures with boundary scan. A topology such as that shown in Figure 6.50 has been proposed by Robinson [115], which claims to minimize the

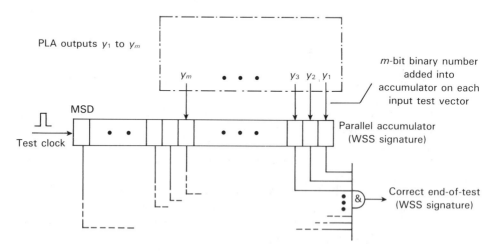

Figure 6.49 The generation of a binary-weighted syndrome summation WSS from the y_i outputs of a PLA under test. Note $y_1, ..., y_m$ need not be routed in the strict power-of-two order shown.

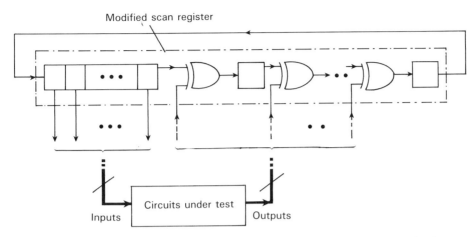

Figure 6.50 The circular BIST organization proposed by Robinson, wherein the initial stages of the scan register provide input test vectors and the later stages act as a multiple-input signature register in the overall feedback loop.

sequence length necessary for test and to give a reduced silicon area overhead in comparison with previous proposals.

These and other developments [60], [86], [116] will undoubtedly continue in order to refine this increasingly necessary form of DFT as chip complexity grows.

6.9.5 *Arithmetic and spectral coefficients*

Although the binary values of 0 and 1 are the normally accepted way of defining the logic states in 2-valued digital logic, there are other mathematical relationships which can be employed. These alternatives have been widely researched, and offer certain academic advantages for digital design and test in comparison with our discrete {0, 1} Boolean representation. We introduce them here for completeness, although they have not yet secured any firm place in design or test practice.

Let us first define the several symbols which will be used in the following survey; some will be the same as previously used but the majority will be new to our discussions so far.

Symbols used below

$x_i, i = 1, \ldots, n$	independent binary variables, $x_i \in \{0, 1\}$, x_1 least significant
$f(x_1, x_2, \ldots, x_n)$ or merely $f(X)$	binary function of the x_i input variables, $f(X) \in \{0, 1\}$
X]	truthtable column vector for function $f(X)$, entries $\in \{0, 1\}$, in minterm decimal order

Y]	truthtable column vector for $f(X)$, entries $\in \{+1, -1\}$, in minterm decimal order
$a_j, j = 0$ to $2^n - 1$	coefficients in canonic minterm expansion for $f(X)$
A]	coefficient column vector of a_j, in decimal order
$b_j, j = 0$ to $2^n - 1$	coefficients in arithmetic canonic expansion for $f(X)$
B]	coefficient column vector of b_j, in decimal order
$c_j, j = 0$ to $2^n - 1$	coefficients in positive Reed–Muller canonic expansion for $f(X)$
C]	coefficient column vector of c_j in decimal order
$r_j, j = 0$ to $2^n - 1$	spectral coefficients of $f(X)$, transformed from **X]**
R]	coefficient column vector of r_j, in decimal order
$s_j, j = 0$ to $2^n - 1$	spectral coefficients of $f(X)$, transformed from **Y]**
S]	coefficient column vector of s_j, in decimal order
[T^n]	general $2^n \times 2^n$ transform matrix
[Hd]	$2^n \times 2^n$ Hadamard matrix
[RW]	$2^n \times 2^n$ Rademacher–Walsh matrix
[I]	$2^n \times 2^n$ identity matrix

(a) The canonic minterm expansion for any $f(X)$

The coefficients directly related to normal Boolean algebra are those associated with the canonic minterm expansion for any given combinatorial function $f(X)$. For any three-variable function we have

$$f(X) = a_0 \bar{x}_1 \bar{x}_2 \bar{x}_3 + a_1 x_1 \bar{x}_2 \bar{x}_3 + a_2 \bar{x}_1 x_2 \bar{x}_3 + a_3 x_1 x_2 \bar{x}_3 + a_4 \bar{x}_1 \bar{x}_2 x_3 + a_5 x_1 \bar{x}_2 x_3$$
$$+ a_6 \bar{x}_1 x_2 x_3 + a_7 x_1 x_2 x_3$$

$$a_j, j = 0, \ldots, 2^n - 1, \in \{0, 1\} \tag{1}$$

The vector **A]** of the a_j in correct canonic order fully defines $f(X)$.

This is a trivial case, since the a_j are merely defining the zero-valued and one-valued minterms of $f(X)$, with **A]** being identical to the normal output truthtable of $f(X)$ in minterm order. In matrix form we may write (see list of symbols)

$$[\mathbf{I}]\mathbf{X]} = \mathbf{A]}, \text{ giving } \mathbf{A]} = \mathbf{X]} \tag{2}$$

(b) The canonic arithmetic expansion for any $f(X)$

The canonic arithmetic expansion for any function $f(X)$ is also available, but not widely known [117], [118]. For any three-variable function we have

$$f(X) = b_0 + b_1 x_1 + b_2 x_2 + b_3 x_1 x_2 + b_4 x_3 + b_5 x_1 x_3 + b_6 x_2 x_3 + b_7 x_1 x_2 x_3$$
$$b_j, j = 0, \ldots, 2^n - 1 \tag{3}$$

where the addition is arithmetic, not Boolean disjunction. The general expression for any $f(X)$ is given by

$$f(X) = \sum_{j=0}^{2^n - 1} b_j X_j$$

$$X_j \in \{1, x_1, x_2, \ldots, x_n, x_1 x_2, \ldots, x_1 x_2 \ldots x_n\} \tag{4}$$

and where the binary subscripts identifiers of the x_i product terms are read as decimal numbers j. As an example, the parity function $\bar{x}_1 \bar{x}_2 x_3 + \bar{x}_1 x_2 \bar{x}_3 + x_1 \bar{x}_2 \bar{x}_3 + x_1 x_2 x_3$ may be written as

$$f(X) = x_1 + x_2 - 2x_1 x_2 + x_3 - 2x_1 x_3 - 2x_2 x_3 + 4x_1 x_2 x_3$$

Note that b_j is not confined to positive values. The vector **B**] of the b_j coefficients is the arithmetic spectrum of $f(X)$. For any 3-variable function ($n = 3$) it may be determined by the transform

$$\begin{bmatrix} 1 & 0 & 0 & 0 & 0 & 0 & 0 & 0 \\ -1 & 1 & 0 & 0 & 0 & 0 & 0 & 0 \\ -1 & 0 & 1 & 0 & 0 & 0 & 0 & 0 \\ 1 & -1 & -1 & 1 & 0 & 0 & 0 & 0 \\ -1 & 0 & 0 & 0 & 1 & 0 & 0 & 0 \\ 1 & -1 & 0 & 0 & -1 & 1 & 0 & 0 \\ 1 & 0 & -1 & 0 & -1 & 0 & 1 & 0 \\ -1 & 1 & 1 & -1 & 1 & -1 & -1 & 1 \end{bmatrix} \begin{bmatrix} X \end{bmatrix} = \begin{bmatrix} b_0 \\ \\ \\ \\ B \\ \\ \\ b_7 \end{bmatrix} \tag{5}$$

The transform for any n is given by

$$\mathbf{T}^n = \begin{bmatrix} \mathbf{T}^{n-1} & \mathbf{0} \\ -\mathbf{T}^{n-1} & \mathbf{T}^{n-1} \end{bmatrix}$$

$$\mathbf{T}^0 = +1 \tag{6}$$

(c) The canonic positive Reed–Muller expansion for any $f(X)$

The Reed–Muller canonic expansion for $f(X)$ involves Exclusive–OR (addition modulo 2) relationships rather than the Inclusive–OR. There are 2^n different possible Reed–Muller expansions for any given function $f(X)$, involving all possible permutations of each input variable x_i true or complemented; here we will confine ourselves to the 'positive polarity' RM expansion for $f(X)$, which involves all x_is true and none complemented, [117], [119]. For any three-variable function we have

$$f(X) = c_0 \oplus c_1 x_1 \oplus c_2 x_2 \oplus c_3 x_1 x_2 \oplus c_4 x_3 \oplus c_5 x_1 x_3 \oplus c_6 x_2 x_3 \oplus c_7 x_1 x_2 x_3$$
$$c_{j,j} = 0 \text{ to } 2^n - 1, \in \{0, 1\} \tag{7}$$

where the addition is mod_2. The general expression for any $f(X)$ is given by

$$f(X) = \sum_{j=0}^{2^n-1} c_j X_j \tag{8}$$

X_j as previously defined.

The vector $\mathbf{C}]$ of the c_j coefficients is the Reed–Muller positive canonic spectrum of $f(X)$. For any three-variable function we have

$$\begin{bmatrix} 1 & 0 & 0 & 0 & 0 & 0 & 0 & 0 \\ 1 & 1 & 0 & 0 & 0 & 0 & 0 & 0 \\ 1 & 0 & 1 & 0 & 0 & 0 & 0 & 0 \\ 1 & 1 & 1 & 1 & 0 & 0 & 0 & 0 \\ 1 & 0 & 0 & 0 & 1 & 0 & 0 & 0 \\ 1 & 1 & 0 & 0 & 1 & 1 & 0 & 0 \\ 1 & 0 & 1 & 0 & 1 & 0 & 1 & 0 \\ 1 & 1 & 1 & 1 & 1 & 1 & 1 & 1 \end{bmatrix}_{\text{mod}_2} \mathbf{X} = \mathbf{C} \begin{bmatrix} c_0 \\ \\ \\ \\ \\ \\ \\ c_7 \end{bmatrix} \tag{9}$$

where the matrix additions are mod 2, not arithmetic. A signal flow graph equivalent of the matrix of Equation 9 may be found in Besslich [120].

The transform for any n is given by

$$\mathbf{T}^n = \begin{bmatrix} \mathbf{T}^{n-1} & 0 \\ \mathbf{T}^{n-1} & \mathbf{T}^{n-1} \end{bmatrix}_{\text{mod}_2}$$
$$\mathbf{T}^0 = +1. \tag{10}$$

The relationships between the c_j coefficients for the positive canonic case (above) and any other polarity expansion may be given by a further matrix operation; this will not concern us here.

(d) *The spectral coefficients for any function $f(X)$*

It is possible to consider the transformation of any binary-valued output vector of

Further testing concepts 361

a function $f(X)$ into some alternative domain by multiplying the binary-valued vector by a $2^n \times 2^n$ complete orthogonal transform matrix. The transformation produced by such a mathematical operation is such that the information content of the original vector is fully retained in the resulting 'spectral domain' vector, that is, the spectral data is unique for the given function $f(X)$; the application of the inverse transformation enables the original binary data to be unambiguously re-created from the spectral domain data.

The mathematical theory of complete orthogonal transforms may be found in many mathematical texts [121]–[123]. However, the simplest to apply for the transformation of binary data, and the most well-structured, is the Hadamard transform, defined by the recursive structure

$$\mathbf{Hd}^n = \begin{bmatrix} \mathbf{Hd}^{n-1} & \mathbf{Hd}^{n-1} \\ \mathbf{Hd}^{n-1} & -\mathbf{Hd}^{n-1} \end{bmatrix}$$

$$\mathbf{Hd}^0 = +1 \tag{11}$$

For $n = 3$, this gives the following, shown here for the example function

$$f(X) = \bar{x}_1 \bar{x}_2 x_3 + \bar{x}_1 x_2 \bar{x}_3 + x_1 \bar{x}_2 \bar{x}_3 + x_1 x_2 x_3$$

$$\begin{bmatrix} 1 & 1 & 1 & 1 & 1 & 1 & 1 & 1 \\ 1 & -1 & 1 & -1 & 1 & -1 & 1 & -1 \\ 1 & 1 & -1 & -1 & 1 & 1 & -1 & -1 \\ 1 & -1 & -1 & 1 & 1 & -1 & -1 & 1 \\ 1 & 1 & 1 & 1 & -1 & -1 & -1 & -1 \\ 1 & -1 & 1 & -1 & -1 & 1 & -1 & 1 \\ 1 & 1 & -1 & -1 & -1 & -1 & 1 & 1 \\ 1 & -1 & -1 & 1 & -1 & 1 & 1 & -1 \end{bmatrix} \begin{bmatrix} 0 \\ 1 \\ 1 \\ 0 \\ 1 \\ 0 \\ 0 \\ 1 \end{bmatrix} \begin{bmatrix} m_0 \\ m_1 \\ m_2 \\ m_3 \\ m_4 \\ m_5 \\ m_6 \\ m_7 \end{bmatrix} = \begin{bmatrix} 4 \\ 0 \\ 0 \\ 0 \\ 0 \\ 0 \\ 0 \\ -4 \end{bmatrix} \begin{bmatrix} r_0 \\ r_1 \\ r_2 \\ r_3 \\ r_4 \\ r_5 \\ r_6 \\ r_7 \end{bmatrix} \tag{12}$$

[Hd]   X]   R]

For completeness, we identify above the function minterms m_0 to m_7 and the resulting spectral coefficients r_0 to r_7. The spectral coefficients may be re-designated and expressed in a re-arranged order as follows:

	r_0	r_1	r_2	r_{12}	r_3	r_{13}	r_{23}	r_{123}
	(r_0)	(r_1)	(r_2)	(r_3)	(r_4)	(r_5)	(r_6)	(r_7)
Spectrum of $f(X)$:	4	0	0	0	0	0	0	-4

It is equally relevant to transform the output truth-table vector **Y**] representing $f(X)$, where **Y**] is a re-coding of **X**], re-coded logic $0 \to +1$, logic $1 \to -1$. [**Hd**] **Y**] then gives the spectral coefficient vector **S**] for $f(X)$, which for the above function would yield the spectrum in functional order:

$$\begin{bmatrix} 1 & 1 & 1 & 1 & 1 & 1 & 1 & 1 \\ 1 & -1 & 1 & -1 & 1 & -1 & 1 & -1 \\ 1 & 1 & -1 & -1 & 1 & 1 & -1 & -1 \\ 1 & -1 & -1 & 1 & 1 & -1 & -1 & 1 \\ 1 & 1 & 1 & 1 & -1 & -1 & -1 & -1 \\ 1 & -1 & 1 & -1 & -1 & 1 & -1 & 1 \\ 1 & 1 & -1 & -1 & -1 & -1 & 1 & 1 \\ 1 & -1 & -1 & 1 & -1 & 1 & 1 & -1 \end{bmatrix} \begin{bmatrix} +1 \\ -1 \\ -1 \\ +1 \\ -1 \\ +1 \\ +1 \\ -1 \end{bmatrix} \begin{bmatrix} m_0 \\ m_1 \\ m_2 \\ m_3 \\ m_4 \\ m_5 \\ m_6 \\ m_7 \end{bmatrix} = \begin{bmatrix} 0 \\ 0 \\ 0 \\ 0 \\ 0 \\ 0 \\ 0 \\ +8 \end{bmatrix} \begin{matrix} s_0 \\ s_1 \\ s_2 \\ s_{12} \\ s_3 \\ s_{13} \\ s_{23} \\ s_{123} \end{matrix} \quad (13)$$

$$[\mathbf{Hd}] \qquad \qquad \mathbf{Y}] \qquad \mathbf{S}]$$

Spectrum of $f(X)$:
$\begin{matrix} s_0 & s_1 & s_2 & s_{12} & s_3 & s_{13} & s_{23} & s_{123} \\ 0 & 0 & 0 & 0 & 0 & 0 & 0 & +8 \end{matrix}$

Both $[\mathbf{Hd}]\mathbf{Y}$ and $[\mathbf{Hd}]\mathbf{X}$ are used extensively in the literature [101]. The relationships between the r_j and s_j spectral coefficients is linear, being

$$r_0 = \frac{1}{2}(2^n - s_0), \qquad r_j, j \neq 0, = -\frac{1}{2} s_j \qquad (14)$$

There are a number of alternative complete orthogonal transforms to the Hadamard which are row re-orderings [101], [124]. The most prominent is the Rademacher–Walsh variant [**RW**], which directly generates the spectral coefficients in the re-arranged logical order shown above rather than in the Hadamard ordering. It will be appreciated that with such variants the recursive structure of the Hadamard transform is lost, but there is no difference in the information content of the resulting $r_j(s_j)$ spectral coefficients.

It should also be noted that the values of the coefficients in the spectral domain may take both positive and negative values. For the majority of functions the coefficients will be non-zero, for example, the spectrum for $f(X) = x_1\bar{x}_2 + \bar{x}_1 x_2 x_3$ is $\mathbf{R}] = 3, -1, 1, -3, -1, -1, 1, 1$ or $\mathbf{S}] = 2, 2, -2, 6, 2, 2, -2, -2$; the example shown in Equations (12) and (13) is a special case of an odd-parity function.

Since all the above coefficients which can be proposed to define a given function $f(X)$ are obtainable by an appropriate matrix operation on the column vector $f(X)$, appropriate matrix relationships exist between all these alternative sets of coefficients. For example, from the Hadamard spectral coefficients $\mathbf{R}]$ the other coefficient vectors are given by the following relationships.

The r_j to a_j coefficient relationships

$$\mathbf{A}] = \mathbf{X}]$$
$$= [\mathbf{Hd}]^{-1}\mathbf{R}]$$

where

$$[\mathbf{Hd}]^{-1} = \frac{1}{2^n}[\mathbf{Hd}] \qquad (15)$$

Further testing concepts 363

In particular,

$$a_0 = \frac{1}{2^n}\left[\sum_{j=0}^{2^n-1} r_j\right] \tag{16}$$

Note that the syndrome parameter **S** without the normalizing factor $1/2^n$ is merely

$$\mathbf{S} = \sum_{j=0}^{2^n-1} a_j = r_0. \tag{17}$$

The r_j to b_j coefficient relationships

$$\begin{aligned}\mathbf{B}] &= [\mathbf{T^n}]\,\overline{\mathbf{X}]} \\ &= [\mathbf{T^n}]\left[\frac{1}{2^n}\,[\mathbf{Hd}]\,\mathbf{R}]\right] \\ &= \frac{1}{2^n}\left[[\mathbf{T^n}]\,[\mathbf{Hd}]\,\mathbf{R}]\right] \end{aligned} \tag{18}$$

where $[\mathbf{T^n}]$ is the transform given in Equation (6). Evaluation of $[\mathbf{T^n}][\mathbf{Hd}]$, see Equations (6) and (11), will give the resulting conversion matrix $[\mathbf{Trb}]$ from $\mathbf{R}]$ to $\mathbf{B}]$, giving

$$\mathbf{Trb^n} = \begin{bmatrix} \mathbf{Trb}^{n-1} & \mathbf{Trb}^{n-1} \\ 0 & -2\,\mathbf{Trb}^{n-1} \end{bmatrix}$$
$$\mathbf{Trb^0} = +1. \tag{19}$$

Equation (18) may therefore be finally written as

$$\mathbf{B}] = \frac{1}{2^n}\,[\mathbf{Trb}]\,\mathbf{R}] \tag{20}$$

Again, the particular (simple) case of

$$b_0 = \frac{1}{2^n}\left[\sum_{j=0}^{2^n-1} r_j\right] \tag{21}$$

may be observed.

The inverse of $[\mathbf{Trb}]$ will provide the conversion matrix from $\mathbf{B}]$ to $\mathbf{R}]$, i.e.

$$\begin{aligned}\mathbf{R}] &= [\mathbf{Trb}]^{-1}\mathbf{B}] \\ &= [\mathbf{Tbr}]\,\mathbf{B}]\end{aligned}$$

where it readily follows from Equation (19) that

$$\mathbf{Tbr}^n = \begin{bmatrix} 2\mathbf{Tbr}^{n-1} & \mathbf{Tbr}^{n-1} \\ 0 & -\mathbf{Tbr}^{n-1} \end{bmatrix}$$

$$\mathbf{Tbr}^0 = +1 \tag{22}$$

The r_j to c_j coefficient relationships

From Equations (9), (10) and (15), the Reed–Muller positive canonic c_j coefficients are related to the r_j spectral coefficients by

$$\mathbf{C}] = [\mathbf{T}^n]_{\mathrm{mod}\,2} \mathbf{X}]$$

$$= [\mathbf{T}^n]_{\mathrm{mod}\,2} \left[\frac{1}{2^n} [\mathbf{Hd}] \mathbf{R}] \right] \tag{23}$$

where $[\mathbf{T}^n]$ is the modulo 2 transform given in Equation (10).

The modulo 2 (Exclusive-OR) relationships imposed by $[\mathbf{T}^n]$ imply that all odd numbers are expressed as $+1$ and all even numbers as 0. Hence it is permissible to combine $[\mathbf{T}^n]$ $[\mathbf{Hd}]$ into a single conversion matrix $[\mathbf{Trc}]$ giving

$$\mathbf{C}] = \frac{1}{2^n} [\mathbf{Trc}] \mathbf{R}] \tag{24}$$

where the final vector product summations are written as 1 or 0 depending upon odd or even summation values respectively. From Equations (10) and (11) the conversion matrix $[\mathbf{Trc}]$ is given by

$$\mathbf{Trc}^n = \begin{bmatrix} \mathbf{Trc}^{n-1} & \mathbf{Trc}^{n-1} \\ 2\mathbf{Trc}^{n-1} & 0 \end{bmatrix}_{\mathrm{mod}\,2}$$

$$\mathbf{Trc}^0 = +1. \tag{25}$$

The inverse of $[\mathbf{Trc}]$ will provide the conversion matrix $[\mathbf{Tcr}]$ from $\mathbf{C}]$ to $\mathbf{R}]$, i.e.

$$\mathbf{R}] = [\mathbf{Trc}]^{-1} \mathbf{C}]$$
$$= [\mathbf{Tcr}] \mathbf{C}]$$

where it readily follows from Equation (25) that

$$\mathbf{Trc}^n = \begin{bmatrix} 0 & \mathbf{Tcr}^{n-1} \\ 2\mathbf{Tcr}^{n-1} & -\mathbf{Tcr}^{n-1} \end{bmatrix}$$

$$\mathbf{Tcr}^0 = +1 \tag{26}$$

Further information on the relationships between these coefficient vectors, and between the very many variants of complete orthogonal transforms which may be proposed instead of the Hadamard transform, may be found in [12], [101], [124].

However, to revert to digital testing considerations: looking back at the developments from Equation (2) to Equation (12) it will be seen that the coefficients defining $f(X)$ in Equation (2) give *discrete information* about the function, merely listing whether each individual minterm of $f(X)$ is logic 0 or logic 1. Looking at any one coefficient gives no information whatsoever about any of the other coefficients, and all 2^n have to be read to define $f(X)$. On the other hand, each of the spectral coefficients in Equation (12) is a *global parameter* of $f(X)$, since all the entries in $f(X)$ influence the value of each coefficient. Between these extremes the other coefficients contain an information content concerning $f(X)$ ranging from (i) discrete, (ii) some 'window' on $f(X)$, through to (iii) some global parameter.

The interpretation of the meaning of the spectral coefficients as correlation coefficients has been widely published [101]. This interpretation is more readily appreciated when the $\{0, 1\}$ function vector $\mathbf{X}]$ is re-coded into the alternative $\{+1, -1\}$ vector $\mathbf{Y}]$ (see Table 6.8). It is then apparent that each resulting coefficient value is numerically equal to the number of agreements between the given function and a matrix row function minus the number of disagreements. However, the linear relationship between the two encodings (see Equation (14)) means that either $\mathbf{X}]$ or $\mathbf{Y}]$ produces a set of unique correlation coefficients between the given function vector and the function which the $\{+1, -1\}$ coding of each row of the transform represents. Further, any functional fault in a given function will of course change one or more of the spectral coefficient values.

The same considerations can be applied to all other coefficient vectors which represent a given function vector $\mathbf{X}]$. All may therefore be regarded as a correlation value between the network output $f(X)$ and some further function of the primary inputs, the mathematical relationships between the actual values being as developed above.

Table 6.8 Illustration of the correlation meaning of the resulting spectral coefficient values of a function $f(X)$. The sixth row of a $n = 3$ Hadamard matrix is shown operating on the function $f(X) = x_1x_2 + x_1\bar{x}_3 + \bar{x}_1x_3$. Note that this sixth row of the transform itself represent the function $x_1\bar{x}_3 + \bar{x}_1x_3$.

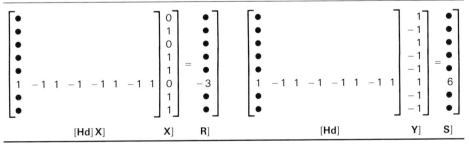

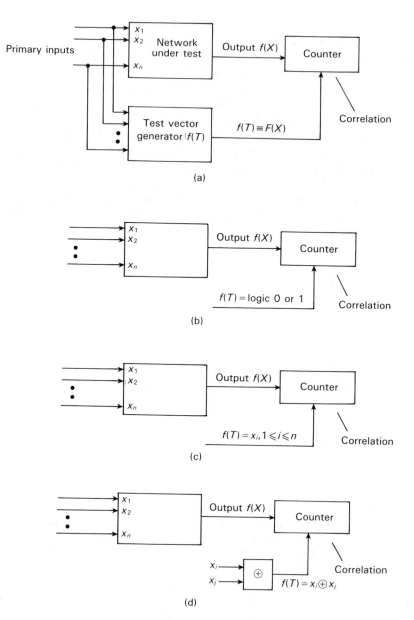

Figure 6.51 Test signatures based upon the correlation between the output $f(X)$ and a test vector generator $f(T)$ over the exhaustive test cycle of 2^n: (a) the ultimate situation of $f(X)$ compared against a copy of itself; (b) $f(X)$ compared against logic 0 or logic 1, = syndrome testing; (c) $f(X)$ compared against a single primary input x_i = spectral coefficient r_i; (d) $f(X)$ compared against the Exclusive-OR of two primary inputs, = spectral coefficient r_{ij}. (Note that in all cases a maximum count of 2^n or a minimum count of zero is the ideal objective.

Hence the commonality between many previously disclosed test signature proposals may be regarded as a search for the simplest test signature for a given $f(X)$ covering all fault conditions (see Figure 6.51). At one extreme we have the situation where $f(X)$ is correlated with itself, see Figure 6.51(a); clearly, this is not a practical proposition except in exceptional circumstances, since the test generator is then a duplicate of, and therefore of the same complexity as, the network under test. At the other extreme we have the situation where $f(X)$ is correlated with logic 1 (or 0) (see Figure 6.51(b)), which is the syndrome count. The syndrome count is therefore the most simple (and the weakest) of all the 2^n correlation coefficients available in the spectral domain data. The two remaining examples shown in Figure 6.51 represent the use of higher-order spectral coefficients as signatures.

A great deal of research has been, and continues to be, carried out on the possibilities of using some of these alternatives for test purposes. Briefly:

1. Whilst the use of *arithmetic coefficients* has been considered [118], there appears to be little attraction in their use.
2. Unless the network is designed using Exclusive-OR primitives there appears to be little attraction in the *Exclusive-OR coefficients*; further work and additional references may be found in [126], [127], [128].
3. Because of the richness of the information content of the *spectral coefficients*, a vast amount of work has been pursued on their possible use for both synthesis and fault detection. In particular, the use of spectral fault signatures for specific fault models, for output compression, and for the testing of PLA and other strongly structured architectures has been considered [12], [95], [101], [129]–[131], but industrial application of such methods has not yet emerged.

Undoubtedly the spectral domain with its global information content parameters provides a much deeper insight into the logical structure of a combinatorial network than is available from discrete Boolean-domain data. Possibly if spectral techniques were adopted for network synthesis [101] then some sub-set of the spectral design data could be acceptably formulated for test purposes. For further details of these areas of research see specifically [12], [101] and [126]; [12] in particular contains a further extensive biography.

6.9.6 *On-line checking*

As has previously been noted, some form of continuous on-line checking of a digital network avoids all the difficulties of having to generate test patterns, or re-configure the normal working mode of the circuit into an alternative test mode. Also, the on-line tests are automatically carried out at the normal working speed of the system. However, the penalty to be paid is a certain increase in silicon area to accommodate the additional circuitry.

Duplication of networks has already been mentioned, see Figure 6.12 for example. Clearly, this involves more than 100% additional silicon overhead, but it may be appropriate under critical circumstances. (Usually, networks will be at least triplicated (triple-modular-redundancy or TMR) so that majority voting to determine the correct output state is possible [13], [17].) However, here we will briefly reference some on-line checker proposals which have a minimal additional silicon overhead, and which merely provide a flag output of fault-free/faulty.

On-line (otherwise known as 'concurrent' or 'implicit') testing involves some form of information redundancy wherein more output lines are present than are required for normal system operation. The digital signals on the additional output line(s) are related to the expected (healthy) output signals by some chosen relationship, and any disagreement between them will be detected as a circuit fault. This is illustrated in Figure 6.52(a).

One of the commonly proposed coding schemes for the check bits is the Berger code [13], [132], [133]. Standard Berger coding involves the use of $[\log_2(m+1)]$ check bits for m data output bits, where the value of the check bits is the binary count of the number of zeros in the data output, for example as follows:

Healthy data output	Check bit value	Berger code
0 0 0 0 0 0 0 0	8	1 0 0 0
0 1 0 1 0 1 0 1	4	0 1 0 0
1 1 1 1 1 1 0 1	1	0 0 0 1

The Berger code generator is designed to generate the appropriate code from the primary inputs, thus making the output check bits continuously available alongside the normal system output bits. The checker circuit also calculates this number by simply counting the number of zeros in the data output, which it then compares with the check bits for agreement. Any disagreement indicates some internal circuit fault [13], [17], [132]–[134].

However, additional security has been proposed by including a parity check on the primary inputs (see Figure 6.52(b)). The Berger circuit now generates an output code which represents the number of zeros in the data output plus the input parity, which the checker circuit continuously compares against the $m + 1$ data bits [135].

The particular strength of the Berger code as an on-line checker is for what are termed 'unidirectional' faults in the circuit under test, that is, additional 1-valued minterms are either added to or subtracted from blocks of 1s in the y_i outputs under fault conditions. These will be recognized as the classic failure characteristics of PLAs (growth faults, shrinkage faults, etc.) and hence the Berger code on-line checking has been advocated particularly for PLAs [134]–[136]. The Berger code generator is made using additional product and sum lines in the PLA, with the checker circuit as a separate circuit at the PLA output.

There are, however, many other checker codes which may be proposed, some of which may be more appropriate for, say, arithmetic or signal processing systems. Among the proposals are the following:

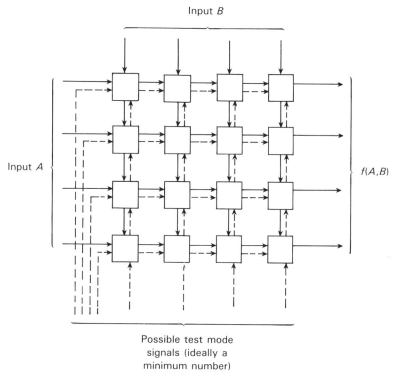

Figure 6.53 Two-dimensional cellular arrays with parallel data flow through this array structure, and the possible test mode for testing purposes.

where every x_i input variable is complemented in the right-hand side of the above expression, together with overall complementation. This holds true for all input combinations. A simple example of a self-dual function is $x_1 x_2 + x_1 x_3 + x_2 x_3$, as may readily be shown on a Karnaugh map.

The particular attraction of the self-dual primitive is that if *any* input vector such as $\bar{x}_1 x_2 \bar{x}_3$ is applied to the primitive followed by the complemented vector $x_1 \bar{x}_2 x_3$, then by definition the function output will switch from 0 to 1 or vice versa. Further, if a network of self-dual functions is made, with any fan-out or any reconvergent topology, it is an easy exercise to show that if any complete input vector $\dot{x}_1 \dot{x}_2 \ldots \dot{x}_n$ is applied followed by the complemented vector $\bar{x}_1 \bar{x}_2 \ldots \bar{x}_n$, then all the self-dual outputs throughout the network including the final output(s) will switch. Hence, if a combinatorial network were composed of self-dual primitives, (see Figure 6.54) then full controllability and observability of internal nodes would be present.

A self-dual function such as $f(X) = \overline{x_1 x_2 + x_1 x_3 + x_2 x_3}$ may therefore be suggested as a universal logic building-block, since 2-input NOR or NAND relationships can be obtained by setting a third variable to 0 or 1. In test mode the

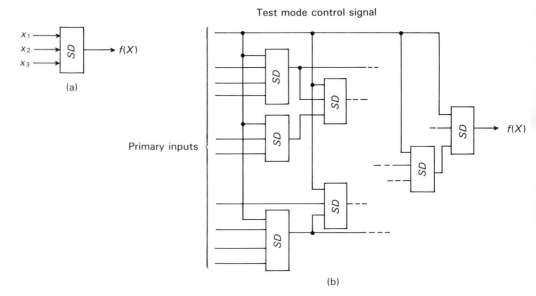

Figure 6.54 The concept of using self-dual functions for test purposes: (a) example self-dual function, say $\overline{x_1 x_2} + \overline{x_2 x_3} + \overline{x_1 x_3}$; (b) network of self-dual functions, where the logic value of all internal nodes and the network output changes when the dual of any input vector is applied.

third variable would become a test mode signal to give the full self-dual property. Other self-dual functions may likewise be suggested as universal primitives. The advantage of this concept is that it has eliminated the need to determine test vectors to give stuck-at-fault cover, since any two complementary vectors will suffice; the disadvantage is the silicon area penalty for the primitives. Details of this concept may be found [141], and is representative of ongoing interests in the controllability and observability characteristics of logic networks.

6.9.8 *Cellular automaton logic-block observation (CALBO)*

The use of linear-feedback shift registers (LFSRs) to provide both pseudo-random test-vector generation and signature analysis has been covered in Section 6.6.3, being a part of the built-in logic block observation (BILBO) method of self-test.

Some further developments of the methods described in this section have been developed, in particular the interesting use of two LFSRs of dissimilar length to act as a test-vector generator and signature analyser; the lengths of the two LFSRs are chosen in accordance with a co-prime rule in order to maximize the testing efficiency, giving what has been termed a 'quasi-exhaustive self-test', which claims to be superior to that provided by a single LFSR with the equivalent total number of stages.

Details of this method of self-test, which has been used in production and which may find wider acceptance, may be found in [142]. Some other alternative work on parallel LFSRs has also been reported [143].

However, research has been carried out on a completely dissimilar means of pseudo-random test-vector generation and signature analysis, not using LFSRs. This uses one-dimensional cellular automata configurations, and has given rise to the proposal known as 'cellular automaton logic block observation', or CALBO, as an alternative to BILBO. A four-stage CALBO circuit, corresponding to the four-stage BILBO circuit of Figure 6.34, is shown in Figure 6.55, this being a one-dimensional row of cells rather than a two- (or more) dimensional matrix of cells which may be present in multidimensional cellular automata [144], [145].

As will be seen from Figure 6.55, the logic to control the next stage of any stage Q_k is some chosen combination of data taken from near-neighbour stages, for example from $k-1, k$, and $k+1$. The input signal to D of the stage is therefore

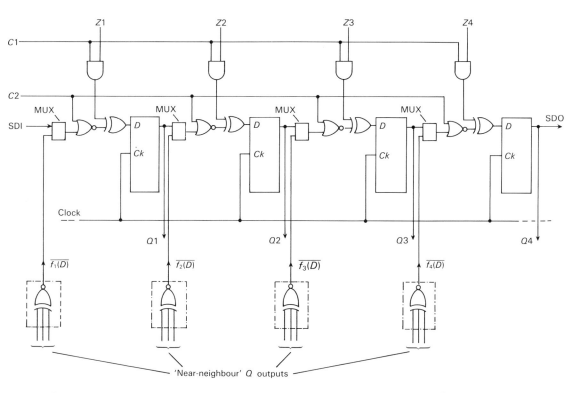

Figure 6.55 The CALBO circuit details with $C1$ and $C2$ as in Figure 6.34. The feedback to each D input is some locally generated Exclusive-OR function $f_k(D)$.

given by some function $f_k(D)$, where $f_k(D)$ may be defined by a truthtable:

Q_{k-1}	Q_k	Q_{k+1}	$f_k(D)$
0	0	0	.
0	0	1	.
0	1	0	.
0	1	1	.
1	0	0	.
1	0	1	.
1	1	0	.
1	1	1	.

With three inputs there are $2^{2^3} = 256$ possible functions $f_k(D)$. However, the majority of these will not be relevant functions for generating useful sequences.

The most useful sequences are generated by linear (Exclusive-OR) relationships between Q_{k-1}, Q_k and Q_{k+1}, since no information content is then lost in the resulting function $f_k(D)$ and hence in the next state of Q_k.

It has been found [145]–[147], and is an intuitive first choice, that the two most useful functions are as follows:

Q_{k-1}	Q_k	Q_{k+1}	$f_k(D) = Q_{k-1} \oplus Q_{k+1}$	$f_k(D) = Q_{k-1} \oplus Q_k \oplus Q_{k+1}$
0	0	0	0	0
0	0	1	1	1
0	1	0	0	1
0	1	1	1	0
1	0	0	1	1
1	0	1	0	0
1	1	0	1	0
1	1	1	0	1

Reading these two output function truthtables as eight-bit binary numbers with the top entry as the least-significant digit, the function $Q_{k-1} \oplus Q_{k+1}$ is termed function '90', and the function $Q_{k-1} \oplus Q_k \oplus Q_{k+1}$ is termed function '150'.

It has been shown that a pseudo-random maximum length sequence of $2^n - 1$ states from an n-stage CALBO register can always be generated by an appropriate assembly of 90 and 150 circuits, with a null (logic 0) input completing the inputs to $f_1(D)$ and $f_n(D)$. This is shown in Figure 6.56. The 'forbidden' state of all outputs zero exists as in an LFSR, and hence a CALBO circuit has likewise to be seeded out of this state when used as a test generator. The configuration of 90/150 circuits which produces a maximum-length sequence has been documented; like the LFSR there are multiple choices for $n > 2$; given the generating polynomial for any maximum-length LFSR the corresponding maximum-length CALBO configuration can be calculated [148]–[150] (see p. 377).

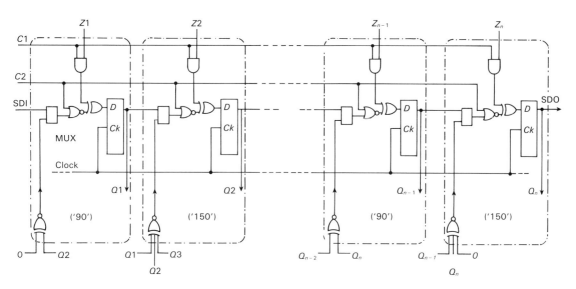

Figure 6.56 A maximum-length CALBO pseudo-random generator composed of 90 and 150 circuits. The exact sequence of 90 and 150 circuits for maximum length depends upon the total number of stages and will not necessarily always be 90, 150, 90, 150, ... Note that a 90 circuit may be made from a 150 circuit by connecting logic 0 to the middle input of the 3-input Exclusive-OR.

The pseudo-random nature of the sequences generated by CALBO configurations is clearly dissimilar to those generated by BILBO LFSRs. They have, however, been investigated, and it has been claimed that they possess a more random nature and lower cross-correlation than LFSR sequences. Hence for use as test-vector generators and signature analysers for built-in test purposes, CALBO circuits may have equal if not preferable testing characteristics, with reduced aliasing compared with BILBO [146], [147], [149]. The other potentially significant advantage of CALBO compared with the BILBO is that all the feedback taps to generate a maximum length sequence are local, unlike the LFSR where they have to run the whole length of the shift register. Hence the length of the CALBO generator may be much more readily increased or decreased by just adding stages and arranging the local 90, 150 connections accordingly; with an LFSR any modification in shift register length involves a new generating polynomial and hence a change of taps and routing along the shift register.

This flexibility may be the practical advantage of CALBO over BILBO. However, there may be an additional silicon area penalty to pay, since the circuits of Figure 6.56 are marginally more complex than those of a BILBO shift register stage. Boundary scan with CALBO has also recently been considered [151], and thus continuing developments may be expected in these areas.

6.10 The silicon area overheads of DFT

Because of the commercial difficulties in quoting actual financial costs, the 'cost' of DFT is usually measured in silicon area overhead for the test features. Only the IC vendor can translate this into true money terms, and then it may vary widely depending upon a range of commercial factors.

Most of the published material on silicon overheads deals with scan-path and BIST techniques; *ad-hoc* techniques are not amenable to formal analysis since by definition they may vary with each design. Thus overheads associated with additional devices (transistors) and/or additional gates form the basis of most overhead estimations.

One of the first papers in this area was by Ohletz, Williams and Mucha [152]. Like the majority of subsequent papers it considers that the combinatorial logic in a network does not increase with the introduction of DFT, but the sequential (storage) part of the circuit becomes more complex. The usual calculations for scan and self-test silicon overheads therefore introduce the factor K, where K is defined as

$$K = \frac{\text{No. of combinational logic gates}}{\text{No. of latches (flip-flops)}}$$

in the circuit. The value of K generally varies between 5 for latch-intensive designs and 12–20 for less latch-intensive circuits. Alternative calculations for the silicon area overhead, specifically in PLA architectures, may also be found; we will consider these later.

The % overhead for scan path or BILBO may be calculated as follows. Assuming CMOS technology, let

- P = number of transistor-pairs per D-type circuit
- L = number of additional transistor-pairs per D-type circuit to form a scan path or BILBO register stage
- F = average number of transistor-pairs per combinational gate, where a gate is conventionally taken as a 2-input NAND or NOR circuit.

Then the total number of transistor-pairs per D-type circuit without DFT is $P + KF$, and with the introduction of DFT is increased by L to $P + L + KF$. Therefore the % transistor-pair overhead, which may be taken as roughly equivalent to the silicon area overhead, as given by

$$\frac{L}{P + KF} \times 100\%$$

(Some publications count the number of transistors rather than transistor-pairs, which does not affect the overhead calculations, but may refer to a master–slave

D-type register as *two* latches when considering K. Here we are taking a master–slave circuit as one circuit. The precise values taken for P and L strongly affect the final percentage values.) The count of the number of transistors or transistor-pairs will vary with the technology and perhaps with the detailed gate and macro design. However, as an example, Figure 6.57 shows a CMOS realization for a D-type master–slave circuit and for the additional circuitry of a BILBO stage (see Figure 6.34), from which we determine the following numbers:

Circuit	No. of transistor pairs
D-type circuit (Figure 6.57(a))	10
Additional BILBO complexity per stage (Figure 6.57(b))	8
Additional one-off BILBO items: first-stage multiplexer (see Figure 6.34)	5
m-input LFSR XOR feedback = $(m - 1)$ 2-input XORs in cascade	5(m − 1)

Thus the total transistor-pair count for n D-type circuits is $10n$, and for n LFSR stages is $\{10n + 8n + 5 + 5(m - 1)\} \approx 18n$ as m is small in comparison with the remaining terms. A similar calculation for n CALBO stages (see Figure 6.56), assuming an equal number of 90 and 150 circuits, gives a count of $26n$. These results are plotted in Figure 6.58 for increasing K. (Recent work has shown that for assemblies of up to 150 stages in length, a maximum-length CALBO sequence can always be generated by using *not more than two* 150 circuits, the remainder being the simpler 90 circuits [160]. The CALBO results plotted in Figure 6.58 are therefore too pessimistic in the light of this development.)

These results should not be taken as exact, as detailed circuit design and the technology will influence the figures. However, they do indicate the form of analysis which may be carried out, and show that there is usually a noticeable silicon area penalty involved in DFT techniques. Other figures may be found in [152] for scan path DFT, and a frequent general quote is that DFT may involve a 10–30% silicon area overhead penalty.

The overheads in incorporating some form of DFT in PLA structures has also been extensively considered. These usually involve a consideration of how many additional columns are necessary in the AND array and how many additional rows in the OR array, plus any separate check circuitry. Results such as those shown in Figure 6.59 may be found [43], [153]–[155].

However, there has also been some development work on the provision of knowledge-based CAD tools which can examine all aspects of test and design-for-test at the early stages of design, and make forecasts of the costs of adopting any chosen design strategy. Such test-strategy planning tools are not yet commercially available, and require vendors to supply figures to be included in the very many

378 Test-pattern generation and design-for-testability

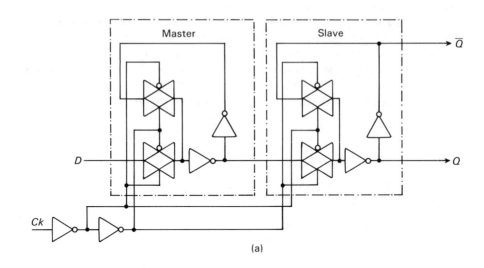

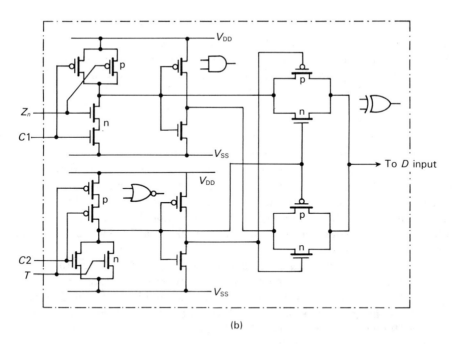

Figure 6.57 CMOS circuit realizations: (a) master–slave D-type flip-flop; (b) the addition to (a) necessary in a BILBO LFSR stage.

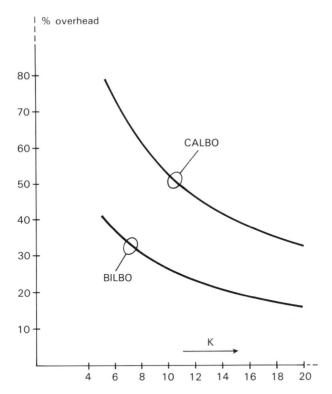

Figure 6.58 Typical transistor-pair overheads necessary in both BILBO and CALBO compared with the use of D-type circuits in a non-DFT architecture. Scan-path DFT may give somewhat lower figures than BILBO.

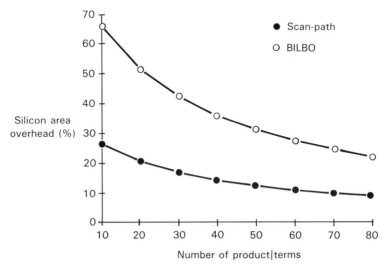

Figure 6.59 Area overhead predictions for scan-path and BILBO DFT applied to PLA architectures.

comprehensive cost equations involving yield, mask costs, test-generation costs, fault coverage, and other parameters which can be incorporated. Without doubt the OEM would appreciate the availability of such a tool in a suite of CAD software in order to be able to quantify the cost and efficiency of design choices at an early stage, but the availability of such a resource outside the vendor's premises depends heavily upon the vendor's co-operation. Details of this area of cost planning may be found in [156]–[158].

6.11 Summary

It will be apparent from the preceding sections of this chapter that testing is a critical consideration for VLSI circuits, due to the limited accessibility provided by the available I/Os. The higher the ratio of internal gate count per I/O the more difficult will be the testing procedure unless some form of design-for-test is incorporated.

Figure 6.60 attempts to summarize the various methods which can be used to test combinatorial networks. For sequential networks the only currently available

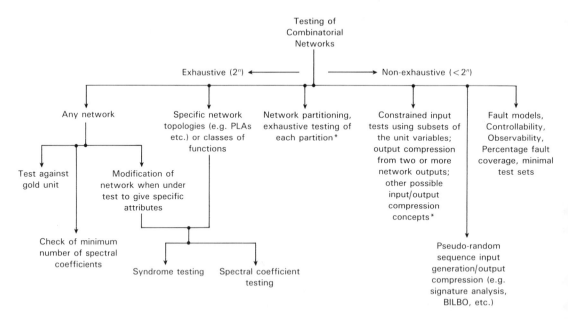

Figure 6.60 A summary of the general scenario of testing combinatorial networks. In the partitioned and constrained methods marked * the total number of tests should be less than 2^n or else there may be little advantage over a single exhaustive global test, unless the tests are performed in parallel to minimize test time.

successful techniques are (i) partitioning to reduce the sequence lengths to acceptable values, and (ii) scan test using the particular attractions of the shift register configuration. PLAs and other strongly structured architectures can use test strategies which capitalize upon their regular structure, and in general may be tested more easily than non-regular architectures.

Test-pattern generation for large digital networks has received extensive consideration, but remains a difficult area to ensure 100% fault coverage without extensive computer time. Thus many recent developments have been pursued in order to avoid having to undertake any ATPG; instead, they use exhaustive or quasi-exhaustive test patterns applied to partitions of the complete circuit under test. Built-in self-test techniques come under this heading, with the further advantage that they do not require any expensive commercial test equipment.

Analogue and mixed analogue/digital testing is perhaps the weakest area of development at present, but the relatively small volume of analogue circuitry has usually allowed conventional analogue testing methods to be used, always assuming accessibility of the analogue inputs and outputs.

The OEM system designer who may be involved in custom IC design activities will encounter many of these factors, and should therefore be knowledgeable on the subject. It is clearly a very broad area; here we have only been able to survey the subject and have not considered the details and underlying mathematics of many of the topics. For further information the several texts which specifically cover testing should be consulted [8], [12]–[14], [16], [17], [31], [138], and continuing attention should be paid to journals such as *IEEE Design and Test of Computers*, *Computer Design*, the *Journal of Electronic Testing: Theory and Applications* and the *IBM Journal of Research* [159] for on-going developments.

6.12 References

1. Parker, K. P., *Integrating Design and Test: Using CAE tools for ATE programming*, IEEE Computer Society Press, Los Angeles, 1987.
2. Bateson, J., *In-Circuit Testing*, Van Nostrand Reinhold, NY, 1985.
3. Bennetts, R. G., *Introduction to Digital Board Testing*, Crane Russak, NY, 1982.
4. Naish, P. and Bishop, P., *Designing ASICs*, John Wiley, NY, 1988.
5. Trontelj, J., Trontelj, L. and Shenton, G., *Analog Digital ASIC Design*, McGraw-Hill, London, 1989.
6. The Open University, *Design for Testability*, Microelectronics for Industry Publication PT505 DFT, The Open University, UK, 1988.
7. Hurst, S. L., *Custom-Specific Integrated Circuits: Design and fabrication*, Marcel Dekker, NY, 1985.
8. Wilkins, B. R., *Testing Digital Circuits*, Van Nostrand Reinhold, UK, 1986.
9. Richardson, N., 'E-beam probing for VLSI circuit debug', *VLSI System Design*, Vol. 8, No. 9, 1987, pp. 24–9, 58.
10. Bennetts, R. G., *Design of Testable Logic Circuits*, Addison-Wesley, London, 1984.
11. Bate, J. A. and Miller, D. M., 'Exhaustive testing of stuck-open faults in CMOS combinational circuits', *Proc. IEE*, Vol. 135, 1988, pp. 10–16.

12. Miller, D. M. (ed.), *Developments in Integrated Circuit Testing*, Academic Press, London, 1987.
13. Russell, G. and Sayers, I. L., *Advanced Simulation and Test Methodologies for VLSI Design*, Van Nostrand Reinhold, London, 1989.
14. Williams, T. W. (ed.), *VLSI Testing*, North-Holland, Amsterdam, 1986.
15. Roth, J. P., 'Diagnosis of automata failure: a calculus and a method', *IBM J. Research and Development*, Vol. 10, 1986, pp. 278–91.
16. Miczo, A., *Digital Logic Testing and Simulation*, Harper and Row, NY, 1986.
17. Lala, P. K., *Fault Tolerant and Fault Testable Hardware Design*, Prentice-Hall, London, 1986.
18. Kirkland, T. and Mercer, R., 'Algorithms for automatic test pattern generation', *IEEE Design and Test of Computers*, Vol. 5, No. 3, 1988, pp. 43–55.
19. Goel, P., 'RAPS test pattern generation', *IBM Technical Disclosure Bulletin*, Vol. 21, 1978, pp. 2787–91.
20. Goel, P. and Rosales, B. C., 'PODEM-X: an automatic test generation system for VLSI logic structures', *IEEE Proc. 18th Design Automation Conf.*, 1981, pp. 260–8.
21. Fujiwara, H. and Shimona, T., 'On the acceleration of test generation algorithms', *IEEE Trans. Computer*, Vol. C32, 1983, pp. 1137–44.
22. Funatsu, S. and Kawsai, M., 'An automatic test generation system for large digital circuits', *IEEE Design and Test of Computers*, Vol. 2, No. 5, pp. 54–60, 1985.
23. Ghosh, S., 'Behavioral-level fault simulation', *ibid.*, Vol. 5, No. 3, 1988, pp. 31–42.
24. McClusky, E. J., 'Comparing causes of IC failures', Centre for Reliable Computing Report No. 86-2, Stanford University, CA, 1986.
25. Breur, M. A. and Freedman, A. D., 'Functional level primitives in test generation', *IEEE Trans. Computers*, Vol. C29, 1983, pp. 223–35.
26. Kawai, M., Shibano, M., Funatso, S., Kato, S., Kurobe, T., Oakawa, K. and Sasaki, T., 'A high level test pattern generation algorithm', *Proc. IEEE Int. Test Conf.*, 1983, pp. 346–52.
27. Lin, T. and Su, S. Y. H., 'The S-algorithm: a promising approach for systematic functional test generation', *IEEE Trans. Computer Aided Design*, Vol. CAD4, 1985, pp. 250–63.
28. Levendel, Y. H. and Menon, P. R., 'Test generation algorithms for non-procedural computer hardware description languages', *Digest IEEE Int. Sym. on Fault Tolerant Computers*, 1981, pp. 200–5.
29. Sasaki, T., Yamada, A., Kato, S., Nakazawa, T., Tomita, K. and Namizu, N., 'MIXS: a mixed level simulator for large digital system verification', *Proc. IEEE 17th Design Automation Conf.*, 1980, pp. 626–33.
30. Miczo, A., 'A sequential ATPG: a theoretical limit', *Proc. IEEE Int. Test Conf.*, 1983, pp. 143–7.
31. Bardell, P. W., McAnney, W. H. and Savir, J., *Built-in Test for VLSI: Pseudorandom techniques*, Wiley, NY, 1987.
32. Jain, S. K. and Stroud, C. E., 'Built-in self-testing of embedded memories', *IEEE Design and Test of Computers*, Vol. 3, No. 5, 1986, pp. 27–37.
33. Sun, Z. and Wang, L-T., 'Self-testing embedded RAMs', *Proc. IEEE Int. Test Conf.*, 1984, pp. 148–56.
34. Nair, R., Thatte, S. M. and Abraham, J. A., 'Efficient algorithms for testing semiconductor random access memories', *IEEE Trans. Computers*, Vol. C.27, 1978, pp. 572–6.
35. Suk, D. S. and Reddy, S.M., 'A march test for functional faults in semiconductor random access memories', *ibid.*, Vol. C.30, 1981, pp. 982–5.

36. Marinescu, M., 'Simple and efficient algorithms for functional RAM testing', *Proc. IEEE Int. Test Conf.*, 1982, pp. 236–9.
37. Bardell, P. H. and McAnney, W. H., 'Built-in self test for RAMs', *IEEE Design and Test of Computers*, Vol. 5, No. 5, 1988, pp. 29–36.
38. Grnodis, A. J. and Hoffman, D. E., '250 MHz advanced test systems', *ibid.*, Vol. 5, No. 2, 1988, pp. 24–35.
39. Smith, J. E., 'Detection of faults in programmable logic arrays', *IEEE Trans. Computers*, Vol. C.28, 1979, pp. 845–53.
40. Ramanatha, K. S. and Biswas, N. N., 'A design for testability of undetectable crosspoint faults in programmable logic arrays', *ibid.*, Vol. C.32, 1983, pp. 551–7.
41. Ostapka, D. L. and Hong, S. J., 'Fault analysis and test generation for programmable logic arrays (PLAs)', *Proc. IEEE Fault Tolerant Computing Conf.*, 1978, pp. 83–9.
42. Eichelberger, E. B. and Lindbloom, E., 'Heuristic test-pattern generator for programmable logic arrays', *IBM Journal of Research and Development*, Vol. 24, No. 1, 1980, pp. 15–22.
43. Zhu, X-A. and Breur, M. A., 'Analysis of testable PLA designs', *IEEE Design and Test of Computers*, Vol. 5, No. 4, 1988, pp. 14–28.
44. Agrawal, V. K., 'Multiple fault detection in programmable logic arrays', *IEEE Trans. Computers*, Vol. C.29, 1980, pp. 518–22.
45. Jacob, J. and Biswas, N. N., 'Further comments on "Detection of Faults in Programmable Logic Arrays"', *IEEE Trans. Computers*, Vol. C.39, 1990, pp. 155–6.
46. Thatte, S. M. and Abraham, J. A., 'Test generation for microprocessors', *IEEE Trans. Computers*, Vol. C.29, 1980, pp. 429–41.
47. Abraham, J.A. and Parker, K. P., 'Practical microprocessor testing: open and closed loop approaches', *Proc. IEEE COMPCON*, 1981, pp. 308–11.
48. Bellon, C., 'Automatic generation of microprocessor test patterns', *IEEE Design and Test of Computers*, Vol. 1, No. 1, 1984, pp. 83–93.
49. Daniels, R. G. and Bruce, W. C., 'Built-in self-test trends in Motorola microprocessors', *ibid.*, Vol. 2, No. 2, 1985, pp. 64–71.
50. Sridhar, T. and Hayes, J. P., 'A functional approach to testing bit-sliced microprocessors', *IEEE Trans. Computers*, Vol. C.30, 1981, pp. 563–71.
51. Perry, T. S., 'Intel's secret is out', *IEEE Spectrum*, Vol. 26, No. 4, 1989, pp. 21–7.
52. Moore, E. F. (ed.), *Sequential Machines: Selected papers*, Addison-Wesley, MA, 1964.
53. The Open University, *Introduction to Devices and Circuits*, Microelectronics for Industry Publication PT505 IDC, The Open University, UK, 1988.
54. Eichelberger, E. B. and Williams, T. W., 'A logic design structure for LSI testability', *IEEE Proc. Design Automation Conf.*, 1977, pp. 462–8.
55. Williams, T. W. and Parker, K. P., 'Design for testability – a survey', *Proc. IEEE*, Vol. 71, 1983, pp. 98–112.
56. Fumatsu, S., Wakatsuki, N. and Yamada, A., 'Easily-testable design of larger digital networks', *NEC Journal of Research and Development*, No. 54, 1979, pp. 49–55.
57. Stewart, J. H., 'Future testing of large LSI circuit cards', *IEEE Digest Semiconductor Test Symposium*, 1977, pp. 6–15.
58. Ando, H., 'Testing VLSI with random access scan', *IEEE Digest COMPCON*, Spring 1980, pp. 50–2.
59. Agrawal, V. D., Cheng, K. T., Johnson, D. D. and Lin, T., 'Designing circuits with partial scan', *IEEE Design and Test of Computers*, Vol. 5, No. 2, 1988, pp. 8–15.

60. Martlett, R. A. and Deer, J., 'RISP (Reduced Intrusion Scan Path) methodology', *J. Semicustom ICs*, Vol. 6, No. 2, 1988, pp. 15–18.
61. Anderson, A. and Lee, P., *Fault-tolerance: Principles and practice*, Prentice-Hall, NJ, 1980.
62. von Neumann, J., 'Probabilistic logic and synthesis of reliable organisms from unreliable components', *Annals of Mathematical Studies*, No. 34, 1956, pp. 43–98.
63. Breur, M. A. and Friedman, A. D., *Diagnosis and Reliable Design of Digital Systems*, Pitman, NY, 1977.
64. Hewlett-Packard, *A Designer's Guide to Signature Analysis*, Application Note 222, Hewlett-Packard Co., CA, 1977.
65. Williams, T. W., Daehn, W., Gruetzner, M. and Stark, C. W., 'Aliasing errors in signature analysis registers', *IEEE Design and Test of Computers*, Vol. 4, No. 2, 1987, pp. 39–45.
66. Beucler, F. and Manner, M. J., '"HILDO" a highly integrated logic device observer', *VLSI Design*, Vol. 5, No. 6, 1984, pp. 88–96.
67. Cosgrave, B., 'The UK 5000 gate array', *Proc. IEE*, Vol. 132.G, 1985, pp. 90–2.
68. Gelsinger, P. P., 'Built-in self-test of the 80386', *Proc. IEEE Conf. on Computer Design*, 1986, pp. 169–73.
69. Paraskeva, M., Knight, W. G. and Burrows, D. F., 'A new test structure for VLSI self-test: the structured test register (STR)', *IEE Electronic Letters*, Vol. 21, 1985, pp. 856, 857.
70. Williams, T. W., Walther, R. G., Bottorff, P. S. and DasGupta, S., 'Experiment to investigate self-testing techniques in VLSI', *Proc. IEE*, Vol. 132.G, 1985, pp. 105–7.
71. Maunder, C. M., 'The status of IC design-for-testability', *J. Semicustom ICs*, Vol. 6, June 1989, pp. 25–9.
72. Stroud, C. E., 'An automated BIST approach for general sequential logic synthesis', *Proc. IEEE 25th Design Automation Conf.*, 1988, pp. 3–8.
73. Daehn, W. and Mucha, J., 'A hardware approach to self-testing of large programmable logic arrays', *IEEE Trans. Computers*, Vol. C.30, 1981, pp. 829–33.
74. Tamir, Y. and Sequin, C. H., 'Design and application of self-testing comparators implemented in MOS technology', *ibid.*, Vol. C.33, 1984, pp. 493–506.
75. Khakbaz, J. and McClusky, E. J., 'Concurrent error detection and testing for large PLAs', *IEEE Trans. Electronic Devices*, Vol. Ed.29, 1982, pp. 756–64.
76. Chen, C. Y., Fuchs, W. K. and Abraham, J. A., 'Efficient concurrent error detection in PLAs and ROMs', *Proc. IEEE ICCD*, 1985, pp. 525–9.
77. Fujiwara, H. and Kinoshita, K., 'A design of programmable logic arrays with universal tests', *IEEE Trans. Computers*, Vol. C30, 1981, pp. 823–8.
78. Saluja, K. K., Kinoshita, K. and Fujiwara, H., 'An easily testable design of programmable logic arrays for multiple faults', *ibid.*, Vol. C.32, 1983, pp. 1038–46.
79. Hong, S. J. and Ostapko, D. L. 'FITPLA: a programmable logic array for function-dependent testing', *Proc. IEEE 10th Fault Tolerant Computing Symp.*, 1980, pp. 131–6.
80. Khakbaz, J., 'A testable PLA design with low overhead and high fault coverage', *IEEE Trans. Computers*, Vol. C.33, 1984, pp. 743–5.
81. LSI Logic, *Databook and Design Manual*, LSI Logic Corporation, CA, 1986.
82. Institute of Electrical and Electronics Engineers, *Standard P.1149.1*, NY, 1988.
83. Maunder, C., 'Boundary scan: a framework for structural design-for-test', *Proc. IEEE Int. Test Conf.*, 1987, pp. 714–23.
84. Pradhan, M. M., Tulloss, R. E., Beenker, F. P. M. and Bleeker, H., 'Developing a standard for boundary-scan implementation', *Proc. Int. Conf. of Computer Design*, New York, 1987, pp. 462–6.

85. Wagner, P. T., 'Interconnect testing with boundary-scan', *Proc. IEEE Int. Test Conf.*, 1987, pp. 52–7.
86. Scholz, H. N., Tulloss, R. E., Yau, C. W. and Wach, W., 'ASIC implementations of boundary-scan and built-in self-test (BIST)', *J. Semicustom ICs*, Vol. 6, No. 4, 1989, pp. 30–8.
87. Plessey Semiconductors, *JTAG/BIST in ASICs*, Application Note AN87, GEC/Plessey Semiconductors Ltd., UK, October 1987.
88. Dettmer, R., 'JTAG: setting the standard for boundary-scan testing', *IEE Review*, Vol. 32, February 1989, pp. 49–52.
89. Andrews, W., 'JTAG works to standardise chip, board and system self-test', *Computer Design*, Vol. 28, No. 13, 1989, pp. 28–31.
90. Birkner, J., Coli, V. and Lee, F., 'Shadow register architecture simplifies digital diagnosis', *Application Note AN.123*, Monolithic Memories Inc., CA, 1983.
91. Akers, S. B., 'In the use of linear sums in exhaustive testing', *Proc. IEEE Design Automation Conf.*, 1985, pp. 148–53.
92. Tang, D. T. and Cheng, C. L., 'Logic test patterns using linear codes', *ibid.*, C.33, 1984, pp. 845–50.
93. McClusky, E. J., 'Verification testing: a pseudo exhaustive test technique', *IEEE Trans. Computers*, Vol. C.33, 1984, pp. 541–6.
94. Saluja, K. K. and Karpovsky, M., 'Testing computer hardware through data compression in space and time', *Proc. IEEE Int. Test Conf.*, 1983, pp. 83–8.
95. Hurst, S. L., 'Use of linearisation and spectral techniques in input and output compaction testing of digital networks', *Proc. IEE*, Part E, Vol. 136, 1989, pp. 48–56.
96. Agrawal, V. K., 'Increasing effectiveness of built-in testing by output data modification', *Proc. IEEE 13th. Int. Symp. on Fault Tolerant Computing*, 1983, pp. 227–33.
97. Savir, J., 'Syndrome testable design of combinational circuits', *IEEE Trans. Computers*, Vol. C.29, 1980, pp. 442–51, and correction pp. 1012, 1013.
98. Savir, J., 'Syndrome testing of "syndrome-untestable" combinational circuits', *ibid.*, Vol. C.30, 1981, pp. 606–8.
99. Markowsky, G., 'Syndrome-testability can be achieved by circuit modification', *ibid.*, Vol. C.30, 1981, pp. 604–6.
100. Barzilai, Z., Savir, J., Markoswky, G. and Smith, M. G., 'The weighted syndrome sums approach to VLSI testing', *ibid.*, Vol. C.30, 1981, pp. 996–1000.
101. Hurst, S. L., Miller, D. M. and Muzio, J. C., *Spectral Techniques in Digital Logic*, Academic Press, NY, 1985.
102. Hayes, J. P., 'Transition count testing of combinational logic circuits', *Trans. IEEE Computers*, Vol. C.25, 1976, pp. 613–20.
103. Hayes, J. P., 'Generation of optimal transition count tests', *ibid.*, Vol. C.27, 1978, pp. 36–41.
104. Reddy, S. M., 'A note on logic circuit testing by transition counting', *ibid.*, Vol. C.26, 1977, pp. 313–14.
105. Fujiwara, H. and Kimoshita, K., 'Testing logic circuits with compressed data', *Proc. IEEE 8th Int. Symp. on Fault Tolerant Computing*, 1978, pp. 108–13.
106. Saxena, N. R. and Robinson, J. P., 'Accumulator compression testing', *IEEE Trans. Computers*, Vol. C.35, 1986, pp. 317–21.
107. Robinson, J. P. and Saxena, N. R., 'A unified view of test compression methods', *ibid.*, Vol. C.36, 1987, pp. 94–9.
108. Savir, J. and McAnney, W. H., 'On the masking probability with one's count and transition count', *Proc. IEEE Int. Conf. on Computer Aided Design*, 1985, pp. 111–13.

109. Serra, M. and Muzio, J. C., 'Testing programmable logic arrays by sum of syndromes', *IEEE Trans. on Computers*, Vol. C.36, 1987, pp. 1097–1101.
110. Yamada, T., 'Syndrome-testable design of programmable logic arrays', *Proc. IEEE Int. Test. Conf.*, 1983, pp. 453–8.
111. Ha, D. S. and Reddy, S.M., 'On the design of pseudo exhaustive testable PLAs', *IEEE Trans. Computers*, Vol. C.37, 1988, pp. 468–72.
112. Daehn, W. and Mucha, J., 'A hardware approach to self-testing of large PLAs', *ibid.*, Vol. C.30, 1981, pp. 829–33.
113. Hassan, S. Z. and McClusky, E. J., 'Testing PLAs using multiple parallel signature analysers', *Digest IEEE Int. Symp. on Fault Tolerant Computing*, 1983, pp. 422–5.
114. Treuer, R., Fujiwara, H. and Agarwal, V. K., 'Implementing a built-in self-test PLA design', *IEEE Design and Test of Computers*, Vol. 2, 1985, pp. 37–48.
115. Robinson, J. P., 'Circular BIST test generation', *J. Semicustom ICs*, Vol. 7, September 1990, pp. 12–16.
116. Zorian, Y. and Agarwal, V. K., 'Optimizing error-masking in BIST by output data modification', *J. Electronic Testing: Theory and Applications*, Vol. 1, February 1990, pp. 59–71.
117. Davio, M., Deschamps, J. P. and Thayse, A., *Discrete and Switching Functions*, McGraw Hill, NY, 1978.
118. Heidtmann, K. D., 'Arithmetic spectrum applied to stuck-at fault detection for combinational networks', *IEEE Trans. Computers*, Vol. C.40, 1991, pp. 320–4.
119. Wu, X., Chen, X. and Hurst, S. L., 'Mapping of Reed–Muller coefficients and the minimisation of Exclusive-OR switching functions', *Proc. IEE*, Vol. 129, Part E, 1982, pp. 15–20.
120. Besslich, P. W., 'Efficient computer method for ExOR logic design', *ibid.*, Vol. 130, Part E, 1983, pp. 203–6.
121. Birkhoff, G. and MacLane, S., *A Survey of Modern Algebra*, Macmillan, NY, 1977.
122. Ayres, F., *Theory and Problems of Matrices*, Schaum's Outline Series, McGraw-Hill, NY, 1962.
123. Hohn, F. E., *Elementary Matrix Algebra*, Macmillan, NY, 1964.
124. Beauchamp, K. G., *Applications of Walsh and Related Functions*, Academic Press, London, 1984.
125. Beauchamp, K. G., *Transforms for Engineers*, Oxford University Press, Oxford and NY, 1987.
126. Karpovsky, M. G. (ed.), *Spectral Techniques and Fault Detection*, Academic Press, NY, 1985.
127. Damarla, T. R. and Karpovsky, M. G., 'Fault detection in combinational networks by Reed–Muller transforms', *IEEE Trans. Computers*, Vol. C.38, 1989, pp. 788–97.
128. Muzio, J. C., 'Stuck fault sensitivity of Reed–Muller and arithmetic coefficients', *3rd International Workshop on Spectral Techniques*, Dortmund, Germany, 1988, pp. 36–45.
129. Miller, D. M. and Muzio, J. C., 'Spectral fault signatures for single stuck-at faults in combinational networks', *IEEE Trans. Computers*, Vol. C.33, 1984, pp. 765–9.
130. Lui, P. K. and Muzio, J. C., 'Spectral signature testing of multiple stuck-at faults in irredundant combinational networks', *IEEE Trans. Computers*, Vol. C.35, 1986, pp. 1088–92.
131. Muzio, J. C. and Serra, M., 'Spectral criteria for the detection of bridging faults', *ibid.*, *Report No. 12-1986*, Fault Detection Research Group, Dept. of Computer Science, University of Victoria, Canada, 1986.
132. Johnson, B. W., *Design and Analysis of Fault Tolerant Systems*, Addison-Wesley, Reading, MA, 1989.

133. Berger, J. M., 'A note on error detection codes for asymmetric channels', *Information and Control*, Vol. 4, 1961, pp. 68–73.
134. Serra, M., 'Some experiments on the overhead for concurrent checking using Berger codes', *Record of 3rd Technical Workshop: New Directions of IC Testing*, Canada, 1988, pp. 207–12.
135. Wessels, D. and Muzio, J. C., 'Adding primary input error coverage to concurrent checking for PLAs', *Record of 4th Technical Workshop: New Directions for IC Testing*, Canada, 1989, pp. 135–53.
136. Abrahams, J. A. and Davidson, E. S., 'The design of PLAs with concurrent error detection', *Proc. IEEE Fault Tolerant Computing Symp.*, 1982, pp. 303–10.
137. Pradham, D. K. (ed.), *Fault Tolerant Computing: Theory and Techniques*, Prentice-Hall, NJ, 1986.
138. Massara, R. E. (ed.), *Design and Test Techniques for VLSI and WLSI Circuits*, IEE Peter Peregrinus, London, 1989.
139. Moore, W. R., Maly, W. and Strojaw, A., *Yield Modelling and Defect Tolerance in VLSI*, Adam Hilger, Bristol, 1988.
140. Marnane, W. P. and Moore, W. R., 'Testing of VLSI regular arrays', *IEEE Int. Conf. on Computer Design*, 1988, pp. 145–8.
141. Taylor, D., *Design of Certain Semicustomised Structures Incorporating Self-Test*, PhD Thesis, Huddersfield Polytechnic, UK, 1990.
142. Illman, R. and Clarke, S., 'Built-in self-test of the Macrolan chip', *IEEE Design and Test of Computers*, Vol. 7, April 1990, pp. 29–40.
143. Bardell, P. H., 'Design considerations for parallel pseudo random pattern generators', *J. Electronic Testing: Theory and Applications*, Vol. 1, No. 1, 1990, pp. 73–87.
144. Wolfran, S., 'Random sequence generation by cellular automata', *Advances in Applied Mathematics*, Vol. 7, 1986, pp. 127–69.
145. Thanailakis, O. O. and Card, H. C., 'Group properties of cellular automata and VLSI applications', *IEEE Trans. Computers*, Vol. C.35, 1986, pp. 1013–24.
146. McCleod, R. D., Hortensius, P. D., Pries, W. and Card, H. C., 'Design for testability using logic block observers based upon cellular automata', *Proc. IEEE Trans. Computers*, Vol. C.38, 1989, pp. 1466–73.
147. Serra, M., Slater, T., Muzio, T. C. and Miller, D. M., 'The analysis of linear cellular automata and their aliasing properties', *IEEE Trans. CAD*, Vol. 9, 1990, pp. 769–78.
148. Serra, M., Miller, D. M. and Muzio, J. C., 'Linear cellular automata and LFSRs are isomorphic', *Record of 3rd Technical Workshop New Directions for IC Testing*, Canada, 1988, pp. 213–23.
149. Serra, M., Slater, T., Muzio, J. C. and Miller, D. M., 'The analysis of one-dimensional linear cellular automata and their aliasing properties', *IEEE Trans. Computer Aided Design*, Vol. 9, 1990, pp. 767–78.
150. Slater, T. and Serra, M., *Tables of Linear Hybrid 90/150 Cellular Automata*, University of Victoria, Dept of Computer Science Report No. DCS-105-IR, January 1989.
151. Gloster, C. S. and Brglez, F., 'Boundary scan with built-in self-test', *IEEE Design and Test of Computers*, Vol. 6, June 1989, pp. 36–48.
152. Ohletz, M. J., Williams, T. W. and Mucha, J. P., 'Overhead in scan and self-test designs', *Proc. IEEE Int. Test Conf.*, 1987, pp. 460–70.
153. Goel, P., 'Test generation cost analysis and projections', *Proc. IEEE 17th Design Automation Conf.*, 1980, pp. 77–84.
154. Miles, J. R., Ambler, A. P. and Totton, K. A. E., 'Estimation of area and performance overheads for testable VLSI circuits', *Proc. IEEE Int. Conf. on Computer Design*, 1988, pp. 213–23.

155. IBM, 'Embedded array test with ECIPT', *IBM Technical Disclosure Bulletin*, Vol. 28, 1984, pp. 2376–8.
156. Varma, P., Ambler, A. P. and Baker, K., 'An analysis of the economics of self-test', *Proc. IEEE Int. Test Conf.*, 1984, pp. 20–30.
157. Zhu, X. and Breur, M. A., 'A knowledge-based system for selecting design methodologies', *IEEE Design and Test of Computers*, Vol. 5, October 1988, pp. 41–58.
158. Dislis, C., Dear, I. D., Miles, J. R., Lau, S. C. and Ambler, A. P., 'Cost analysis of test method environments', *Proc. IEEE Int. Test Conf.*, 1989, pp. 402–7.
159. Special issue, *Electrical Testing, IBM Journal of Research*, Vol. 34, No. 2/3, 1990, pp. 137–448.
160. Zhang, S., Miller, D. M. and Muzio, J. C., 'The determination of minimum cost one-dimensional linear hybrid cellular automata', awaiting publication (1991).

7 The choice of design style: technical and managerial considerations

Every original-equipment manufacturing company has to make a number of initial decisions when considering the introduction of a new or enhanced product relating to the technology in which the product is to be designed and manufactured. Decisions involve managerial, technical and economic aspects, and the final choice may not be a clear-cut decision.

The main diversification of design and manufacture is shown in Figure 7.1. Obviously, there are physical products which cannot incorporate any microelectronics, for example fabrics, mechanical fixtures, adhesives, and so on, and equally there are products which, by definition, must be built around microelectronics, for example pocket calculators and radar; but between the two extremes is a vast range of products which already do or (more importantly) could incorporate electronics in order to improve their performance or sales appeal.

The breadth of present-day electronics is shown in Figure 7.2, from which it is clear that microelectronics generally covers the low-voltage, low-power areas, with high voltage, high power or special applications being covered by discrete components. However, the microelectronics itself is invariably only part of a complete product: there must always be some further devices or fixtures associated with the microelectronics to make the product useful for its intended use. Frequently, weaknesses in these peripheral items make a product unreliable or unsuitable for the end-user, and emphasizes that the non-microelectronic design activities are *equally if not more important than the microelectronics*. It is no good having a superb IC if the input and output peripherals are cheap and unsatisfactory.

The remaining sections of this chapter will deal largely with those items contained within the dotted rectangle of Figure 7.2, and will consider the factors and costs involved in choosing a particular microelectronic design style. However, there is one parameter which the majority of digital ICs have in common, that is,

390 *The choice of design style*

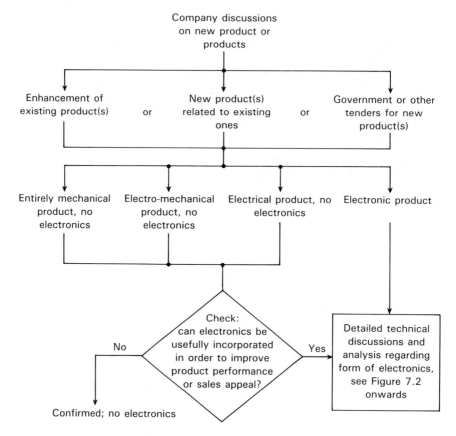

Figure 7.1 Initial OEM company considerations relating to the introduction of any new engineering product.

the d.c. supply voltage. The most common voltage at present is 5 V, which was initially established in the early days of SSI and MSI TTL circuits and has remained enshrined ever since. There is no particular merit in 5 V — indeed, it would not be the preferred choice now — and there are moves to reduce the value to 3 V for VLSI circuits in order to reduce total IC power dissipation. Custom microelectronics does allow a variation from 5 V to be chosen, particularly for MOS technologies where low-voltage battery operation is desired.

On the other hand, analogue circuits have not standardized on 5 V. Instead, a number of voltage ratings will be found in off-the-shelf devices, for example ±9 V and +15 V. This is yet another problem confronting the manufacturer of a mixed analogue/digital product, and may require the use of two or more power supplies. Custom microelectronics may simplify this mixed supply requirement, but will not eliminate it entirely since it is not practical to run sensitive analogue circuits from

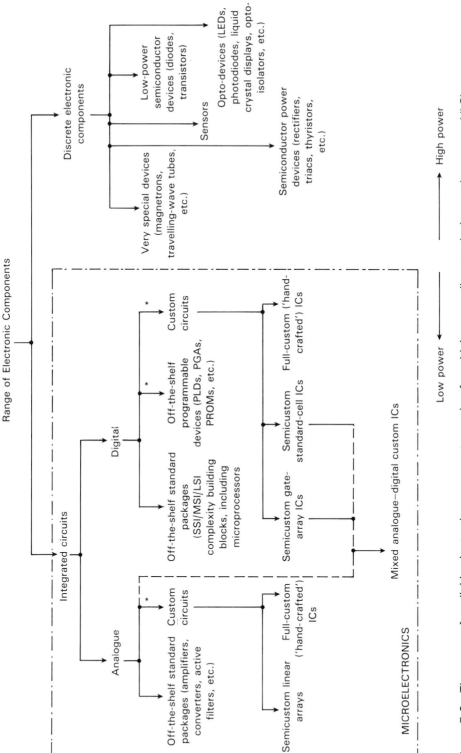

Figure 7.2 The range of available electronic components ranging from high-power discrete devices to low-power VLSI microelectronics. The areas of specific concern in this text are those marked with an asterisk.

392   The choice of design style

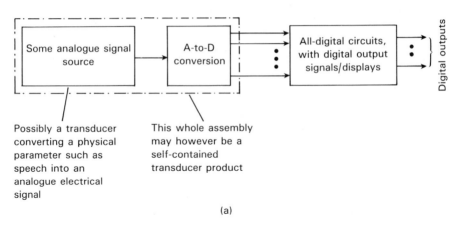

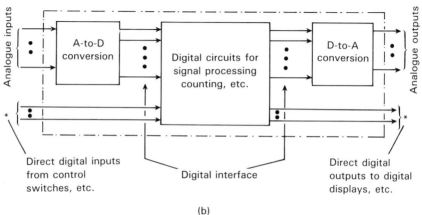

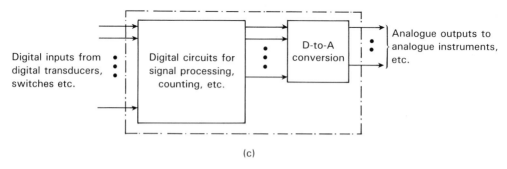

Figure 7.3   The general structure of a mixed analogue/digital product: (a) analogue source signal(s) with some A-to-D conversion; (b) analogue inputs and analogue outputs, with internal digital signal processing; (c) digital inputs, with final analogue outputs.

the same supply rails as digital circuits due to the interference which can be caused by fast-changing logic signals.

Unless otherwise specified, the following sections will assume that no unusual power supply voltages are required. Power dissipation, however, may be a crucial factor, and will be referred to where relevant.

7.1 The microelectronic choices

The first choice, which automatically arises from the definition of a new microelectronic-based product, is whether the circuits are:

- entirely digital, with binary input and binary output signals;
- entirely analogue, with no digital signals present but, instead, with a need for accurate voltage amplification or other signal processing; or
- mixed analogue/digital.

The last category almost invariably involves some form of analogue-to-digital (A-to-D) and/or digital-to-analogue (D-to-A) conversion, for example as is shown in Figure 7.3, with the digital part performing some logic processing on the digital data.

In the following sections we will consider the decisions which face the OEM designer in each of the above categories. As will be seen the last category often has to be regarded as a mixture of separate analogue and digital parts, with the best design style having to be chosen for each part independently.

7.1.1 Digital-only products

The categories of digital circuits available to the OEM designer for a new product design range from full-custom ICs, designed in every detail to match the required specification, through the various standard-cell and gate-array choices, to programmable logic devices and, finally, standard off-the-shelf parts. The principal characteristics of these various groups have been considered in preceding chapters, but in order to summarize them Table 7.1 gives the design features as far as the OEM is concerned.

Which type of circuit to use for a new product depends upon the anticipated production quantity, since it is clearly uneconomic to incur the expense of designing a custom circuit if only very small quantities are required [1], [2]. There will be exceptions to this generalization, for example where product weight or size or absolute performance takes precedence over all economic factors, as may arise in avionics, military or other specialist fields. But for the majority of products involving microelectronics the initial choice of design style is likely to be influenced by the broad guide shown in Figure 7.4. Further considerations of all the economic

Table 7.1. The control which the OEM product designer has over the detailed IC design and fabrication

Category of circuit	Control over the chip design and fabrication	CAD availability
1 Hand-crafted Full custom All mask levels unique, no fixed chip size	Full control; anything can be done within the limits of the technology and budget	Yes
2 Standard cells, parameterized (or 'soft') cells/macros All mask levels unique, no fixed chip size	Can only use the functional cells in the CAD library, but can (i) optimize cell performance for the particular application and (ii) place and route in any manner	Yes
3 Standard cells, fixed (or 'hard') cells/macros All mask levels unique, no fixed chip size	Can only use the pre-designed cells in the CAD library but can place and route in any manner	Yes
4 Uncommitted arrays, component-level cells Final interconnection mask(s) only unique, fixed chip sizes	Fixed chip layout of components. Full flexibility to interconnect components in each cell and route cells in any manner, subject to constraint of available wiring space	Yes
5 Uncommitted arrays, functional cells Final interconnection mask(s) only unique, fixed chip size	Fixed chip layout of fixed functional cells (gates or other primitives). Full flexibility to route cells in any manner, subject to constraint of wiring space	Yes
6 Programmable devices, field-programmable Completely prefabricated, no unique mask, fixed chip sizes	Fixed chip layout and circuit topology. Can only destroy existing internal connections or active wanted connections to dedicate the chip	Yes
7 Programmable devices, mask-programmable Final interconnection mask only unique, fixed chip sizes	A vendor's version of (6) above should volume production of a field-programmable device be required, still the same fixed chip layout and circuit topology as (6)	Same as for (6) above
8 Fixed standard functions Completely prefabricated standard off-the-shelf products, e.g. microprocessor, memory. TTL 7000 series, CMOS 4000 series, etc.	Absolutely standard products, with optimized chip layouts. Customer/user has no control or knowledge of the detailed chip design or fabrication	Not specifically available

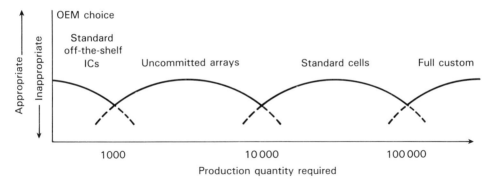

Figure 7.4 A repeat of Figure 4.37: a general guide to the most economic form of digital electronics for varying production quantities.

factors, including non-recurring engineering (NRE) costs, will be made in Section 7.4.

The product's technical requirements must, however, play the dominant role in the choice of design style. The off-the-shelf microprocessor or microcontroller IC, with its selling price of a few dollars, is a strong contender in the choice of production method for a new product, and must enter into this initial decision-making. Microprocessors and microcontrollers for OEM use range from 4-bit, such as the Texas Instruments 1000 series, 8-bit such as the Zilog Z8* and Motorola 68** series, to 16-bit such as the Intel 80** and the Motorola 68*** series. 32-bit and 64-bit µPs are also available, but the smaller circuits are entirely appropriate for very many industrial applications [3].

The general decision tree of when or when not to use a microprocessor is given in Figure 7.5. In comparison with later decision routes, factors such as space, power constraints and, to some extent, cost are not deciding factors in the decision to adopt a microprocessor-based design. Rather, the dominant features are as follows:

- the architecture and computational power of the processor;
- the flexibility whereby several versions of the same product can be made by changing the software only, leaving the hardware standardized for all the variants.

The last item is the supreme advantage of software-based equipments.

Outside product areas such as word processors, etc., it is unlikely that production quantities of software-based products will be high, possibly thousands rather than tens of thousands. However, if very large production requirements are anticipated, then the possibility of using a standard-cell design route, with the microprocessor/microcontroller as an on-chip macro, may be justified. This will be seen later in connection with Figure 7.8. It is unlikely that any OEM would reach this level of sophistication without appreciable prior in-house experience in the design and manufacture of microelectronic-based products.

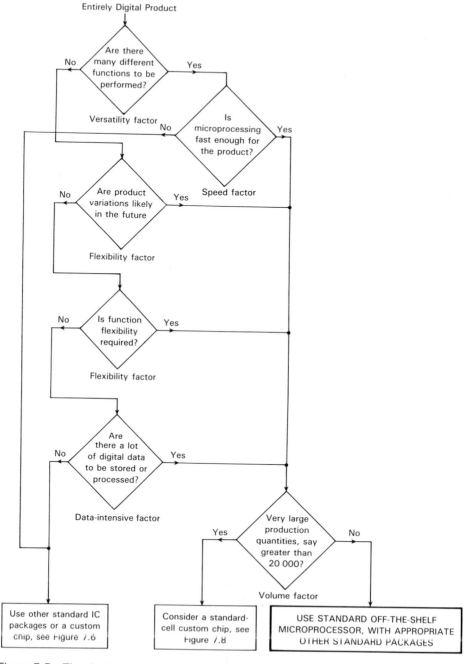

Figure 7.5 The decision tree of when to use an off-the-shelf microprocessor or microcontroller. (Source: based upon [2].)

For other digital-only applications where small production quantities are anticipated, other widely available off-the-shelf standard parts, including programmable logic devices, may be relevant. The general decision tree leading to the choice of standard parts is given in Figure 7.6.

Further choices between:

- standard digital logic ICs or PLDs and
- bipolar technology or CMOS technology

are not clear-cut decisions: standard ICs allow ready second-sourcing, but PLDs can reduce the number of IC packages and hence the size of the printed-circuit board(s) in the final equipment. Also, PLDs can provide some degree of design security unavailable with standard parts, since the exact device programming is hidden from the end-user. On the technicalities of bipolar vs. CMOS, advanced low-power Schottky TTL 74ALS series ICs will provide a higher speed, but for lower power the 74HC or 74HCT CMOS series are good all-round choices, and also provide a wider operating temperature range than does TTL. For wide voltage requirements the older 4000 series CMOS ICs provide a working supply voltage range of $+3$ to $+15$ V. Price per IC now generally favours the 74HC or 74HCT series over the others [2]. Costs of CAD resources for PLD design duties will be referred to in Section 7.4.2.

From the production quantity viewpoint the next choice of possible design style is the gate array. For very-high-speed applications this may require ECL or GaAs products, but for the majority of applications it will currently be CMOS, usually a channelled array for general-purpose applications, but perhaps channel-less for more sophisticated needs. However, here we will concentrate upon the decisions leading to the adoption of a relatively simple CMOS product.

Figure 7.7 illustrates the general decision tree leading to the choice of a gate array. This tree should be read in conjunction with both the preceding and the subsequent diagrams, since there is no absolute division between them. However, a number of additional practical considerations are involved when considering the normal gate-array route, not shown in Figure 7.7; among these are the following:

- the number of available gates and I/Os on the vendor's products compared with the OEM's needs;
- I/O interface requirements;
- power dissipation if critical;
- the number of custom mask levels necessary to customize the array;
- the comprehensiveness of the vendor's standard library of cell interconnections and the supporting CAD tools;
- how much of the custom design is to be carried out in-house and how much by the vendor or independent design house.

398 *The choice of design style*

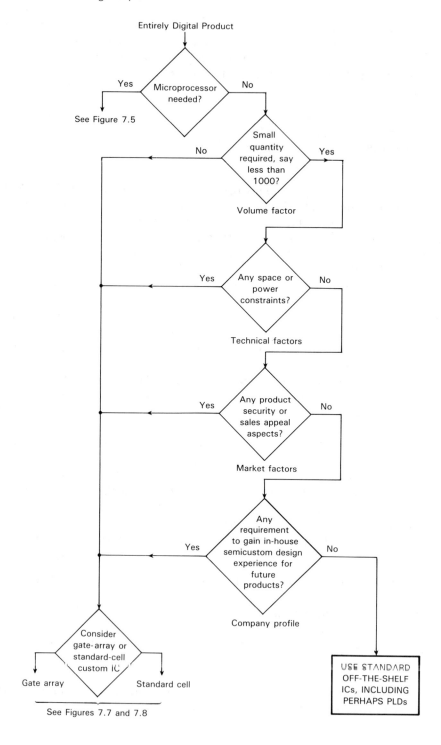

These are all factors requiring close consideration when it seems that a gate-array choice may be appropriate for the new product.

IC design and manufacturing costs will inevitably be involved in the adoption of a gate-array solution which are not present when using standard parts. These cost relationships (see Section 7.4) must therefore be considered, unless the particular attributes of a custom design such as design security or minimum size and power dissipation are overriding considerations.

If large production needs without any product design change are envisaged, then a standard-cell choice may be preferable to the gate array; or if complex macros and memory requirements are present, this choice may provide these needs more readily.

Figure 7.8 illustrates the decision routes to a standard-cell design. However, design costs, time to market, the possibility of product design revisions, and hence iteration of the custom design and the full mask set, makes this a choice which perhaps needs more careful attention than a gate-array design decision.

The availability of microprocessor and RAM and ROM macros in a number of vendors' standard-cell libraries may be a deciding factor in choosing the standard-cell design solution, since, as we have seen in preceding chapters, it is not practical to compile these on a channelled architecture gate-array IC. However, this decision is itself not without alternatives, the following in particular:

- the use of cheap off-the-shelf microprocessor and memory ICs, leaving the custom IC to accommodate the remaining digital logic; or
- the use of channel-less gate arrays, which may be able to provide complex macro requirements if the vendor has included them in his cell library.

The last has not yet found a well-defined place in the range of design styles open to the OEM, but may do so during the 1990s. They are, however, likely to be used for the more sophisticated and high-speed product requirements rather than for the average OEM needs.

A full-custom design solution has not been listed in the preceding decision charts; the increasing availability of efficient high-level macros in standard-cell form and the forecast increasing availability of more digital signal processing (DSP) macros diminishes the need for full-custom design. Hence, it is not until extremely large production quantities of a new digital product are envisaged – for example in video recorders or other high-volume home entertainment products with global sales – that full-custom becomes appropriate, or unless complex DSP requirements cannot readily be achieved with existing standard parts or standard-cell libraries.

Figure 7.6 The decision tree of when to use off-the-shelf standard digital ICs including programmable logic devices if the latter can provide the required complexity. (Source: based upon [2].)

400   The choice of design style

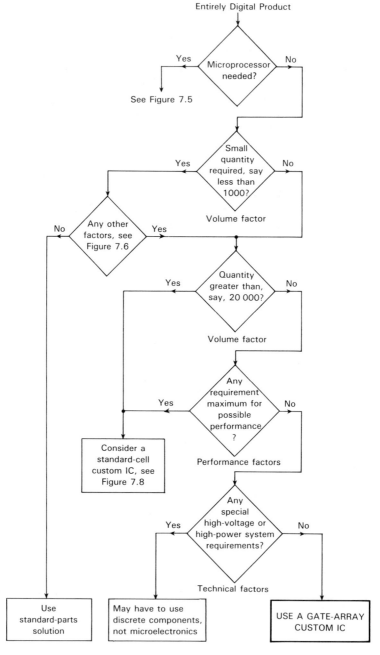

Figure 7.7   The decision tree leading to the choice of a channelled-architecture gate-array IC for digital applications. The place of the channel-less gate array, however, is currently less sure. (Source: based upon [2].)

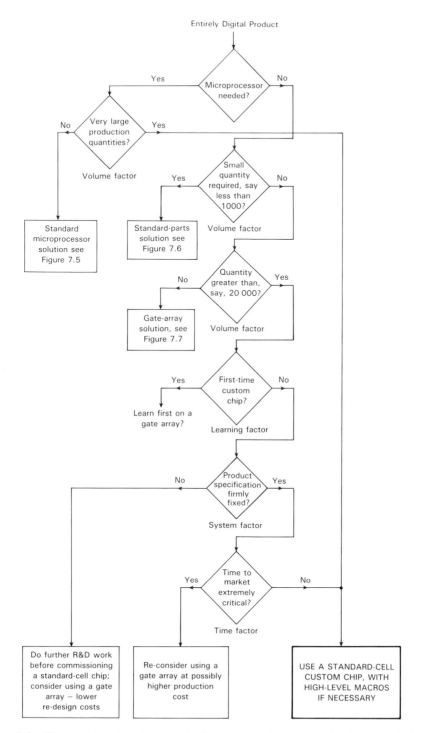

Figure 7.8 The decision tree leading to the choice of a standard-cell IC for digital applications. (Source: based upon [2].)

402   The choice of design style

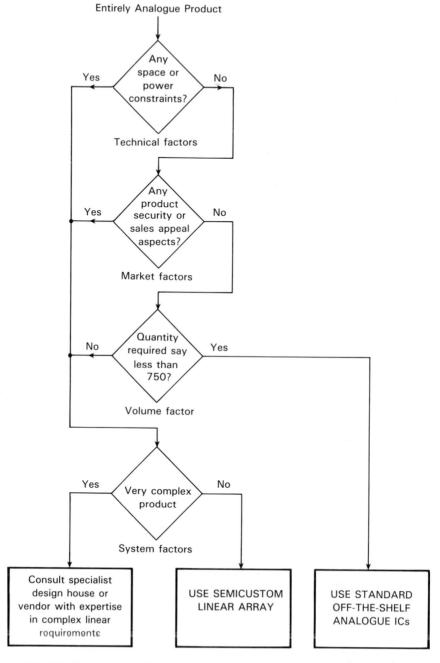

Figure 7.9 The decision tree for purely analogue applications. (Source: based upon [2].)

7.1.2 *Analogue-only products*

In contrast to VLSI digital circuits where tens of thousands of transistors may be involved, analogue-only systems tend to remain relatively small, with the whole system design capable of being undertaken by one system designer. The individual building-blocks – the operational amplifiers, comparators, active filters, etc. – may be much more complex than individual logic gates, so that the majority of OEM system designers would not have the necessary skills to undertake their detailed design, and hence analogue-only systems are almost invariably built up using pre-designed circuits in one form or anther.

Figure 7.9 shows the decision tree for purely analogue systems. The majority are still produced using standard off-the-shelf parts due to their wide availability and very low cost, so that it is extremely difficult for any form of custom design to compete on purely financial terms. Both bipolar and CMOS analogue products are available, although bipolar generally gives better performance.

However, if an appreciable quantity of a product is planned, or if space, design security or even sales appeal are significant factors, then, as indicated in Figure 7.9, a custom IC may be a suitable design choice. A further advantage not specifically listed in Figure 7.9 is that a wider range of packaging options, see Section 7.2, may be available in the custom route than with standard parts, which may have advantages.

The most widely available custom product for purely analogue purposes is the uncommitted component array, with an increasing range of tile-based architectures becoming available from vendors. Standard-cell products for analogue purposes tend to be combined with digital standard cells, and are therefore most relevant for

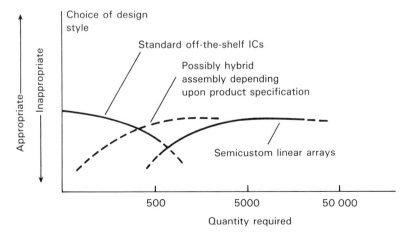

Figure 7.10 A general guide to the most economic design style for analogue-only microelectronics.

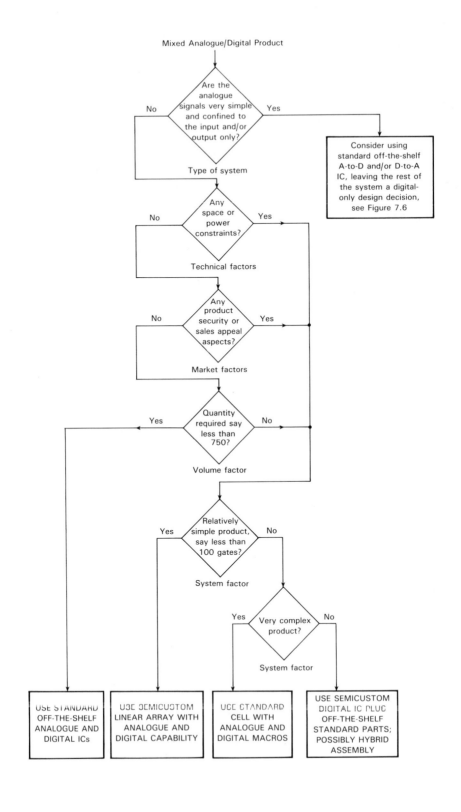

mixed analogue/digital purposes (see Section 7.1.3). Perhaps the greatest risk of adopting the custom IC path rather than using standard parts is that it is clearly more expensive and time-consuming to correct or modify a custom IC, and since analogue system design is prone to require final 'fine-tuning' or other last-minute circuit modifications, it is much more important to have a right-first-time design with the custom route than when using standard parts. The saving factor is that SPICE simulation is usually capable of performing a complete custom circuit simulation before any manufacturing costs are incurred, and should therefore highlight errors or omissions in the design specification.

Ignoring these practical considerations, on purely financial grounds and omitting any NRE costs (see Section 7.4.1), the choice between standard parts and custom is generally summarized in Figure 7.10. The dotted line indicating hybrid assemblies (see Section 7.2) is not an area that we have been able to detail in this text, but it is nevertheless one which may offer both financial and technical attractions.

7.1.3 *Mixed analogue/digital products*

The design style for mixed analogue/digital products (see Figure 7.3) is perhaps most difficult to summarize, since it depends heavily upon the size of the analogue part(s) and the size of the digital part(s). If the main design component is the digital part with the analogue merely, say, A-to-D input and/or a D-to-A output, then it is most easy to purchase the A-to-D and D-to-A parts as multi-sourced, off-the-shelf standard components, leaving the digital design style to be decided on its own merits.

However, if the digital part is small and the analogue part dominates, then an uncommitted linear array which can also provide some digital capability may be appropriate. For complex digital and complex analogue requirements, then a mixed standard-cell approach may be considered. These possibilities are summarized in Figure 7.11, but it must be emphasized that the final decision is heavily dependent upon the actual circuit requirements.

7.2 Packaging

The assembly of semiconductor dice into appropriate packages is normally necessary in order that the IC may be handled by the OEM and assembled into the final product. The only exception to this is in forms of 'direct-die-mounting', when individual dice ('naked dice') scribed from a tested wafer are used in thick-film or thin-film hybrid assemblies, each assembly containing one or more IC plus further

---

Figure 7.11 The decision tree for mixed analogue/digital applications. (Source: based upon [2].)

components if appropriate; each individual die is fixed to the hybrid substrate, usually with a gold/eutectic solder, the die I/Os then being wire-bonded to the substrate to complete the interconnections. Other forms of die-to-substrate interconnect, such as 'bump-on-die' soldering, may be used. The whole assembly is then usually given some hermetic protective covering [4], [8].

Such assembly methods are only undertaken by companies who specialize in hybrid assembly techniques. For the OEM it can offer a very compact assembly for new products, but the work has to be sub-contracted to a specialist vendor who is not usually the IC manufacturer. Hence, an additional outside agency becomes involved in the OEM's project management.

The most usual situation is for the OEM to use standard ICs or custom ICs pre-assembled upon receipt into appropriate packages, with one IC per package. The OEM has no choice in the packaging when using standard ICs, the majority of circuits being packaged in the familiar plastic encapsulated dual-in-line ('DIP' or 'DIL') packages. Analogue or other special off-the-shelf ICs may have alternative forms of packaging, but the DIL is the most common. However, for custom microelectronics the situation is completely different: the OEM now comes into the decisions on packaging, and, in consultation with the custom IC vendor or outside design agency, has to agree upon the type of packaging to be used for his or her circuit. This is, therefore, a further technical and financial aspect to be actively considered by the OEM during the custom adoption procedure.

The technology of packaging is complex. The package into which an IC is mounted has to provide many features, including the following:

- mechanical robustness, so that it may be handled in automatic testers and final OEM product assembly without damage;
- thermal properties so that the heat dissipated on the dice can be removed without unacceptable temperature stresses;
- electrical properties, so that the on-chip I/O signals can be connected to the package I/O pins without unacceptably degrading the available performance.

The thermal design [6] is complicated by the need to ensure that expansion stresses are minimized as the package temperature varies, which usually involves some small degree of flexibility between package and printed-circuit board (PCB) or other mounting. The electrical design is equally, if not more, difficult [7], since die-to-package lead inductances and capacitances mean that the fastest on-chip performance now available cannot be brought out without some loss of performance; also the electrical characteristics of *all* die-to-package interconnections should be identical, which implies that all such connections must be of identical length within the package.

Package design is thus a complex specialist field [5]–[11]. However, the OEM can only accept what is available and make an informed choice between alternatives.

The two broad categories of IC packaging are as follows:

(a) through-hole mounting packages, in which the package I/O pins pass through holes in the PCB or other mounting, being soldered on the underside; and
(b) surface-mounting packages, which do not use any through-holes but are mounted directly on the face of the PCB, etc.

The former is the most common, but the latter has many electrical and other advantages at the cost to the OEM of mastering more difficult attachment/connection/inspection techniques, with possibly higher tooling costs for the product assembly line. Removal of faulty or incorrectly assembled surface-mounted ICs may be particularly awkward.

The range of available packages is shown in Table 7.2. Figure 7.12 illustrates some of these possibilities. The overall dimensions and pin spacings are defined by internationally accepted JEDEC standards with $0.100 \pm 0.010''$ pin spacing being common in DIL packages, although other types of packaging employ contact

Table 7.2. Details of the principal range of IC packages. Mounting sockets are available for many of these types, but this introduces additional interconnection risks in the final product

Type of package P = plastic encapsulation C = ceramic	Designation	Through-hole mounting (T) or surface mounting (S)	Connector pitch	No. of I/Os[a]
Standard dual-in-line P	DIL or DIP	T	0.1"	16–64
Shrink-size dual-in-line P	SDIP	T	0.07"	40–64
Small-outline dual-in-line P	SOP	S	0.5"	8–32
Quad-in-line P	QUIP	T	0.1" (two rows each side)	16–40
Flat plastic package P	FPP	S	1 mm or less	54–120
Plastic leaded chip carrier P	PLCC	S	0.05"	24–156
Standard dual-in-line C	DIL	T	0.1"	16–64
Ceramic leadless chip carrier C	LCC	S	0.5" OR 0.4"	16–156
Pin-grid arrays C (also available in plastic encapsulation = PPGA)	PGA	T	0.1" (nested squares of pins)	64–324

[a] Minimum and maximum number of pins may vary from different vendors.

408 The choice of design style

Figure 7.12 Some of the range of IC packaging methods. Reading from top to bottom: (a) dual-in-line (DIL) plastic packaging; (b) flat plastic packaging (FPP); (c) plastic and ceramic leadless chip carriers (PLCC and LCC); and (d) pin-grid array and plastic encapsulated pin-grid array (PGA and FPGA) packaging. (Courtesy of NEC Electronics (UK) Ltd.)

spacings of 0.050″, 0.040″, 1.0 mm, 0.8 mm, 0.65 mm, etc. (JEDEC = Joint Electronic Devices Engineering Council, now part of the Electronic Industry Association (EIA), which is itself a full member of the International Electrotechnical Commission (IEC). However, JEDEC is still the term normally used in connection with IC packaging standards.)

As far as custom ICs are concerned there are several factors which strongly influence the most appropriate type of packaging [12]–[14]. These include the following:

1. The volume of custom ICs required may not be large, and therefore the set-up cost and possible delay in plastic encapsulation may not be acceptable; ceramic packages which can be held in stock and 'hand assembled' may be more economic and give a much faster turn-around time.
2. Custom ICs often require a large number of I/Os in comparison with standard off-the-shelf ICs, and hence the relatively small number of pins on dual-in-line and other standard packages is insufficient. (A typical provision is 25 pins per 1000 gates, but some custom designs require substantially more than this.)
3. One of the reasons for adopting custom microelectronics may be the space factor, and thus a package with the smallest 'footprint' area (or footprint area per I/O) may be particularly significant – in this respect also the dual-in-line package is inefficient.
4. There may be circumstances where the circuit dissipation is particularly high, and thus the thermal parameters of the package may become critical.
5. The size of the custom die may require a larger die size cavity, and hence a larger package in which to mount it, than is necessary from the actual number of I/Os required – thus spare I/Os may be available which could perhaps be profitably used for additional test purposes if appreciated at an early design stage.

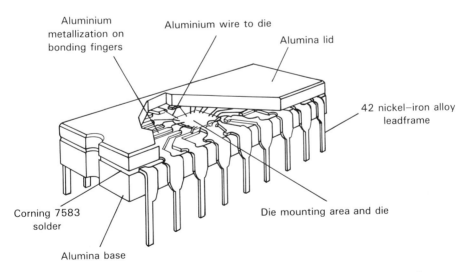

Figure 7.13 The CERDIP ceramic dual-in-line package which is appropriate for many OEM applications. A higher reliability ceramic DIL package, the ceramic sidebraze package, is also available, which is an accepted industry standard for high reliability applications.

Details of the many packaging parameters may be found in vendor's literature and elsewhere [5]–[14]. The OEM should be aware of these technicalities and implications as part of the choice of custom microelectronics. It is probable that the final choice will be between ceramic dual-in-line for relatively simple USICs (see Figure 7.13) or LCCs or PGAs for more sophisticated requirements, moving to plastic encapsulation if production volumes subsequently justify the change. (Ceramic packaging may still be required for very high reliability and/or military, avionics, space and similar products.)

7.3 Time to market

The time to market, that is, the design, manufacture and testing time for a new product, has been mentioned in preceding chapters as an important parameter, particularly so in microelectronic-based products where there is continuous evolution and hence rapid obsolescence of products. This is particularly important in designs involving custom ICs, since any error or omission in design may require a re-run of the fabrication and hence an appreciable delay to market of the final product.

The typical market life cycle for a microelectronic product is shown in Figure 7.14(a). The time scale on the x-axis varies with product, but may typically be as shown. However, market research has indicated that if a microelectronic-based product suffers an unforeseen delay of six months in its time to market, then there will be a 30% loss of sales over the lifetime of the product [15]. This is indicated in Figure 7.14(b).

To attempt to put some illustrative figures on this loss, consider a product with a selling price of $1000 giving a 20% profit margin, with a target production of 10 000 units over three years. With the correct time to market, this gives a company profit of $1000 \times 0.2 \times 10\,000 = \$2$ million.

However, in the event of an unforeseen six months' delay to market there will be a loss of sales of 3000 units, which represents a loss of company income of $600 thousand over the product lifetime. This factor is discussed in several publications [2], [16]–[18], and necessitates consideration of the design styles which provide:

- the shortest design-to-prototype time;
- the shortest time to correct initial design errors or omissions; and
- the lowest risk factor of encountering any delays.

To attempt to quantify the above factors would need a full analysis of each particular product and company expertise. However, the general trends are as shown in Figure 7.15, taking the adoption of off-the-shelf standard parts as 1.0 in each case. These generalizations also assume the following:

- The OEM has experience in the use of standard parts before adopting a PLD

Time to market 411

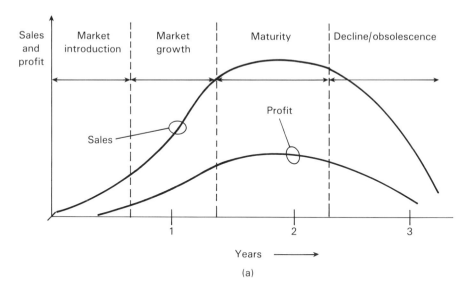

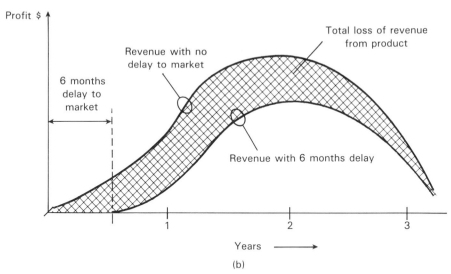

Figure 7.14 Product life cycle and product revenue: (a) typical life cycle of a microelectronic-based product, with possibility a two-year useful market life before obsolescence; (b) the general effect of a six-month delay in market introduction on product revenue.

412 The choice of design style

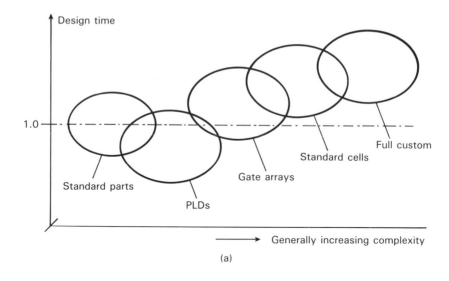

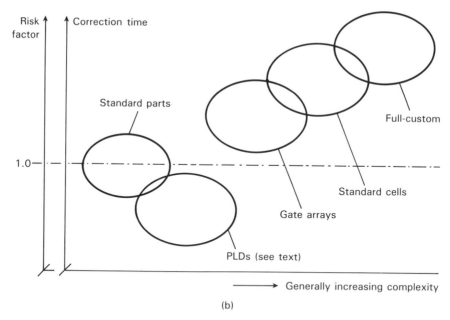

Figure 7.15 A relative comparison of design and correction times and the risk factor with differing design styles. Exact quantification is product, company, vendor and cost dependent: (a) broad comparison of design-to-prototype times; (b) broad comparison of risk and correction times, which are roughly related.

or custom design style, and would not attempt complex circuits without considerable company experience — if this is not the case, then the risks and times may escalate considerably.
- Full in-house supporting CAD is available for the PLD design choice, which can provide comprehensive design and simulation resources and automatic down-loading into the PLD programming unit.

The greatest risk with the gate-array, standard-cell and full-custom design styles remains that of some error at the design or specification stages unnoticed by the OEM at the simulation approval stage, which then necessitates a re-run of the mask and wafer fabrication, with possible queueing for these outside re-run activities. There is, therefore, considerable attraction in being able to implement any design corrections in-house, as with a standard-parts or PLD design approach, but it may not always be possible to meet complexity, performance, space or cost requirements.

7.4 Financial consideration

Financial considerations are possibly the most difficult parameters to survey, since they are clearly dependent upon moving technical and market factors. All we can attempt here is to illustrate certain underlying principles and provide broad information. Any financial figures quoted are for illustrative purposes only, and should not be read as detailed information which remains continuously valid. In real terms, the cost of silicon and CAD/CAE tools is likely to continue falling, but manpower costs may increase. Periodically updated publications such as *ASIC Outlook* [1] give current guidelines.

7.4.1 *NRE costs*

The design of all products involves non-recurring engineering (NRE) costs, including designers' time and overheads and other factors which are ideally completely paid for by the time the product is on the assembly line.

In the case of PLDs and custom microelectronics the OEM may have to consider costs arising from the factors listed in Table 7.3. Some of these may not arise, and others, such as training, may not be considered by some companies to be NRE expenditure directly costed to a specific new product. Nevertheless, they are costs which must be covered somewhere within the company finances. Note that external PLD costs should not arise since the whole object of PLDs is to undertake all design activities in-house so as to give maximum control over the design and prototype manufacture.

We will examine certain of these NRE costs more closely in the following sections, but first let us illustrate how such costs ought to be considered at the early

Table 7.3. The microelectronic NRE costs which may be incurred by the OEM when using PLD or custom microelectronic design styles. The individual item costs are not necessarily the same between the two columns

NRE item	PLD	Custom IC
In-house NRE costs		
Purchase or hire of CAD tools	✓	✓
Designer training	✓	✓
Design time and overheads	✓	✓
Special assembly and test resources	✓	✓
Purchase of PLD programmer	✓	–
Outside NRE costs		
Subcontract part or all of the IC design work	–	✓ ⎫
Cost of mask-making	–	✓ ⎬ [a]
Cost of custom test jigs if necessary to interface with standard tester	–	✓ ⎭
Iteration costs	–	✓

[a] Possibly a single lump sum NRE cost to OEM.

product planning stages. Let

E_1 = the total microelectronic NRE costs involved with a particular design style
E_2 = the remaining OEM NRE costs for the rest of the product, e.g. mechanical design, PCB design, tooling, etc.
A = the per-unit cost to the OEM of the PLD or custom IC
B = the per-unit cost of all the remaining product components and
C = the per-unit assembly and test costs.

Then for a production run of N units the total OEM costs are

$$E_1 + E_2 + N(A + B + C)$$

giving a per-unit cost of

$$\frac{E_1 + E_2}{N} + (A + B + C)$$

Clearly, if E_1 is large and/or N is small the microelectronic NRE costs will have a profound effect on per-unit cost, and the OEM should consider a design style which reduces this factor as much as possible. On the other hand, NRE costs become much less significant for large quantities, as the $E_1 + E_2$ costs can be amortized over a much larger production number N, with A and B now becoming the significant parameters.

These are simple equations to give the most economic form of design style; the often insuperable difficulty, however, is to obtain accurate numbers to put into the equations for different design styles with different suppliers and cost structures.

7.4.2 *CAD costs*

The CAD costs possibly represent the largest NRE expenditure when considering a design style other than one using standard off-the-shelf parts, particularly for first-time adoption by the OEM. Chapter 5, Section 5.6 indicated some typical hardware costs which may be incurred for both IC design and PCB design activities. Software costs are more elusive, since vendors may often supply software relating to their particular technology at preferential rates, and also the OEM may elect to purchase a minimum resource or a more comprehensive software suite.

A broad generalization of CAD costs which the OEM may consider is shown in Figure 7.16, to which software costs and hardware and software maintenance costs should be added. Hence, a total of at least twice the initial hardware costs may be involved when an OEM builds up in-house resources for design purposes. (This is broadly true for all computer-based activities within the company.)

Thus the major CAD decisions confronting the OEM when a design style other

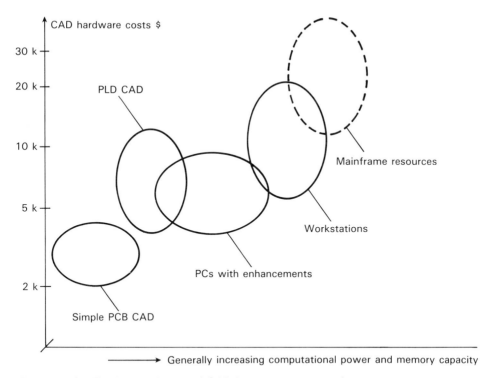

Figure 7.16 The broad picture of CAD hardware costs; software costs are additional except, possibly, with the PLD CAD which may be marketed as a complete vendors' package. See also Table 5.7 for additional workstation costings.

than one using off-the-shelf standard parts is under consideration are as follows:

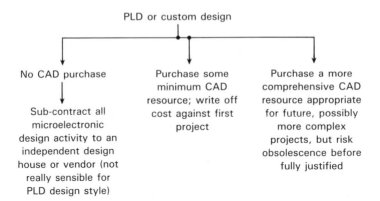

For first-time company adoption of non-standard parts, it is perhaps most appropriate for an OEM to budget for the middle course, which enables some in-house design experience to be built up without the risk of purchasing too much too soon. However, these are decisions for informed managerial judgements.

7.4.3 Learning costs

The learning time required by OEM designers to become conversant with new CAD tools is not inconsiderable. Indeed, for a manufacturing company embarking on microelectronic design activities for the first time it may be a traumatic experience, with considerable company expenditure in salaries during this period over and above that incurred in profitable design work.

Because of this above there is a school of thought which says that the OEM designer can never be as efficient as a design engineer in a specialist design office who spends 100% of his or her time on custom microelectronic design [19]. In terms of the productivity parameter 'number of gates or transistors committed per hour' this is probably true, but if the OEM sub-contracts all this design activity, then the savings in in-house learning costs have to be set against the commercial rates charged by the outside agency. In purely financial terms, the latter may be cheaper, but this does not give the OEM any increased in-house expertise for future design activities, and there are always the problems of potentially incomplete or misunderstood interface communications and staff mobility [19]–[23].

In broad terms, it takes perhaps two years for an OEM designer to become fully conversant and 'up-to-speed' with CAD tools for IC design activities, of which time perhaps one-quarter is spent in non profitable activities. With a salary plus company overhead figure of around $80 thousand per annum (£50 thousand), this implies a company learning cost per designer of $40 thousand. For new CAD tools the subsequent learning times will be considerably less, perhaps $5–10 thousand per

Financial considerations 417

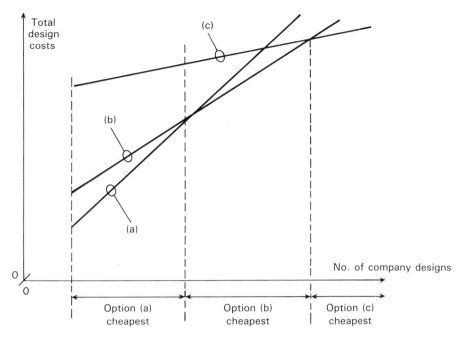

Figure 7.17 A broad picture of custom microelectronic design costs, which include capital cost of CAD tools and OEM learning costs, where: (a) sub-contract all the design work to outside agencies; (b) minimum in-house CAD resources purchased, say up to schematic capture and netlist preparation; (c) full in-house CAD capability. (The cross-over points of these three categories are difficult to define without detailed product analyses.)

new system. There are also the even more intangible learning costs for the company as a whole if no previous microelectronic-based products have been marketed: in this case it may well take five years for a company to build up expertise in source of components, reliable and unreliable suppliers and sub-contractors, new shop floor assembly and testing skills, and new marketing requirements.

The general trend of design costs is thus as shown in Figure 7.17. As before, the difficulty is to put real figures into this picture, although some figures may be extracted from certain publications [24]–[26].

7.4.4 *Testing costs*

The total testing costs involved in a microelectronic-based product includes the following:

(a) OEM cost of incoming IC testing (if done);

(b) OEM cost of final product testing; and
(c) vendor's cost of wafer probing and packaged IC testing if custom circuits are adopted.

The first two costs occur whatever the chosen microelectronic design style, but the last is particularly significant for custom circuits.

We have discussed the significance of test in many sections of preceding chapters. It is one of the most critical technical factors to discuss with the vendor once the custom IC design style is chosen. However, it is likely that the vendor will not charge the OEM separately for testing unless very long or special test sequences are required, but will instead include any necessary test jigs, etc., in an overall NRE charge (see Table 7.3) and unit test costs in the per-unit price of the packaged custom circuits. Some details on test-cost modelling may be found in [27], but these relate more to volume tests and yield than to small-quantity situations.

There is, however, one practical difference between the test of a gate-array USIC and a standard-cell USIC. In the former case the die size and the number and positions of all the I/Os are fixed, and d.c. supply inputs always use the same I/Os regardless of custom dedication. In the standard-cell case, however, both the die size and the number and allocation of the I/Os are flexible, and may (probably will) vary between each custom design.

Figure 7.18 Example test set interface boards required to interface a standard computer-controlled test resource with a particular IC under test. (Courtesy of MTL Microtechnology Ltd., UK.)

For the vendor this means that standard wafer-probe cards and test set interface boards can be used for each size of gate-array product, but a new probe card and a new interface card may have to be manufactured for each standard-cell design. (A probe card is a small card containing a number of spring-loaded probes which bear down on the I/O pads of a die during the wafer test, the wafer being stepped and each die probed in turn. The interface board provides the necessary signal processing between the circuit under test, whether at wafer-probe or subsequent packaged stage, and a standard test set such as that shown in Figure 6.4.) Figure 7.18 illustrates the types of interface boards which may be involved, and, clearly, if this and the probe card are unique, then their cost will have to be passed on to the OEM in one form or another.

The OEM may not be advised of the cost breakdown involved in the vendor's testing, but the additional cost involved in standard-cell testing will be reflected somewhere in the vendor's pricing structure.

7.4.5 *Overall product costings*

Finally, let us attempt to bring these factors together to illustrate the costs facing an OEM considering different design styles. The following figures should be taken as relative or illustrative only, exact figures being dependent upon circuit complexity, existing OEM expertise and CAD/CAM resources, and vendors' needs to secure new orders. More detailed sources of costing may be found in [1], [2], [16], [17], the first having the advantage of being annually updated. Also, a very complex series of analyses of the economics of custom microelectronics has been pursued by Fey and Paraskevopoulos who have developed extremely detailed cost models [24]–[26], but the usefulness of these analyses in pointing an OEM to the 'best' choice of design style for a first-time project is perhaps limited. Reference [26] will be found re-printed and commented upon in [2].

Perhaps the most widely publicized costings relate to the PLD vs. gate-array choice, since PLD vendors are strongly attacking the product area previously held by gate arrays. Taking 2000-gate and 5000-gate circuit requirements, cost such as that shown in Table 7.4 may be compiled, which gives the total costs for a production run of N units as follows:

1. 2000-gate array
   $\$30k + (2000 \times 0.0015 \times N)$
2. 5000 gate gate array
   $\$50k + (5000 \times 0.002 \times N)$
3. 2000-gate PLD
   $\$10k + (2000 \times 0.004 \times N)$
4. 5000-gate PLD
   $\$15k + (5000 \times 0.005 \times N)$

The choice of design style

Table 7.4. Comparative NRE and manufacturing costs for gate array and PLD designs

	2000-gate gate array	5000-gate gate array	2000-gate PLD	5000-gate PLD
Total NRE (including design) costs ($)	30 k	50 k	10 k	15 k
Manufacturing cost per gate (cents)	0.15	0.2	0.4	0.5

Assuming a linear relationship with N (which is not necessarily valid), the economics of the gate-array vs. PLD choice is shown in Figure 7.19.

Care should always be taken to ensure that like-for-like is being compared in such comparisons. For example, in some figures it may be assumed that for the gate-array choice all the circuit design costs are lumped into a global NRE cost, but for the PLD choice, as there is no sub-contracted design work, the NRE design

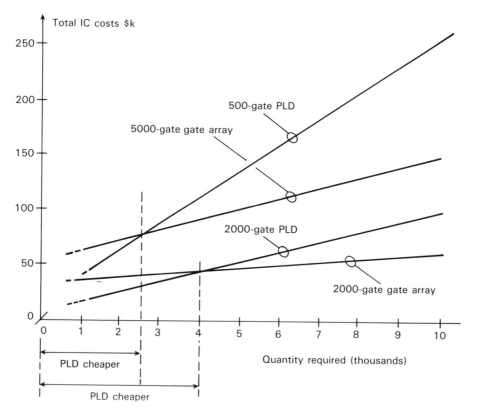

Figure 7.19 Total PLD and gate-array costs for 2000-gate and 5000-gate applications. Note, falling PLD pricing will move the cross-over points to higher production values.

costs are zero. Similarly, the IC package cost may be very significant, at least as much as an untested die, which will tend to smooth out differences in the cost per gate in different design styles [1]. And, finally, there may be a propensity to 'adjust' values in order to produce neat graphical representations. Nevertheless, there is good evidence that a PLD may be more cost-effective than a gate array over a wide band of quantity requirements.

Cost comparisons between gate arrays and off-the-shelf standard parts usually assume that some specific CAD resource must be available for in-house use for the former, but not for the latter. Hence higher NRE charges are involved in first-time adoption of the gate-array design style than when using standard parts, but it is then often assumed that this capital expenditure on CAD is freely available for subsequent new product design activities.

A typical breakdown of NRE and other costs is given in Table 7.5. The significant feature here is that, due to higher initial NRE costs, the first custom IC product is not cheaper than when using standard parts. Subsequent designs, however, when the CAD tools are available and the OEM has gained experience in custom-design activities, now begin to show a cost advantage.

Irrespective of how realistic are the costs put into these calculations, it is almost always true that the first-time adoption of custom microelectronics does not save money where an alternative off-the-shelf design style is technically possible. This law is illustrated in Figure 7.20, and is a natural result of new CAD costs and OEM learning time.

Finally, we consider standard-cell vs. gate-array design styles. While it is

Table 7.5. Design and manufacturing costs for a typical digital product using off-the-shelf parts and custom ICs, showing no saving to the OEM in first-time adoption of the latter

Activity or expenditure	Off-the-shelf standard parts ($)	First custom circuit ($)	Second custom circuit ($)
Global NRE costs up to the approval of prototype stage			
Total CAD costs	–	40 000	–
OEM design time costs	40 000	50 000	30 000
Simulation costs	–	5000	4000
Prototype costs	5000	10 000	10 000
Test costs	7500	7500	7500
Design iteration costs	5000	10 000	3000
Non-IC design and documentation costs	10 000	7500	7500
Company overheads	15 000	15 000	15 000
Sub-totals	82 500	145 000	77 000
Manufacturing costs of 2000 units	140 000	106 000	106 000
Total costs	222 500	251 000	183 000
Ratio	1.0	1.12 (12% increase)	0.82 (18% saving)

422   *The choice of design style*

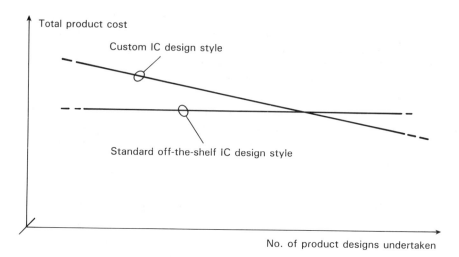

Figure 7.20 The usual cost characteristic between using standard parts and a custom IC when either is technically feasible for a product. The cross-over point where OEM expertise and resources makes the custom choice less expensive is difficult to quantify since it depends upon many company and product factors, but may be the second or third company design provided no new CAD tools have to be purchased and learnt.

generally true that more powerful CAD resources may be required for standard-cell design activities compared with corresponding gate-array design, design time may possibly be shorter due to less constraints in placement and routing. Thus, if we assume that, overall, NRE design costs roughly balance, the major cost difference between the two design styles is in the mask-making and fabrication. In particular, the *full mask set* of the standard-cell IC compared with the reduced mask set of the gate-array IC will be the major cost distinction.

Since masks cost up to $5000–10 000 per mask depending upon area and resolution, a complex standard-cell USIC requiring fifteen or more mask levels involves a considerable initial expenditure, being the major component of the vendor's NRE charges. If a design alteration should unfortunately be necessary, then this cost will be repeated.

Considering an average 2 μm custom circuit, mask costs may be:

Gate-array choice, two commitment masks @ $5000 per mask = $10 000

Standard-cell choice, ten masks @ $5000 per mask = $50 000

Hence a saving of $40 000 has to be made on the production cost of the latter circuits compared with the former before the standard-cell approach becomes financially advantageous. The position of the break-even point is very vendor-dependent, but could be in the region of 20 000 or more production circuits. Notice also that if, as some vendors suggest, an original gate-array design is moved over to a standard-cell design because of increasing product sales, the full mask set costs have to be recovered before the change becomes financially worth while; in practice, due to the limited product life cycle (see Figure 7.14), it is unlikely that an OEM would finance and risk such a change unless exceptionally good market circumstances prevail.

A final rule-of-thumb of all these considerations is that on financial grounds the optimum choice of design style is generally as shown in Figure 7.21. However, technical and other factors, including time to market, are in the end more likely to determine the OEM's choice for any given new project than are purely financial figures.

Analogue and mixed analogue/digital custom IC finances cannot be analysed in the same detail as the above digital-only circuits, since vendors do not publish the costs of such products widely. This is largely because the broad breadth of custom requirements make it difficult to give general design, manufacturing and

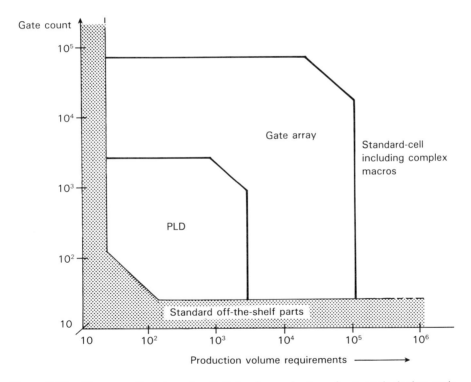

Figure 7.21 The overall picture for digital-only products, where each design style has its band of gate count/number required which is financially advantageous. Full custom may be considered as a higher continuation of the standard-cell design choice.

testing costs, plus the fact that the OEM is heavily involved in the detailed design of the analogue parts.

Little further information can thus be given on these finances. The financial equations for analogue USICs may be found in [28], but few hard figures can be quoted.

7.5 Summary

From all the preceding discussions it will be apparent that it is difficult to give firm guidelines on the optimum choice of microelectronic design style for a new company product. The technical decisions on what is *not* appropriate for a given product are much easier than deciding on the 'best' choice; financial considerations are even more diffuse and often indeterminate when all costs, both OEM's and vendor's, are taken into account.

Analyses and cost models are still being developed to assist in design style choice. Mention has previously been made of the definitive considerations of Fey and Paraskevopoulos [24]–[26]; further recent work in cost modelling is that of Edward [29], but the difficulty of finding accurate values representing the current state of the market remains. The cost of design-for-test was mentioned in the preceding chapter (see Chapter 6 [151]–[154]), which is another decision which the OEM has to face for other than simple USICs, but here again values are difficult to obtain for all the individual parameters built into the models and cost equations.

Management and design engineers in original-equipment manufacturing companies should be aware of all these aspects in order to be able to take informed decisions on design styles. Material such as that covered in [1], [2], [9], [24], [27], [30], [31], which include some case studies, should be appreciated. However, in the end it is most probable that if a USIC design style is chosen, it will be for technical and other reasons rather than cost (see the list of potential advantages at the beginning of Chapter 4, p. 110; cost is unlikely to be an overriding factor unless extremely high production volumes are envisaged. Time to market perhaps remains the most crucial parameter for the success of a new microelectronic product, see Figure 7.14, and must be very high on the list of factors requiring detailed consideration.

7.6 References

1. ICE, *ASIC Outlook 1992*, Integrated Circuit Engineering, Scottsdale AZ, 1991 (periodically republished).
2. The Open University, *Microelecrtonic Decisions*, Microelectronics for Industry Publication PT505MED, The Open University, UK, 1988.
3. Wilson, R., '8-bit MCUs wage counterattack', *Computer Design*, Vol. 29, No. 11, 1990, pp. 73–82.

4. Haskard, M. R., *Thick Film Hybrids: Manufacturing and design*, Prentice-Hall, NJ, 1989.
5. Griffiths, G. W., 'A review of semiconductor packaging and its role in electronics manufacture', *J. Semicustom ICs*, Vol. 7, No. 1, 1989, pp. 31–9.
6. DiGiacomo, J. (ed.), *VLSI Handbook*, McGraw-Hill, NY, 1989.
7. Bakaglu, H. G., *Circuits, Interconnections and Packaging for VLSI*, Addison-Wesley, Reading, MA, 1990.
8. Sinnadurai, F. N. (ed.), *Handbook of Microelectronic Packaging and Interconnect Technologies*, Electrochemical Publications, Scotland, 1985.
9. BEP Data Services, *Semicustom IC Yearbook*, Elsevier Scientific Publications, Oxford, UK, 1988.
10. VLSI Support Technologies: *Computer-Aided Design, Testing and Packaging*, IEEE Computer Society Press, CA, 1982.
11. Booth, W., *VLSI Era IC Packaging*, Electronic Trend Publications, CA, 1987.
12. National Semiconductor, *ASIC Packaging*, Booklet No. 101185, National Semiconductor Corporation, W. Germany, 1987.
13. Freeman, E., 'Cost, device speed, size, and reliability determine the best package for an ASIC', *EDN*, Vol. 32, No. 9, 1987, pp. 77–82.
14. Fahr, G. K., 'Logic array packaging', *Application Report No. A23*, LSI Logic Corporation, CA, 1982.
15. Reinertsen, D. G., 'Whodunit? The search for the new product killers', *Electronic Business*, Vol. 9, July 1983, pp. 62–6.
16. Xilinx, *The Programmable Gate Array Data Book*, Xilinx Corporation, CA, 1989.
17. Altera, Product Information Bulletin No. 6, *PLDs vs. ASIC Economics*, Altera Corporation, CA, 1988.
18. Hurst, S. L., 'Custom microelectronics: the risk factor', *Microprocessors and Microsystems*, Vol. 14, No. 4, 1990, pp. 197–203.
19. Kutzin, M., 'Advantages of ASIC design centres', *J. Semicustom ICs*, Vol. 5, No. 1, 1987, pp. 12–15.
20. Heberling, C., 'The modern silicon foundry and CAD software', *ibid.*, Vol. 6, No. 1, 1988, pp. 19–24.
21. Horne, N. W., 'The case for strategic alliances', *ibid.*, pp. 25–8.
22. Shenton, G., 'ASIC procurement: is getting married a good idea?', *ibid.*, pp. 29–32.
23. White, A., 'The painless path to ASIC design', *ibid.*, pp. 33–6.
24. Fey, C. F. and Paraskevopoulos, D. E., 'Studies in LSI technology: a comparison of product cost using MSI, gate arrays, standard cells and full custom VLSI', *IEEE J. Solid State Circuits*, Vol. SC21, 1986, pp. 297–303.
25. Fey, C. F. and Paraskevopoulos, D. E., 'A model of design schedules for application specific ICs', *J. Semicustom ICs*, Vol. 4, No. 1, 1986, pp. 5–12
26. Fey, C. F. and Paraskevopoulos, D. E., 'A techno-economic assessment of application-specific integrated circuits: current status and future trends', *Proc. IEEE*, Vol. 75, 1987, pp. 829–41.
27. Teradyne, *The Economics of VLSI Testing*, Teradyne, Inc., CA, 1984.
28. Trontelj, J., Trontelj, L. and Shenton, G., *Analog Digital ASIC Design*, McGraw-Hill, UK, 1989.
29. Edwards, L. N. M., 'USIC cost simulation: a new solution to an old problem', *J. Semicustom ICs*, Vol. 8, No. 2, 1990, pp. 3–12.
30. The Open University, *Microelectronic Matters*, Microelectronics for Industry, publication No. PT504 MM, The Open University, UK, 1988.
31. Altera, *A Manager's Perspective*, Product Information Bulletin No. 5, Altera Corporation, CA, 1990.

8 Conclusions

8.1 The present status

From the information which we have attempted to survey in the preceding chapters, it is apparent that the present status of custom microelectronics is one of considerable expertise and achievement, particularly in the following respects:

- Silicon fabrication technology has reached a very high level of maturity, with 2 μm geometry being a widely available standard process capability.
- Extensive gate-array and standard-cell developments have provided the OEM with a wide range of products and choice of vendors for digital applications.
- CAD resources have also developed to the stage where most digital applications can be readily designed and simulated for a target gate-array or standard-cell realization.
- Costs of both design and fabrication have fallen in real terms over the past two decades to the point where − assuming no start-up/learning-time costs − a custom realization may become financially viable for production quantities, possibly as low as 5000 units.
- The problems of testing large digital networks are now much more widely appreciated and freely discussed between OEMs and vendors; and lastly:
- Recent increases in the capability of programmable logic devices means that they must now be regarded as a viable alternative to other custom design styles.

However, as we have also seen, there are still shortcomings and risks which the OEM should appreciate when deciding to use or not to use custom circuits. Among these factors are the following:

- Analogue and mixed analogue/digital custom applications are not as widely served by vendors' available products and CAD as the digital-only area − this is partly a reflection of the much greater flexibility of requirements found in analogue applications.

- CAD resources for the very highest levels of design, the behavioural and architectural levels, have yet to reach the maturity and widespread acceptance of the lower-level CAD tools.
- First-time adoption of custom-design techniques by an OEM inevitably involves a learning period and possible risk of errors and omissions.
- Obsolescence of products and CAD tools, difficulties of second-sourcing without incurring further NRE costs, and take-overs and mergers between vendors are further factors which an OEM has to consider if long-term product availability is involved.

Silicon is and will remain the dominant semiconductor technology. The availability of gallium-arsenide is increasing, with GaAs custom products becoming more comprehensive, but the market for GaAs performance is likely to remain small in comparison with silicon.

8.2 Future developments

Improvements in:

- silicon and GaAs fabrication capability,
- custom products for OEM use, and
- supporting CAD/CAE tools including networking and data-exchange standards,

will continue, with emphasis on smaller geometries, higher performance from a given technology, and more user-friendly and intelligent CAD resources.

Current surveys [1], [2], indicate that global semiconductor production will continue to increase by around 10% per annum, with an increase in co-operative agreements between major IC vendors in order to spread the escalating costs of development and fabrication lines. Figure 8.1 shows the anticipated increase in the custom microelectronics segment, which is forecast to grow from the mid-1980s figure of about 10% of the total semiconductor market to about 20% of the market by the mid-1990s, although it is likely that the major increase in value will be due to the output of a few very large product manufacturers rather than the result of a large number of small companies adopting custom microelectronics for short production runs. However, the wider adoption of programmable logic devices for small-quantity production runs may modify these predictions.

CMOS technology will continue to be the major technology for custom microelectronics for the early-1990s, with geometries reducing to the sub-micron level for the highest performance requirements. But as geometries shrink we find

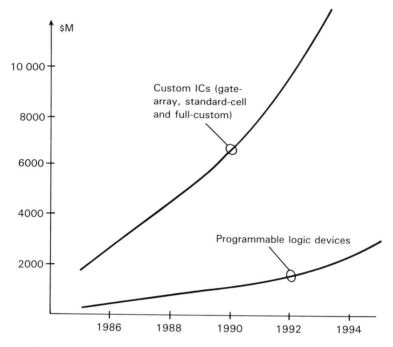

Figure 8.1 Global sales of custom microelectronic ICs, forecast into the 1990s.

that:

- the ohmic resistance of via connections to the silicon becomes more significant and may cause limitations in performance; and
- the limits of optical lithography are being approached, see Figure 8.2.

Probably 2–3 μm will remain the most relevant and cost-effective for the majority of OEM and product applications.

However, it is possible that the present R and D in small-geometry, low-power bipolar technology may challenge the dominance of CMOS in the 1990s, providing equal if not better speed performance without the excessive power dissipation that currently characterizes very-high-speed bipolar technologies such as ECL. The advantages of bipolar developments will also be significant for analogue and mixed analogue/digital applications, and may mean that the present BiCMOS developments become unnecessary if bipolar VLSI digital capability becomes available.

For very-high-speed requirements it is possible that GaAs may challenge ECL successfully on both price and performance as the technology becomes more mature. A natural advantage of GaAs over ECL is that fewer mask levels are normally involved, possibly ten to twelve masks compared with twenty or more

Future developments 429

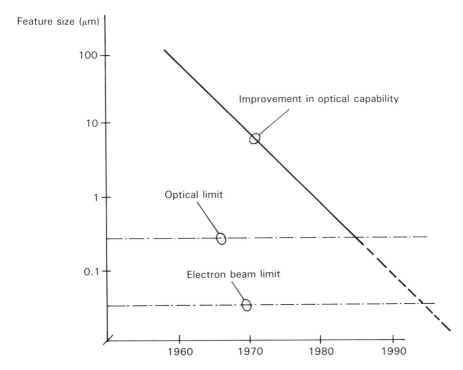

Figure 8.2 The limits of optical lithography, with electron-beam lithography providing about a further order of magnitude reduction in limiting capability. (Note: mask alignment and etching accuracy are separate limitations.)

which may be used on very advanced ECL processes [1], and hence if base material costs and defect densities can be reduced, then a market challenge to ECL will occur. However, this top end of the performance market is unlikely to be relevant to other than specialist OEMs.

Absolute fabrication capabilities will continue to be driven by memory circuits, which will always utilize as much complexity as can be fabricated on a single die, and will demand still more. The utilization of this capability for analogue and digital networks has always tended to lag behind, as shown in Figure 8.3. In spite of the promise of CAD resources of increasing power, this is likely to continue, testing as well as design capability being a limitation on the volume of circuitry that a designer can realistically compile on a single die.

Thus, future developments in the CAD area will have the greatest impact for the OEM and company adoption of custom microelectronics. Depending upon the level of company requirements, some or all of the following may be anticipated:

- more powerful hardware platforms providing more power per $, possibly with RISC architectures and UNIX operating systems [3], [4];

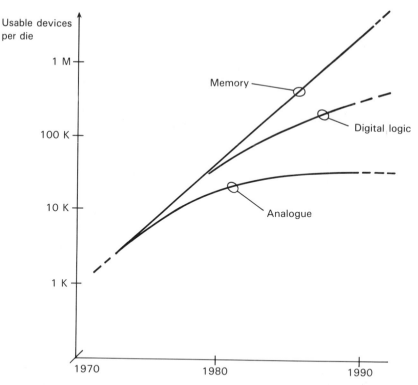

Figure 8.3 The trends in capability of memory, digital logic and analogue circuits per die, with design and test being the principal limitations in the latter two cases rather than fabrication capabilities. (The value for the latter two cases should not be read too closely.)

- increased standardization of data exchange formats for the design data – possibly EDIF;
- improved busses to link CAD resources and handle the increased data transfer requirements of complex resources and/or RISC architectures, etc. – possibly Futurebus+ [5];
- better graphics capabilities, with extensive availability of icons and windows [6];
- increased use of distributed CAD resources linked by local area networks (LANs), promoted by the Open Software Foundation (OSF) which is dedicated to the development of distributed CAD/CAM/CAE [4], with the possible use of Fibre-Distributed Data Interface (FDDI) links [7];
- increased availability of hardware description languages (HDLs) for behavioural- and architectural-level digital design activities, spanning or linking to the lower levels of design activity [8];

- increased availability of knowledge-based CAD tools, which incorporate some of the factors and decision-making which an experienced designer uses when involved in a new IC design;
- increased availability of design-for-testability features in the design tools so that scan-test, built-in self-test, concurrent test, or any other means to ease the final circuit testing, can be built in as the design proceeds; and, finally,
- increased availability of analogue and mixed analogue/digital design tools, including simulation capabilities for mixed analogue/digital designs [9].

Silicon compilation has not been specifically mentioned in the above listings; it has never been easy to define exactly what was meant by 'silicon compilation', but its principal aims and objectives are essentially covered by HDLs and hierarchical design tools. Further information and discussions may be found in [3]–[15].

The ideal CAD resource for future digital design activities may therefore be as shown in Figure 8.4, where several possible data entry routes are available before optimization and targeting on the final form of realization. Standardized data interfaces are implicit in such a hierarchical system. However, it will be noted that analogue needs do not yet enter this comprehensive picture, which emphasizes the still separate nature of much of the analogue design requirements.

Finally, in packaging assembly and test we may find the following:

- the development of new forms of packaging which match the increasing on-chip performance capabilities, and provide an easier means of assembly on to PCBs;
- increasing use of die mounted directly on to PCB or other substrates in order to circumvent any packaging shortcomings; and
- the increasing use of on-chip self-test or concurrent on-line checking in order to minimize the amount of test equipment and set-up necessary to test custom circuits.

Concurrent on-line test has not yet received very much attention in the custom microelectronics field, but this may be one of the growth features of the next decade.

All the above factors represent the continuing evolution of existing concepts and expertise. In terms of concepts which are not yet near the usage stage, we may see some of the following being turned into practice in the longer term:

- opto-electronics, with on-chip photonic logic elements and interconnect, a great attraction of optical signals being their freedom from interfering with each other;
- new devices based upon thin layers of different semiconductor materials fabricated as one continuous 3-D structure; and
- new computing architectures to supersede the binary-based von Neumann and parallel processing techniques of the present.

432   Conclusions

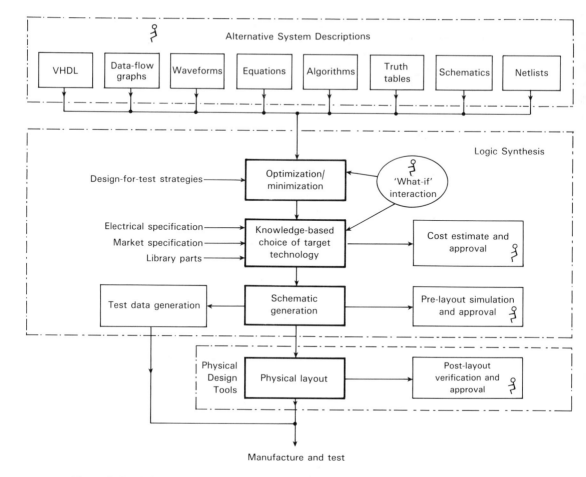

Figure 8.4 A future CAD resource for digital system design, incorporating knowledge-based tools to choose the optimum design style and provide cost and test data. Some feedback loops (not shown) must be present to accommodate non-approval or designer's interactive requirements.

Exactly when these developments will impact upon the well-established semiconductor and computer industries remains to be seen.

8.3 Final summary

To return from the future to the present, for the majority of OEMs the present status and capabilities of silicon fabrication can provide all their performance requirements. The current most important needs are concerned with design and test aspects, with time-to-market being a crucial factor in company profitability.

Thus the existing status provides a solid basis for the future of custom microelectronics. The principles which we have covered in the preceding chapters will continue to remain valid, with no revolutionary concepts being anticipated in the near future, but increasing emphasis may be found on improving existing CAD tools and making them more 'intelligent' and easy for OEM designers to use, by providing more digital signal processing (DSP) and analogue design resources, by providing more ready means of moving from one form of realization to another (e.g. PLD ↔ gate array), and by smoothing the data transfer between OEM and custom microelectronic vendors.

8.4 References

1. ICE, *ASIC Outlook 1992*, Integrated Circuit Engineering, Scottsdale, AZ, 1991 (periodically republished).
2. Elsevier, *Semicustom IC Yearbook*, Elsevier Scientific Publications, UK, 1988 (periodically republished).
3. Wilson, R., 'RISC chips', *Computer Design*, Vol. 29, No. 9, 1990, pp. 80–98.
4. CMC, *User's Guide to Design Automation*, CMC Publications, NY, 1988 (periodically recompiled).
5. Andrews, W., 'Open buses broaden foothold at all levels', *Computer Design*, Vol. 29, No. 9, 1990, pp. 55–64.
6. Bond, J., 'TIGA and 8514/A vie for dominance in graphics', *ibid.*, Vol. 29, 1990, No. 5, pp. 55–64.
7. Wilson, R., 'Second generation FDDI chips', *ibid.*, Vol. 29, No. 9, 1990, pp. 38–40.
8. Harding, W., 'HDLs: a high-powered way to look at complex designs', *ibid.*, Vol. 29, No. 9, 1990, pp. 74–84.
9. Meyer, E., 'Analog synthesis: how soon?', *ibid.*, Vol. 29, No. 9, 1990, pp. 30–4.
10. Special issue, 'Towards 2000', *Electronic Product Design*, Vol. 11, April 1990.
11. Special issue, 'Technology directions', *Computer Design*, Vol. 27, No. 22, 1988 (periodically recompiled).
12. Special issue, 'Technology '89', *IEEE Spectrum*, Vol. 26, No. 1, 1989, (periodically recompiled).
13. Dettmer, R., 'User-programmable logic chasing the gate array', *IEE Review*, Vol. 36, No. 5, 1990, pp. 181–4.
14. Meyer, E., 'Logic synthesis fine tunes abstract design descriptions', *Computer Design*, Vol. 29, No. 11, 1990, pp. 84–97.
15. Meyer, E., 'BiCMOS, ECL and GaAs chips fight for sockets in 1990s', *Computer Design*, Vol. 29, No. 11, 1990, pp. 39–43.

Appendix A: The elements and their properties

To supplement the specific information covered in the main chapters of this text, this and the following appendices give further information which may provide additional background material. This first appendix gives the familiar periodic table of the elements, followed by some factors of significance for microelectronic fabrication or circuit performance.

The periodic table is detailed in Table A.1, with the atomic number of each element given in parentheses. The elements between Group II and Group III are the transition elements, while the elements from lanthanum (La), atomic number 57, to lutecium (Lu), atomic number 71, are the lanthanons or rare earth elements. The higher end of the table constitutes the naturally radioactive and increasingly unstable elements.

The arrangement of the electrons and their energies determines virtually all the mechanical, chemical and electrical properties of the elements. Table A.2 lists some of the elements with the number of atoms in the various electron shells surrounding the nucleus in their lowest energy state (0 K). Note that each shell has a principal quantum number 1, 2, ..., and may have several orbital quantum numbers s, p, d, ... For example, within the N-shell there may be four orbital quantum numbers 4s, 4p, 4d and 4f.

Good electrical conductors such as copper (Cu, 29), silver (Ag, 47), platinum (Pt, 78) and gold (Au, 79) are characterized by a single loosely bound electron in their outermost (valency) orbit, with inner orbits full or almost full. Poor conductors (insulators), on the other hand, such as the inert gases neon (Ne, 10), argon (Ar, 18), etc., are characterized by an atomic structure having no electrons easily available for conduction purposes. Insulating compounds such as silicon dioxide (SiO_2) likewise have no readily-available electrons for conduction purposes in their molecular structure.

The semiconductor elements and semiconductor compounds fall between these two extremes. Table A.3 gives a section of the periodic table which is of interest for microelectronic purposes, specifically the elements which have four electrons in their valency band and which can be made in a crystal lattice structure, and the elements in the immediately adjoining valency bands.

The elements and their properties

Table A.1. The periodic table of the elements with the atomic number of each element given in parentheses

Group I	Group II											Group III	Group IV	Group V	Group VI	Group VII	Group VIII
H (1)																	He (2)
Li (3)	Be (4)											B (5)	C (6)	N (7)	O (8)	F (9)	Ne (10)
Na (11)	Mg (12)				TRANSITION ELEMENTS							Al (13)	Si (14)	P (15)	S (16)	Cl (17)	Ar (18)
K (19)	Ca (20)	Sc (21)	Ti (22)	V (23)	Cr (24)	Mn (25)	Fe (26)	Co (27)	Ni (28)	Cu (29)	Zn (30)	Ga (31)	Ge (32)	As (33)	Se (34)	Br (35)	Kr (36)
Rb (37)	Sr (38)	Y (39)	Zr (40)	Nb (41)	Mo (42)	Tc (43)	Ru (44)	Rh (45)	Pd (46)	Ag (47)	Cd (48)	In (49)	Sn (50)	Sb (51)	Te (52)	I (53)	Xe (54)
Cs (55)	Ba (56)	La (57*)	Hf (72)	Ta (73)	W (74)	Re (75)	Os (76)	Ir (77)	Pt (78)	Au (79)	Hg (80)	Tl (81)	Pb (82)	Bi (83)	Po (84)	At (85)	Rn (86)
Fr (87)	Ra (88)	Ac (89†)															

Lanthanons *(Rare earths)	Ce (58)	Pr (59)	Nd (60)	Pm (61)	Sm (62)	Eu (63)	Gd (64)	Tb (65)	Dy (66)	Ho (67)	Er (68)	Tm (69)	Yb (70)	Lu (71)
Actinons †(Radioactive elements)	Th (90)	Pa (91)	U (92)	Np (93)	Pu (94)	Am (95)	Cm (96)	Bk (97)	Cf (98)	Es (99)	Fm (100)	Md (101)	No (102)	Lr (103)

The two naturally occurring elements for practical semiconductor purposes are silicon and germanium, both of which can be made in a crystal lattice form and which have an intrinsic conduction midway between conductors and insulators. Carbon in its crystal form (diamond) has too high an electron energy gap in its valency band to be useful – it is effectively an insulator – whilst tin and lead have too low an energy gap and are effectively electrical conductors at normal ambient temperatures. This leaves silicon and germanium as the useful elements, with silicon having the preferable fabrication and performance properties under normal operating conditions. The possible n-type dopants are given by the group V elements, contributing an additional electron for conduction purposes when introduced into a silicon or germanium crystal structure, while the possible p-type dopants are from the group III elements. Semiconductor compounds such as GaAs will be seen to be formed from Group III/V elements. The values of significant physical parameters for silicon, germanium and gallium-arsenide are given in Table A.4.

The effect on the intrinsic resistivity of silicon with n-type and p-type doping is shown in Figure A.1. It will be seen that an impurity concentration of, say, one atom of impurity per 10^6 atoms of silicon will give an electrical resistivity of about

Appendix A

Table A.2. An extract from the periodic table giving the electron shell arrangements

Atomic number	Element	K	L		M			N				O					P				Q	
		1s	2s	2p	3s	3p	3d	4s	4p	4d	4f	5s	5p	5d	5f	5g	6s	6p	6d	6f	7s	7p
1	H	1																				
2	He	2																				
3	Li	2	1																			
4	Be	2	2																			
5	B	2	2	1																		
6	C	2	2	2																		
7	N	2	2	3																		
8	O	2	2	4																		
9	F	2	2	5																		
10	Ne	2	2	6																		
11	Na	2	2	6	1																	
12	Mg	2	2	6	2																	
13	Al	2	2	6	2	1																
14	Si	2	2	6	2	2																
15	P	2	2	6	2	3																
16	S	2	2	6	2	4																
17	Cl	2	2	6	2	5																
18	Ar	2	2	6	2	6																
19	K	2	2	6	2	6		1														
20	Ca	2	2	6	2	6		2														
21	Sc	2	2	6	2	6	1	2														
22	Ti	2	2	6	2	6	2	2														
23	V	2	2	6	2	6	3	2														
24	Cr	2	2	6	2	6	5	1														
25	Mn	2	2	6	2	6	5	2														
26	Fe	2	2	6	2	6	6	2														
27	Co	2	2	6	2	6	7	2														
28	Ni	2	2	6	2	6	8	2														
29	Cu	2	2	6	2	6	10	1														
30	Zn	2	2	6	2	6	10	2														
31	Ga	2	2	6	2	6	10	2	1													
32	Ge	2	2	6	2	6	10	2	2													
33	As	2	2	6	2	6	10	2	3													
34	Se	2	2	6	2	6	10	2	4													
35	Br	2	2	6	2	6	10	2	5													
36	Kr	2	2	6	2	6	10	2	6													
37	Rb	2	2	6	2	6	10	2	6			1										
38	Sr	2	2	6	2	6	10	2	6			2										
39	Y	2	2	6	2	6	10	2	6	1		2										
40	Zr	2	2	6	2	6	10	2	6	2		2										
41	Nb	2	2	6	2	6	10	2	6	4		1										
42	Mo	2	2	6	2	6	10	2	6	5		1										
43	Tc	2	2	6	2	6	10	2	6	6		1										
44	Ru	2	2	6	2	6	10	2	6	7		1										
45	Rh	2	2	6	2	6	10	2	6	8		1										
46	Pd	2	2	6	2	6	10	2	6	10												
47	Ag	2	2	6	2	6	10	2	6	10		1										
48	Cd	2	2	6	2	6	10	2	6	10		2										

Table A.2. (Continued)

Atomic number	Element	Electron arrangement																			
		K		L			M			N				O				P				Q	
		1s	2s	2p	3s	3p	3d	4s	4p	4d	4f	5s	5p	5d	5f	5g	6s	6p	6d	6f	7s	7p
49	In	2	2	6	2	6	10	2	6	10		2	1									
50	Sn	2	2	6	2	6	10	2	6	10		2	2									
51	Sb	2	2	6	2	6	10	2	6	10		2	3									
52	Te	2	2	6	2	6	10	2	6	10		2	4									
78	Pt	2	2	6	2	6	10	2	6	10	14	2	6	9			1					
79	Au	2	2	6	2	6	10	2	6	10	14	2	6	10			1					
80	Hg	2	2	6	2	6	10	2	6	10	14	2	6	10			2					
81	Tl	2	2	6	2	6	10	2	6	10	14	2	6	10			2	1				
82	Pb	2	2	6	2	6	10	2	6	10	14	2	6	10			2	2				
83	Bi	2	2	6	2	6	10	2	6	10	14	2	6	10			2	3				
84	Po	2	2	6	2	6	10	2	6	10	14	2	6	10			2	4				
85	At	2	2	6	2	6	10	2	6	10	14	2	6	10			2	5				
86	Rn	2	2	6	2	6	10	2	6	10	14	2	6	10			2	6				
87	Fr	2	2	6	2	6	10	2	6	10	14	2	6	10			2	6			1	
88	Ra	2	2	6	2	6	10	2	6	10	14	2	6	10			2	6			2	
89	Ac	2	2	6	2	6	10	2	6	10	14	2	6	10			2	6	1		2	
90	Th	2	2	6	2	6	10	2	6	10	14	2	6	10			2	6	2		2	
91	Pa	2	2	6	2	6	10	2	6	10	14	2	6	10	2		2	6	1		2	
92	U	2	2	6	2	6	10	2	6	10	14	2	6	10	3		2	6	1		2	

Table A.3. An extract from the periodic table of the elements closely associated with microelectronics

Group II	III	IV	V	VI
	Boron (5)	Carbon (6)	Nitrogen (7)	Oxygen (8)
Magnesium (12)	Aluminium (13)	Silicon (14)	Phosphorus (15)	Sulphur (16)
Zinc (30)	Gallium (31)	Germanium (32)	Arsenic (33)	Selenium (34)
Cadmium (48)	Indium (49)	Tin (50)	Antimony (51)	Tellurium (52)
Mercury (80)	Thallium (81)	Lead (82)	Bismuth (83)	

Table A.4. Significant physical parameters for silicon, germanium and gallium-arsenide at 300 K

	Si	Ge	GaAs
Energy gap E_g (eV)	1.1	0.7	1.4
Intrinsic carrier density n_i (carriers m^{-3})	1.45×10^{16}	2.5×10^{19}	1×10^{13}
Relative permittivity e_r	12	16	12
Hole mobility μ_P (m^2 V^{-1} s^{-1})	0.05	0.19	0.5
Electron mobility μ_N (m^2 V^{-1} s^{-1})	0.14	0.39	0.85
Hole diffusion constant D_P (m^2 s^{-1})	1.3×10^{-3}	5×10^{-3}	1.3×10^{-3}
Electron diffusion constant D_n (m^2 s^{-1})	3.6×10^{-3}	10×10^{-3}	22×10^{-3}

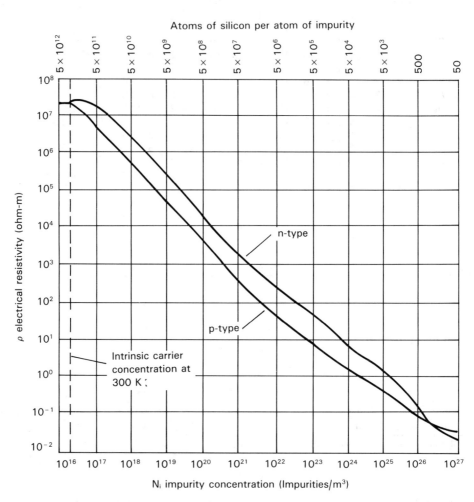

Figure A.1 The variation of electrical resistivity ρ for silicon at 300 K as a function of the impurity doping concentration. Crystalline silicon has approximately 5×10^{22} atoms per cubic metre.

10^2 ohm-metres compared with the intrinsic conductivity of silicon of about 2.3×10^7 ohm-metres, a reduction of over 10^5. However, note how these values compare with conductors and insulators, which at 300 K have values shown as follows.

Material	$\rho(\Omega m)$
Copper	1.7×10^{-8}
Aluminium	2.7×10^{-8}
Gold	11×10^{-8}
Intrinsic silicon	2.3×10^7
Glass, mica, etc.	10^{16} to 10^{22}

Further physical data may be found in many appropriate texts, for example [1] – [8] given in Chapter 2, and elsewhere.

Appendix B: Fabrication and yield

Lithography is currently at the heart of almost all microelectronic fabrication processes. Table B.1 summarizes the steps which are conventionally involved in the production of masks for photolithography or for direct-write-on-wafer; see also Chapter 4, Section 4.4.

The procedures which conventionally take place at each mask level during wafer fabrication are summarized in Table B.2. From this it will be appreciated that at each mask stage a number of separate fabrication and inspection processes must be done, all of which are critical to the production of defect-free final circuits. With E-beam direct-write-on-wafer the steps are similar except that the wafers have to be loaded into and aligned in the E-beam machine for each patterning of the resist material. Further details of these various processing stages may be found in [3] in Chapter 2.

The number of whole dice (chips) which can be obtained from wafers of increasing diameter is given in Table B.3. The extreme edge of a wafer is invariably kept free of circuits, since there is likely to be an increasing number of defects near its circumference due to minor mechanical damage as well as processing inaccuracies/tolerancing at the edges of the wafers as they go through the many stages of fabrication. Table B.3 gives the die size in microns (10^{-6} m) and the wafer diameter in cm (10^{-2} m); imperial measure with die size in mils (10^{-3} inches) and wafer diameter in inches is unfortunately also used in some circumstances. The relationships between these measures is as follows:

$$
\begin{aligned}
1\ \mu\text{m} &= 0.04\ \text{mils} \\
1\ \text{mm} &= 40\ \text{mils} \\
1\ \text{mil} &= 25.4\ \mu\text{m}\ (2.54 \times 10^{-5}\ \text{m}) \\
100\ \text{mils} &= 2.54\ \text{mm} \\
1\ \text{inch} &= 2.54\ \text{cm}\ (2.54 \times 10^{-2}\ \text{m}) \\
1\ \text{mil}^2 &= 645\ \mu\text{m}^2\ (6.45 \times 10^{-10}\ \text{m}^2)
\end{aligned}
$$

Modelling of die yield on a wafer, which is principally dependent upon die size and process 'goodness' and not upon circuit complexity, has been extensively studied, since by accurate modelling the yield, and hence cost, of new circuits may

Table B.1. A summary of the principal alternative ways of patterning resist on silicon wafers

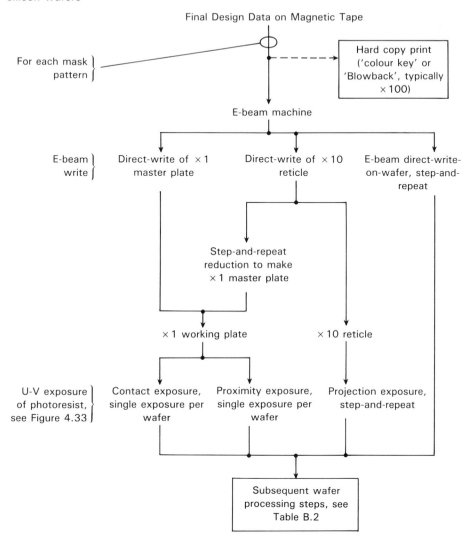

be forecast. Also, once the detailed modelling parameters are known, the quality of production lines can be maintained and improvements made to increase the yield from that production process. However, the available modelling theory and available parameter values usually lag the actual production processing, and as a result the yield of most production lines has historically tended to be higher than that predicted by the available modelling theory.

Wafer yield $Y, 0 \leqslant Y \leqslant 1$, is normally considered to be a function of the

Table B.2. A summary of the steps at each mask level of a wafer fabrication process

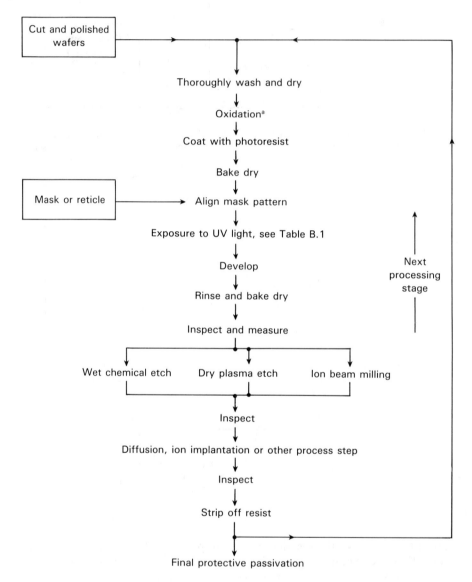

[a] Not at every stage.

Table B.3. The approximate number of whole dice available from increasing diameter wafers, neglecting the presence of any process control monitoring circuits (sometimes termed 'process evaluation devices', PEDs or 'drop-ins')

Square die size ($\mu m \times \mu m$)	Die area (μm^2)	Approximate no. of dice per wafer			
		Wafer diameter			
		7.5 cm	10 cm	12.5 cm	15 cm
2000	4×10^6	1035	1840	2920	4800
2250	5.06×10^6	835	1480	2290	3300
2500	6.25×10^6	655	1180	1800	2600
2750	7.56×10^6	545	980	1490	2200
3000	9×10^6	450	820	1250	1900
3250	10.56×10^6	372	703	1090	1650
3500	12.25×10^6	324	601	935	1400
3750	14.06×10^6	284	510	810	1200
4000	16×10^6	256	448	710	1000
4250	18.06×10^6	214	400	625	880
4500	20.25×10^6	190	350	550	800
4750	22.56×10^6	168	313	490	730
5000	25×10^6	148	282	445	660
5250	27.56×10^6	135	264	395	580
5500	30.25×10^6	123	230	360	500
5750	33.06×10^6	110	212	316	450
6000	36×10^6	98	190	300	410
6250	39.06×10^6	90	172	273	370
6500	42.25×10^6	82	158	250	340
6750	45.56×10^6	75	146	232	310
7000	49×10^6	68	134	217	290
7250	52.56×10^6	67	124	197	270
7500	56.25×10^6	60	115	184	250
7750	60.63×10^6	52	105	173	235
8000	64×10^6	52	100	162	220
8250	68.06×10^6	51	90	152	200
8500	72.25×10^6	43	86	142	185
8750	76.56×10^6	42	81	132	175
9000	81×10^6	40	80	125	160

density of defects per unit area of wafer, D_0, the area of the die, A, and parameters α which are peculiar to the particular model and process. Many equations have been proposed for Y, all of which have a statistical basis. The many variants try to take into account factors such as the defect density D_0 not being uniform across a wafer, but all predict that the yield of fault-free circuits decreases as the die size increases. Further details may be found in [3] in Chapter 2.

One of the modelling equations which has been used is

$$Y = \left\{ \frac{1 - e^{-D_0 A}}{D_0 A} \right\}^2 \times 100\%$$

The difficulty now is to find a realistic value for D_0 which will vary with technology, wafer diameter and production expertise. Vendors rarely release their own in-house figures of yield, so most published information tends not to be up-to-date and usually pessimistic for current practice. Some early 1980 values for the above equation are

$$D_0 = 10^5 \text{ defects per m}^2 \text{ for bipolar technology}$$

and

$$D_0 = 5 \times 10^4 \text{ defects per m}^2 \text{ for MOS technology}$$

These values give the yield characteristics shown in Figure B.1.

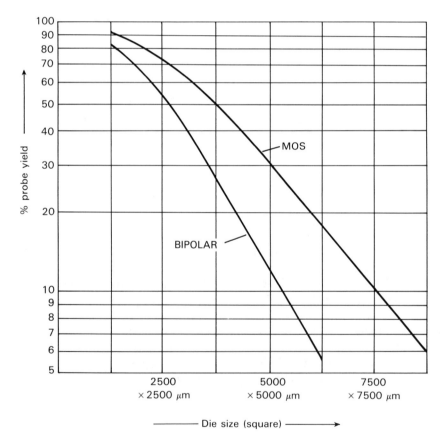

Figure B.1 Statistical wafer yield from a 20 cm diameter wafer as a function of die size.

Whilst the general characteristics for yield shown in Figure B.1 still remain valid, the values for the percentage yield should now be considered pessimistic. For the 1990s' state-of-the-art, a re-scaling of the X-axis by perhaps a factor of 10 may be appropriate to take into account improvements in fabrication and the general pessimistic characteristics of this early modelling operation.

Appendix C: The principal equations relating to bipolar transistor performance

The basic equation which underpins all bipolar transistor performance is the theoretical pn junction diode equation. Considering a simple pn junction such as that shown in Figure C.1(a), the forward direction of conduction is when the p-region is made positive with respect to the n region, and the reverse direction of conduction is when the p-region is negative with respect to the n-region. The expression relating the current I_D through the diode to the voltage V_D across it is given by the non-linear equation

$$I_D = I_S(e^{qV_D/kT} - 1) \tag{C.1}$$

where

I_S = the reverse saturation current
q = electron charge = 1.6×10^{-9} C
k = Boltzmann's constant, 1.38×10^{-23} J K^{-1}
T = temperature, K

The term q/kT has the dimensions of V^{-1} and has a numerical value of between 20 and 40 at 300 K depending upon the detailed properties of the pn junction. However, for the majority of bipolar transistors the value of this term is taken to be 40 V^{-1} for emitter–base junctions. Hence, the diode equation (C.1) may be re-written as

$$I_D = I_S(e^{40V} - 1), \tag{C.2}$$

where V is in volts.

The current I_S for a typical silicon pn junction at room temperature may be less than a picoamp (10^{-12} A). Thus when the reverse voltage V_D across the junction is, say, -1 V, the exponential term in Equation (C.2) is e^{-40} which is negligible, giving $I_D = -I_S$. However, when V_D is made positive the forward current I_D increases rapidly, being about 1 mA when V_D is $+0.63$ V, and continuing to increase by a factor of about 50 (e^4) for every 0.1 V increase in V_D. This is indicated in Figure C.1(b). The -1 term in Equation (C.2) may now be neglected.

In the normal bipolar junction transistor (BJT), it is the emitter–base pn junction which is forward-biased by the transistor input conditions (see Figure C.2(a)). Hence Equation (C.2) can be re-written in terms of the transistor parameters giving

$$I_E = I_S(e^{40 V_{EB}} - 1) \tag{C.3}$$

for pnp transistors, where V_{EB} is positive for forward biasing, and

$$I_E = I_S(e^{40 V_{BE}} - 1) \tag{C.4}$$

for npn transistors where V_{BE} is positive for forward-biasing.

The remaining d.c. current relationships for a transistor operating in common-base configuration are

$$I_C = h_{FB} I_E$$

and

$$I_B = I_E - I_C$$
$$= (1 - h_{FB}) I_E \tag{C.5}$$

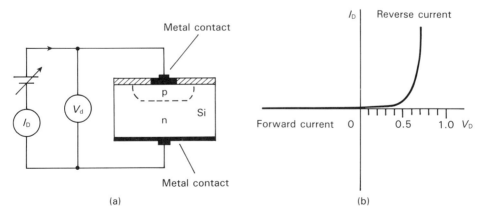

Figure C.1 The silicon pn junction diode: (a) the measuring circuit; (b) the forward and reverse current vs. voltage characteristics.

see Figure C.2(b), where h_{FB}, sometimes termed α, is the large-signal (or d.c.) common-base current gain. Note that as h_{FB} has a value close to, but not exceeding, unity, we may write Equation (C.4) with little error as

$$I_C = I_S(e^{40 V_{BE}} - 1)$$
$$\simeq I_S(e^{40 V_{BE}}) \tag{C.6}$$

when the transistor is conducting.

For the BJT operating in common-emitter configuration where the base current I_B is now regarded as the input current, the above d.c. relationships are re-arranged to become

$$I_C = \frac{h_{FB}}{1 - h_{FB}} I_B$$
$$= h_{FE} I_B \tag{C.7}$$

where h_{FE}, sometimes termed β, is the large-signal common-emitter current gain.

These relationships do not take into account the small but temperature-sensitive leakage current I_{CBO} flowing through the reverse-biased collector–base junction (see Figure C.2(b)). Inclusion of this current modifies the current distribution to

$$I_C = h_{FB} I_E + I_{CBO}$$

and

$$I_B = I_E - (h_{FB} I_E + I_{CBO})$$
$$= (1 - h_{FB}) I_E - I_{CBO}$$

whence

$$I_E = \left(\frac{1}{1 - h_{FB}}\right) I_B + \left(\frac{1}{1 - h_{FB}}\right) I_{CBO} \tag{C.8}$$

Substituting for I_E in the equation for I_C now gives

$$I_C = \left(\frac{h_{FB}}{1 - h_{FB}}\right) I_B + \left(\frac{h_{FB}}{1 - h_{FB}}\right) I_{CBO} + I_{CBO}$$
$$= h_{FE} I_B + (h_{FE} + 1) I_{CBO} \tag{C.9}$$

where h_{FE} is as previously. However, for silicon transistors operating at normal temperatures this leakage term may usually be ignored.

The small-signal (or a.c.) parameters of the junction transistor, which represents the transistor performance about some fixed (quiescent) d.c. operating

Equations relating to bipolar transistor performance   449

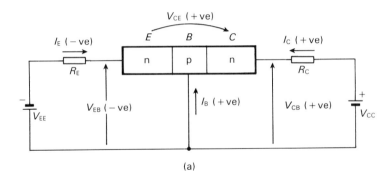

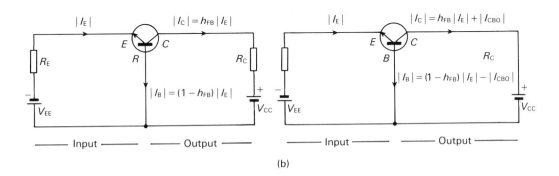

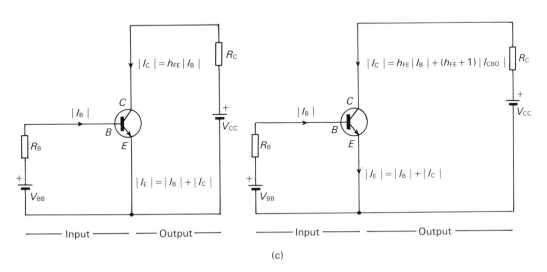

Figure C.2 The silicon npn bipolar junction transistor (BJT): (a) the normal operating polarities; (b) current relationships in common-base working with and without the collector leakage current I_{CBO}; (c) current relationships in common emitter working also with and without I_{CBO}.

point, are determined by evaluating the slope of the various current/voltage relationships. Consider the transconductance g_m of the BJT, where g_m is the rate of change of output current with input voltage. In common-emitter mode this is

$$g_m = dI_C/dV_{BE}$$

which by differentiating Equation (C.6) gives

$$g_m = 40 I_s e^{40 V_{BE}}$$

or

$$g_m = 40 I_C \qquad \text{(C.10)}$$

Thus the g_m of a bipolar transistor is *proportional to the d.c. current flowing in the collector*, and is often quoted as 40 mA V^{-1} per milliamp of collector current. Thus if I_C is 100 μA, $g_m = 4$ mA V^{-1}; if I_C is 10 mA, $g_m = 400$ mA V^{-1}, and so on.

Of the many other small-signal parameters which may be found, the hybrid or h parameters are often used for low/medium frequency modelling. The four h parameters for common-emitter working are

a.c. input impedance $h_{ie} = \dfrac{\Delta V_{BE}}{\Delta I_B} \bigg|_{V_{CE}\text{constant}}$

a.c. output conductance $h_{oe} = \dfrac{\Delta I_C}{\Delta V_{CE}} \bigg|_{I_B\text{constant}}$

a.c. forward current gain $h_{fe} = \dfrac{\Delta I_C}{\Delta I_B} \bigg|_{V_{CE}\text{constant}}$

and

a.c. reverse voltage feedback $h_{re} = \dfrac{\Delta V_{BE}}{\Delta V_{CE}} \bigg|_{I_B\text{constant}}$

The a.c. equivalent circuit using h parameters is shown in Figure C.3 in comparison with the low-frequency hybrid π model.

The significance of the four h parameters is that they can be readily measured individually without requiring inconvenient operating conditions on the transistor. The two parameters h_{oe} and h_{re} are often neglected leaving h_{ie} and h_{fe} as the parameters of major significance. Comparing the two simplified equivalent circuits shown in Figure C.3, the two current generators $g_m v_{be}$ and $h_{fe} i_b$ should give the same output performance, and thus we have the relationships

$$g_m v_{be} = h_{fe} i_b$$

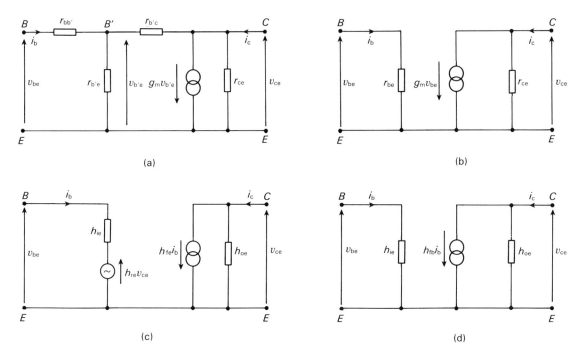

Figure C.3 BJT small-signal a.c. equivalent circuits for common-emitter working: (a) the low-frequency hybrid π model; (b) the simplified hybrid π model with only the dominant parameters; (c) the low-frequency h-parameter model; (d) the simplified h-parameter model.

whence

$$h_{fe} = g_m h_{ie} \tag{C.11}$$

and

$$h_{ie} = \frac{h_{fe}}{g_m}$$

$$= \frac{h_{fe}}{40 I_C}$$

$$= \frac{25 h_{fe}}{I_C} \text{ where } I_C \text{ is in mA} \tag{C.12}$$

This last equation is particularly significant since it states that the common-emitter a.c. input impedance is approximately given by the *current gain times 25 Ω per milliamp of I_C*.

These and the many other BJT relationships may be more rigorously and accurately determined, for example see [1] of Chapter 2, and elsewhere. However, consider a little further the a.c. performance of the bipolar transistor: from Figure C.3(b) it will be noted that the a.c. gain performance of the BJT is very largely governed by the value of g_m, which as we have seen has a pre-determined value proportional to the value of I_C. But from Figure C.3(d) it would appear that the gain performance is largely governed by the value of the forward current gain term h_{fe} — a parameter that is dependent upon the expertise of the vendor in making high-gain devices. To reconcile these two aspects it must be remembered that h_{ie} in the h parameter model is also dependent upon h_{fe} and the transistor g_m value as shown above, and therefore the effect of high current gain h_{fe} is to increase the input impedance term h_{ie}, reduce the input current i_b and hence maintain $h_{fe}i_b$ sensibly constant for any given value of I_C. The mutual conductance term g_m thus remains the dominant parameter which largely determines the a.c. gain performance of a BJT except at high frequencies, being independent of device dimensions. High gain is, however, desirable in order to increase the input impedance in common-emitter working.

Further details of transistor action and small-signal equivalent circuits may be found in standard textbooks, for example

Sedra, A. S. and Smith, K. C., *Microelectronic Circuits*, Holt, Rinehart and Winston, Toronto, 1987.

Glazer, A. B. and Subak-Sharpe, G. E., *Integrated Circuit Engineering*, Addison-Wesley, MA, 1979.

Goodge, M., *Semiconductor Device Technology*, Macmillan, London, 1983.

And many others.

Appendix D: The principal equations relating to unipolar (MOS) transistor performance

Unlike the bipolar junction transistor, most of the MOS device parameters are dependent upon device geometry, particularly the length L and width W of the conducting channel between source and drain. It is this characteristic of MOS devices which allows scaling and the freedom for the system designer to become involved in design at the silicon level in certain circumstances.

Here we will consider only enhancement-mode, insulated-gate field effect transistors (IGFETs), sometimes still ambiguously referred to as 'metal-oxide–silicon field effect transistors' (MOSFETs), even though the gate may be polysilicon, which are non-conducting with $V_{GS} = 0$ V. Junction FETs (JFETs) and depletion-mode types have generally similar equations for their performance; full details may be found from many texts such as [1]–[8] in Chapter 2, [8] in Chapter 4, and elsewhere.

The initial development of FET characteristics classically begins with a consideration of the gate input capacitance, with its dielectric of silicon-dioxide between gate and substrate. This capacitance is given by

$$C_{ox} = \frac{\varepsilon_0 \varepsilon_r A}{x} \tag{D.1}$$

where

C_{ox} = the capacitance of the physical gate-substrate capacitor (F)
ε_0 = absolute permittivity (F m^{-1})
ε_r = relative permittivity of the gate oxide
A = gate area (m^2)
x = gate oxide thickness (m)

This may be given in per unit area, namely

$$C_0 = \frac{\varepsilon_0 \varepsilon_r}{x} \quad \text{F m}^{-2} \tag{D.2}$$

The importance of capacitance in understanding the action of FETs is well illustrated by consideration of the model used in SPICE simulation, which is shown in Figure D.1. (The significance of SPICE for device-level modelling has been covered in Chapter 5.) As will be seen, capacitive terms between gate, drain, source and substrate are very conspicuous; they are, however, together with the drain and source diodes, largely non-linear terms so that considerable computational complexity accrues in undertaking SPICE simulations over a range of operating conditions.

From a consideration of the charge induced by the gate in the channel between source and drain the expression for the drain current I_D may be derived, namely

$$I_D = \frac{W}{L} \mu_N C_0 \left\{ (V_{GS} - V_{TH}) V_{DS} - \frac{V_{DS}^2}{2} \right\} \tag{D.3}$$

where

I_D = drain current (A)
W = width of gate (m)
L = length of gate (m)
μ_N = drift mobility of channel electrons (m^2 V^{-1} s^{-1})
C_0 = gate capacitance per unit area (F m^{-2})
V_{GS}, V_{TH}, V_{DS} = gate-to-source, threshold, and drain-to-source voltages, as previously (V)

The above equation is used for n-channel devices; for p-channel devices μ_N is replaced by μ_P, the drift mobility of channel holes. Further, $\mu_N C_0$ (or $\mu_P C_0$) may be referred to as the process gain factor K', (A V^{-2}), and in turn $W/L\,K'$ may be considered as a single parameter β, termed the transistor gain factor. Hence

$$I_D = \beta \left\{ (V_{GS} - V_{TH}) V_{DS} - \frac{V_{DS}^2}{2} \right\} \tag{D.4}$$

It will be appreciated that because the mobility μ_N for electrons is higher than the hole mobility μ_P (in silicon 0.15 m^2 V^{-1} s^{-1} compared with 0.05 m^2 V^{-1} s^{-1}), β for n-channel devices will be higher than for p-channel devices. Note, however, that the geometric aspect ratio L/W of the transistor is present in the gain factor and therefore in the expression for the resulting drain current I_D.

The above expressions for I_D contain certain simplifications particularly bulk substrate effects which tend to act as a second 'back' gate to modify the drain

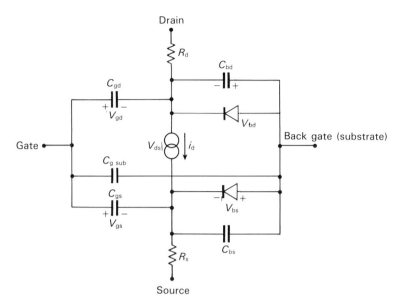

Figure D.1   The SPICE model of an insulated-gate field-effect transistor.

current, but for most practical purposes Equations (D.3) and (D.4) are appropriate. However, if the term $(V_{GS} - V_{TH})V_{DS} - \frac{1}{2}V_{DS}^2$ is examined more closely, it will be seen that for $(V_{GS} - V_{TH})$ constant the resulting current I_D will increase with increasing V_{DS} up to some maximum value. The value for V_{DS} at this point is given by differentiating I_D with respect to V_{DS} and equating to zero, that is,

$$\frac{dI_D}{dV_{DS}} = \left\{ (V_{GS} - V_{TH}) - \frac{1}{2}(2V_{DS}) \right\} = 0$$

whence

$$V_{DS(sat)} = (V_{GS} - V_{TH}) \tag{D.5}$$

The point at which this occurs is known as the saturation (or 'pinch-off') point; below this point the FET is operating in its unsaturated mode, sometimes termed its 'triode' or 'ohmic' mode, where I_D increases rapidly with V_{DS}, V_{GS} constant. The value of I_D at $V_{DS(sat)}$ is given by

$$I_{D(sat)} = \beta \left\{ \frac{(V_{GS} - V_{TH})^2}{2} \right\}$$

$$= \beta \left\{ \frac{(V_{DS(sat)})^2}{2} \right\} \tag{D.6}$$

Appendix D

When V_{DS} is *small* compared with $(V_{GS} - V_{TH})$, the $\frac{1}{2}V_{DS}^2$ term in the expression for I_D may be neglected, and I_D is now given by

$$I_D = \frac{W}{L} \mu_N C_0 \{(V_{GS} - V_{TH})V_{DS}\} \tag{D.7}$$

whence

$$\frac{V_{DS}}{I_D} = \frac{L}{W} \frac{1}{\mu_N C_0 (V_{GS} - V_{TH})}$$

$$= \frac{L}{W} \frac{1}{K'(V_{GS} - V_{TH})} \tag{D.8}$$

The FET is now behaving as a linear voltage-controlled resistor, the resistance value being determined by the aspect ratio L/W of the channel and the gate voltage V_{GS}.

At values of V_{DS} *greater than* $V_{DS(sat)}$, increasing V_{DS} increases the length of the conducting channel which is in the pinch-off condition (known as the channel length modulation effect), a factor which can be accounted for by incorporating a new factor λ in the expression for I_D, giving

$$I_D = I_{D(sat)}(1 + \lambda V_{DS})$$

$$= \beta \left\{ \frac{(V_{GS} - V_{TH})^2}{2} \right\} (1 + \lambda V_{DS}) \tag{D.9}$$

for the current in the saturated region, where λ has a small value typically between 0.01 and 0.05. The classic I_D/V_{DS} characteristics for increasing V_{GS} are therefore as shown in Figure D.2.

The expression for the threshold voltage V_{TH} depends heavily upon the fabrication process and to a negligible extent on the device geometry, as would be expected. The usual equation is

$$V_{TH} = V_{FB} + 2\psi_B + \gamma\{2\psi_B\}^{1/2} \tag{D.10}$$

where

V_{FB} = transistor flat-band voltage

$$\psi_B = \frac{kT}{q} \ln\left\{\frac{N_A}{n_I}\right\},$$

that is a parameter dependent upon the acceptor level concentration N_A in the substrate and the free electron concentration n_I of intrinsic silicon (assuming n-

Equations relating to unipolar transistor performance 457

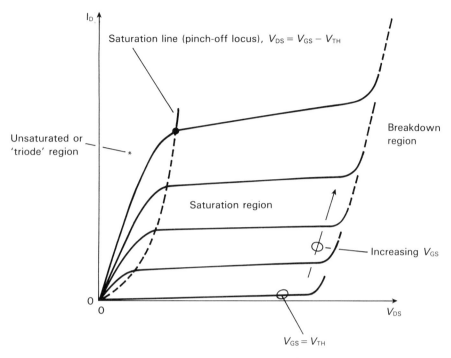

Figure D.2 The I_D vs. V_{DS} output characteristics of an enhancement-type FET for increasing values of gate voltage V_{GS}.

channel device),

$$\gamma = \frac{\{2\varepsilon_0\varepsilon_r q N_A\}^{1/2}}{C_0},$$

i.e. a further parameter dependent upon the acceptor level concentration together with the gate capacitance parameters.

It is not, therefore, a parameter which can be freely varied with device geometry, as can the drain current, and hence is *not a flexible parameter for the system designer's use* at the silicon level. Its value for enhancement-mode FETs may very roughly be taken as 0.2 V_{DD}, so for +5 V operation V_{TH} is about +1 V. For depletion-mode devices V_{TH} may roughly be taken as $-0.7\ V_{DD}$, giving V_{TH} as -3.5 V for +5 V operation.

Turning to the small-signal a.c. performance of the IGFET, because of its very high input impedance, with input leakage currents perhaps as low as 10^{-15} A, the usual low-frequency a.c. equivalent is as shown in Figure D.3. For higher-frequency operation additional capacitive terms have to be added to this simple a.c. equivalent circuit.

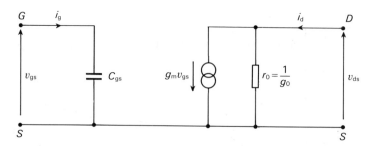

Figure D.3 The low-frequency, small-signal a.c. equivalent circuit for the insulated-gate FET. At higher frequencies additional capacitive terms need to be introduced.

The small-signal gate input capacitance C_{gs} depends upon the gate area $W \times L$ and the gate oxide capacitance per unit area C_0 (see Equation (D.2)). Typically, C_{gs} may be taken as

$$0.7 \, C_0 WL \tag{D.11}$$

The value of the transconductance g_m may be found by differentiating the preceding equations for I_D with respect to V_{GS}, remaining parameters being held constant. This therefore gives

$$g_m = \frac{W}{L} \mu_N C_0 V_{DS}$$
$$= \beta V_{DS} \tag{D.12}$$

from Equations (D.3) and (D.4) for the non-saturated region of operation, that is, when $V_{DS} \leqslant (V_{GS} - V_{TH})$, and

$$g_m = \beta(V_{GS} - V_{TH})(1 + \lambda V_{DS})$$
$$\simeq \beta(V_{GS} - V_{TH})$$
$$= \beta V_{DS(sat)} \tag{D.13}$$

from Equation (D.9) for the saturated region of operation. Hence g_m increases to a maximum value when in saturation.

The small-signal a.c. output conductance g_o is given by differentiating I_D with respect to V_{DS}, remaining parameters being held constant. This therefore gives

$$g_o = \beta\{(V_{GS} - V_{TH}) - V_{DS}\} \tag{D.14}$$

for the non-saturated region of operation, and

$$g_o = \beta \left\{ \frac{(V_{GS} - V_{TH})^2}{2} \right\} \lambda \qquad (D.15)$$

from Equation (D.9) for the saturated region of operation, which is a small-value conductance = high-value impedance.

An alternative expression for g_m in the saturated region can be obtained by substituting for $(V_{GS} - V_{TH})$ in terms of I_D, which yields

$$g_m = \sqrt{(2\mu_N C_0)} \cdot \sqrt{(W/L)} \cdot \sqrt{I_D} \qquad (D.16)$$

which emphasizes the following:

- g_m is proportional to the square root of the d.c. bias current I_D in an amplifier circuit; and
- is proportional to the square root of the channel dimensions W/L.

Similarly, an alternative expression for the output conductance in terms of I_D can be obtained, giving

$$g_o = \frac{\lambda I_D}{1 + \lambda V_{DS}} \qquad (D.17)$$

whence the output impedance r_o is

$$r_o = \frac{1 + \lambda V_{DS}}{\lambda I_D} \qquad (D.18)$$

The above Equation (D.16) for g_m should be contrasted with that for the bipolar junction transistor, where g_m is independent of the geometrical size of the transistor but proportional to the collector current (see Equation (C.10)). From Equation (D.16) an IGFET having $\mu_N C_0 = 20 \; \mu\text{A V}^{-2}$ and equal width and length dimensions, $g_m = 0.2 \text{ mA V}^{-1}$ at $I_D = 1 \text{ mA}$. With $W/L = 100$, which is an extremely wide transistor for a given length, g_m is still only 2 mA V^{-1}. This should be contrasted to the value of 40 mA V^{-1} for a BJT at an operating current of 1 mA. Hence IGFETs are generally characterized by a low value of g_m in comparison with other active devices. Finally, from Figure D.3 it will be seen that the maximum possible voltage gain is g_m/g_o, = $g_m r_o$, which from Equations (D.16) and (D.18) gives

$$A_{max} = \left\{\frac{2\mu_N C_0 W I_D}{L}\right\}^{1/2} \left\{\frac{1 + \lambda V_{DS}}{\lambda I_D}\right\}$$

$$= \frac{k}{\sqrt{I_D}} \tag{D.19}$$

Hence A_{max} is inversely proportional to the square root of the drain current, decreasing with increasing values of I_D. Therefore, for maximum gain performances FET currents should be kept low, which implies a difficulty in using FETs for higher-current, high-amplification duties.

Further details may be found in many standard textbooks, see end of Appendix C.

Index

accelerators *see* simulation accelerators
acceptor, 19
accumulator syndrome count, 354
additive direct-write, 167
algorithmic state machine (ASM), 185, 192
aliasing, 332
all-NAND, 92
all-NOR, 92
analogue, 15, 21–7, 80, 157, 281
analogue cells, 127, 157–61
analogue/digital *see* mixed analogue/digital
analogue simulation, 232–5, 254
analogue synthesis, 209, 210
analogue test, 281–3
analogue-to-digital (A-to-D), 81, 128, 160, 233, 393
AND/OR arrays, 90–8
antifuse, 104–6
appearance fault, 306
application configurable technology (ACT), 106
application-specific IC (ASIC), 18, 106, 107
architecture, 180–3
architectural level design, 180–91
architectural level simulation, 215–18, 236
arithmetic coefficients, 359–63, 367
array *see* gate array and uncommitted array
ASIC design kit, 157
automatic test pattern generation (ATPG), 259–303

back annotation 230, 231
back-end design, 180–3
bandgap *see* energy bandgap
base, 20, 22–5
basic cells, 111, 125, 132, 140–5
behavioural level design, 180–91
behavioural level simulation, 215–18
Berger code, 368
BiCMOS, 5, 68–71, 73, 172
bipolar technology, 3, 5, 19–39, 128, 446–52
block routing, 123, 206
Boolean gates, 91, 140–5

bottom-up design, 183
boundary scan, 342–7
branch-and-bound, 195
bridging fault, 279, 291, 292
bristle-block, 241
buffered FET logic (BFL), 70
building blocks, 13, 117
built-in logic block observation (BILBO), 325–33, 372–9
built-in self-test (BIST), 324–40, 355–7, 376, 431
bump-on-die, 406
buried region/buried layer, 22–38

CAD *see* computer-aided design
CADDIF format, 259
Caltech Intermediate Form (CIF) format, 208, 259
capacitor, 26, 27, 157–9
cell library, 12, 117, 124–6, 128, 132, 137, 139, 142
cellular array, 370
cellular automata built-in self-test (CALBO), 372–5, 377
ceramic package, 407–9
channel-less architecture, 130, 149, 153–6, 171
channel router, 123, 202–8
channel stopper, 43
check bits, 368
checker, 368–70
chip area *see* die size
choice of design style
  analogue, 403–5
  full custom, 399
  gate-array, 397–400
  microprocessor/microcontroller, 395–7
  mixed analogue/digital, 405
  off-the-shelf standard parts, 395–98
  programmable logic devices, 398
  standard-cell, 399, 401
circuit level simulation *see* SPICE
circuit under test, 279

clustering, 196-8
collector, 20, 22, 32, 34-6
collector-diffusion isolation (CDI), 36-8, 83, 137, 138, 160
compacted array, 153
compaction *see* input compaction and output compaction
complementary MOS (CMOS), 49-54, 57, 58, 73, 83, 118, 137-42
compiled cell, 126
compiler *see* silicon compiler
component-level array, 132-40
computer-aided design (CAD), 7-11, 179-256
computer-aided engineering (CAE), 179
computer-aided manufacture (CAM), 179
concurrent testing *see* on-line testing
conductors, 434-7
constructive initial placement, 195
contact exposure, 162
contact window *see* via
controllability, 284-8
correct-by-construction, 7, 215
correlation, 365-7
cost of
    CAD resources, 260-2, 415, 421
    design-for-testability, 376-8, 421
    learning, 416, 417
    manufacture, 419-24
    masks, 422
    non-recurring engineering, 169, 413, 414
    test, 417-19, 421
cost models, 424
crosspoint fault, 306
crossunder, 139, 150
crystal structure, 21, 435
crystal structure defects, *see* defects
current-mode logic (CML), 21, 36-8, 73
custom-specific (or customer-specific) integrated circuit (CSIC), 18, 111

data formats/data standards, 256-60, 263
data bus, 249
D-cubes of failure, 297-300
DMOS, 46
deep-trench isolation, 25, 38, 66
defect level (DL), 274
defects, 440-4
de Morgan's theorem, 90
depletion-mode FET, 41-3
design-for-testability (DFT), 310-41
design rule checking (DRC), 7, 179, 207
design security, 110, 398, 402, 404
design style *see* choice of design style
Deutsch's router, 205
dice per wafer, 443
die size, die area, 112, 169, 443
die size cavity, 409
digital signal processing (DSP), 433

digital test, 277-81
digital-to-analogue, (D-to-A), 81, 128, 160, 233, 393
diode *see* pn junction diode
direct-die-mounting, 405
direct write-on-wafer, 163-7
disappearance fault, 306
discrete devices, 77
dissipation *see* power dissipation
dog-leg, 155, 202, 205
donor, 19
drop-in, 273
double-layer-metal, 54, 129, 149
dual-in-line (DIL) packaging, 407-9
dual guardband, 49
dynamic RAM (DRAM), 89, 90

EAPROM, 57, 87
E-beam lithography, 162-6, 429
EBES format, 258-60
EDIF, 208, 258-60, 430
EEPROM, 57, 87
electrical rule checking (ERC), 8, 179
electron mobility, 1, 70, 437
ELLA, 186
emitter, 20, 22, 32, 34-6
emitter-coupled logic (ECL), 21, 30-2, 58, 73, 80, 129, 132-5, 172-5
energy bandgap, 435, 437
enhancement-mode FET, 41-4
epitaxial fabrication 19, 22, 68
epitaxial (or epi.) layer, 19, 21, 38, 68
EPROM, 57, 59, 87
error masking, *see* aliasing
ESPRESSO, 192
Ethernet, 247, 248
exhaustive test, 208, 274, 278-80

fabrication, 19-69, 440-5
fault cover, fault coverage, 274, 275
fault model, 288-95
ferroelectronic random-access memory (FRAM), 63-8
fibre distributed-data interface (FDDI), 249, 430
field-effect transistor (FET), 39-57
field oxide, 43-53
field programmable, 100, 105
financial factors *see* cost of
fixed cell architectures, 117-23
floating gate, 61, 62, 93
floorplan, 130, 149-54, 157
flush test, 281, 327
footprint area, 409
front-end design, 180-2
full custom, 12, 112-16, 171-5, 412
full mask set, 117
full technical data (FTD), 123
functional-level arrays, 140-5

functional-level design, 180–4, 191–3
functional-level simulation, 215–18

galloping patterns, 304
gallium-arsenide (GaAs), 6, 41, 70–3, 127, 135–7, 172, 437
gate array 15, 116, 128–61, 171–5, 412 see also uncommitted array
gate-level design, 183, 184, 191–3
gate-level simulation, 218–21
gate oxide, 43, 53
GDS II format, 208, 259
general purpose instrumentation bus (GPIB), 283
general routers, 202–4
Gerber format, 208, 258
germanium transistor, 1, 2, 18, 434–7
gigacell, 126
global placement, 195
greedy router, 205
grid expansion router see wavefront router
growth fault, 306

Hadamard matrix, 358, 361–5
hand-crafted, 112–16
hand routing see manual routing
hard array logic (HAL), 97
hard cells/hard-coded cells see fixed-cell architectures
hardware description language (HDL), 185–91, 430
hardware modelling, 236
Hashimoto and Steven's router, 204
hermetic sealing, 2, 406
HHDL, 186
hierarchical designs, 180–3
hierarchical router, 205
Hightower's router, 203, 204
hole mobility, 1, 70, 437
housekeeping, 256
hybrid assembly, 405, 406

IGES format, 259
implant, 23, 43, 44
input compaction, 349, 350
input/output (I/O) cells, 117, 130, 340–2
insulated-gate field-effect transistor (IGFET), 39, 41–57
insulator, 434–9
integrated-injection logic (I^2L), 21, 32–6, 58, 73, 83
integrated-Schottky logic (ISL), 34, 35
interchange formats see data formats
isolation moats, 23, 49
isoplanar I^2L (I^3L), 34
iterative re-placement, 195–9

Joint Test Action Group (JTAG), 343

junction diode see pn junction diode
junction field-effect transistor (JFET), 39, 41, 70
junction isolation, 23

Karnaugh map, 192, 308, 371
Kernighan-Lin algorithm, 199
kit parts, 157

lambda, 113–5
Landmann and Russo, 146
large scale integration (LSI), 3, 77, 78
laser technology, 166–9
laser pantography, 167–9
lasography see laser pantography
lateral transistor, 32
lattice constant, 6
lead-zirconate-titanate (PZT), 65
leaf cell, 125
learning cost/learning curve, 416
Lee's router, 202–4
left-edge router, 204, 205
level-sensitive scan design (LSSD), 318–20
library see cell library
limited technical data (LTD), 120
linear see analogue
linear array, 157–61
linear feedback shift register (LFSR), 330–3
loadmost, 43, 60, 63
local area network (LAN), 247–50, 430
logic cell array (LCA), 102–7
L/W ratio, 56, 453–60

macrocell, 123–7
macros/macro blocks/macro functions, 125
Manhattan distance, 195, 204
manual routing, 150–3
marching patterns, 304
mask, 22, 116, 162
maskless fabrication, 161–9
mask-programmable, 57
maximum length sequence, 330
maze router see wavefront router
Mead and Conway design rules, 112–14
medium scale integration (MSI), 3, 78, 117
megacell, 126
megamacro/megafunction, 126
memory, 21, 57–68, 86–94 see also ROM, RAM, DRAM, SRAM
merged transistor logic (MTL), 32
metal gate, 43, 51, 78
metallurgical junction, 1
metal-oxide-silicon, 3, 41, 78
metal-semiconductor junction field-effect transistor (MESFET) 70, 71
microcell, 125
microcontroller/microprocessor, 81–6, 124, 309
Miller effect, 46
min-cut placement, 198, 199

minimum spanning tree, 200
mixed analogue/digital design, 209, 210
mixed analogue/digital test, 282, 283
mixed-level simulation, 216
mixed-mode simulation, 217, 234, 235
mixed-signal simulation, 216
mobility *see* electron mobility and hole mobility
module generator, 240
Moore's law, 3
MOS technology, 3, 5, 39–57
M-out-of-N codes, 370
MS-DOS, 245
multilevel simulation, 216
multiple-input signature register (MISR), 332
multiproject wafer, 163–5

NAND, 79, 90, 140–4
netlist, 193, 207, 211, 230, 259
networking, 247–250
network under test, 279
nMOS, 41–7, 58, 73
non-exhaustive test, 284, 295, 314
non-recurring engineering (NRE), 169, 395, 413–424
non-volatile memory, 61, 63–7
NOR, 90, 91, 140–4
npn transistor, 19–21, 158
n-type material, 19, 48–54
n-well, 49, 51

observability, 284–8
obsolescence, 78
off-line test, 325
off-the-shelf standard parts, 12, 18, 77–107, 117, 412
on-line testing, 367–70, 431
open-circuit fault, 292–5
operational amplifier, (op.amp.), 80–2
operating point, 225
opto-electronics, 431
original equipment manufacturer (OEM), 393–424
orthogonal transform, 361
output compaction, 350–2
oxide isolation, 22–6, 51–4, 56
oveheads, 333, 376–80, 413–19

packaging, 120, 405–10, 431
packing density, 27, 73, 74
pad oxide, 43, 52

paracell/parametised cell, 126
parity check, 368
partitioning, 312, 317
periodic table of the elements, 434–9
personal computer (PC), 8, 244–7
photolithography, 161–3, 166, 429
photolysis, 167

photoresist, 161, 441
physical design, 180–4, 193–208, 253
pin limitation, 409
placement, 117, 132, 145, 194–200, 253, 263
planar, 19
planar fabrication, 19
PLA test, 306–9
PLD design, 101, 254
pMOS, 47–9, 58
pn junction diode, 19
pnp transistor, 21, 32, 158
polycell, 118, 125
polysilicide, 25
polysilicon, 44
polysilicon connection, 149–51
post-layout simulation, 179, 230
power-delay product, 74
power dissipation, 31, 34, 38, 74, 171
primary inputs, 285, 315–17
primary outputs, 285, 315–17
primitive, 125, 140–5
printed-circuit board (PCB), 117, 255
probe card, 419
process device monitor (PDM), *see* drop-in
process evaluation device (PED), *see* drop-in
process monitoring circuit (PMC), *see* drop-in
production volume, 170, 395, 403, 413–23
programmable array logic (PAL), 96, 100
programmable gate array (PGA), 100, 101
programmable logic array (PLA), 93–6, 123, 306–9, 334–40, 355
programmable logic controller (PLC), 101, 102
programmable logic device (PLD), 90–102, 111, 210–15, 240, 412
programmable logic sequencer (PLS), 100
programmable read-only memory (PROM), 57–61, 305
projection printing, 163
prototype, 120, 165, 421
proximity exposure, 163
pseudo-random sequence, 330, 374
p-type material, 22–39, 41–47
p-well, 49, 51
pyrolysis, 167
PZT *see* lead-zirconate-titanate

Quine-McClusky minimisation, 192

race conditions, 318, 320
rack-and-stack resource, 282
Rademacher-Walsh transform, 362
random access memory (RAM), 57, 61–8, 89, 90, 123, 154, 240, 303
Random Access Scan (RAS), 322
read-only memory (ROM), 57, 59–61, 87, 123, 154, 240, 305
reduced mask set, 117, 129
reduced test set *see* non-exhaustive test

redundancy, 325
Reed-Muller coefficients, 359, 364, 367
Reed-Muller codes, 370
register transfer language (RTL), 191, 192
Rent's rule, 147-8
resist, 161-3
resistivity, 158
resistor, 19, 26
reticle, 163
RISC, 244
risk, 173, 413, 430
routing, 117, 132, 145-53, 159, 200-8, 253, 263
rubyliths, 7
run-out, 166

sales appeal, 390, 398, 402, 404
saturation, 30, 34, 38, 39
Scan-Path, 321
Scan-Set, 322
scan test, 314-24, 431
schematic capture, 192, 193, 253
Schottky junction diode, 30, 34, 70
Schottky diode FET logic (SDFL), 70
Schottky transistor logic (STL), 35
SDIF format, 259, 260
sea-of-gates/sea-of-cells, 153
security *see* design security
seed cells, 199
semiconductor, 434-9
self-alignment, 46, 51, 56
self-dual function, 371, 372
self-isolated, 43
self-test, 324-33
shadow register, 347
shift register, 280, 281, 325-33
shrinkage fault, 306
signature *see* test signature
signature analysis, 330-33
silicon, 1, 2, 18, 434-7
silicon assembler, 239
silicon compiler, 9, 239-41, 253, 431
silicon dioxide (SiO_2), 2, 19, 22, 61
silicon dioxide isolation, 19, 22-55, 61
silicon gate, 42-6, 51
silicon-on-insulator (SOI), 5, 54-6, 83
silicon-on-sapphire (SOS), 6, 55
silicon nitride (Si_3N_4), 23, 43, 53
simulated annealing, 199
simulation, 120, 215-39, 253-5, 263
simulation accelerator, 235-9
single-layer-metal, 129
small scale integration (SSI), 3, 78-80, 117
small-signal equivalent circuit, bipolar, 446-52
small-signal equivalent circuit, MOS, 453-60
soft cells/soft-coded cells, 123, 124, 192
software availability, 250-5
spectrum, 361, 362
spectral coefficients, 360-7

SPICE, 225-33, 253
standard buried collector (SBC), 22-6
standard cell library *see* cell library
standard cells, 13, 116-28, 170, 412
standard parts *see* off-the-shelf standard parts
static read-only memory (SRAM), 58, 89, 90
Steiner tree, 201
stuck-at-0 (s-a-0), 288-302
stuck-at-1 (s-a-1), 288-302
substrate, 19-55, 66-71
sum-of-minterms, 91, 93
sum-of-products, 90-2
supply voltages, 79, 390
supracell, 126
surface mounting, 407
surface protection, 2
switchbox router, 206
switch level simulation, 221-5
syndrome, 352-4
system level design *see* behavioural and architectural level design
system level verification, 235-9

TAU, 192
test, 208, 271-381
tester, 276
test interface card, 419
test mode, 311-24, 343, 348
test overheads *see* cost of test
test pattern, 277, 304
test pattern generation (TPG), 271-309
test set, 277
test signature, 330-3, 375
test vector, 120, 277
thick film, 405
thin film, 405
through-hole mounting, 407
tile, 157, 158
tile array, 157-61
time to market, 410-13, 432
token ring, 247, 248
top-down design, 183, 191
transistor equations, bipolar, 446-452
transistor equations, MOS, 453-460
transistor-pairs, 118, 137-41
transistor-transistor logic (TTL), 12, 21, 28-30, 73
transition count, 353
transmission gates, 222
travelling salesman problems, 200
tunnel, 150
twin-tub, 49, 51, 54
two-layer-metal *see* double-layer-metal

ultra large scale integration (ULSI), 3
uncommitted array/uncommitted gate array, 15, 132-55
uncommitted component array, 132-40

uncommitted logic array (ULA), 132, 137, 160
unipolar technology, 3, 5, 39–57, 453–460
universal logic primitive, 140–4
user-specific integrated circuit (USIC), 18, 111
UV-PROM, 61, 87

vertical transistor, 22
very large scale integration (VLSI), 3, 123
VHDL, 186–91, 217, 218
via, 45, 53, 119, 129, 153
VMOS, 46
volatile memory, 61, 104

wafer probe, 272, 418
wafer scale integration (WSI), 3
wafer stepper, 163

wafer test, 271–3
wafer yield, 441–5
walking pattern, 304
wavefront router, 202–4
weighted syndrome, 355
width (of data bus), 83
window, 19, 22, 45
wiring capacity, 145–9
wiring channel, 118, 130, 145–53
workstation, 8, 243–7

x-ray lithography, 166

yield, 440–5

zener diode, 77

US 17.95